Biotechnological Approaches in Biocontrol of Plant Pathogens

Biotechnological Approaches in Biocontrol of Plant Pathogens

Edited by

K. G. Mukerji and
B. P. Chamola
University of Delhi
Delhi, India

and

R. K. Upadhyay
The Directorate of Plant Protection
Haryana, India

Kluwer Academic / Plenum Publishers
New York, Boston, Dordrecht, London, Moscow

ISBN 0-306-46104-8

233 Spring Street, New York, N.Y. 10013

10 9 8 7 6 5 4 3 2 1

A C.I.P. Catalogue record for this book is available from the Library of Congress.

Printed in the United States of America

PREFACE

According to the International Food Policy Research Institute (IFPRI) report, the world population is expected to grow to a total of 8 to 12 billion between now and 2020. There will be approximately 8 to 12 billion people to feed in the year 2020 which is about twice as many as were in 1990. Together with this population growth, urbanisation and displacement of people have resulted in diminishing the arable area per person from 0.3 hectares in 1981 to 0.18 hectares in 2020. Exponential population growth at the cost of receeding land resources and reduction in cultivated area forced to intensify agriculture by improving tillage, water management, crop varieties, fertilisation and other agronomic practices. Paradoxically these greatly enhanced disease incidence coupled with more and more delibrate use of pesticides/fungicides/bactericides (agrochemicals) etc. These further degraded environment through contamination of air, water, soil and food with these chemicals, and also altered the structure (species richness, biological density and diversity) and functional activities of ecosystem. Eventually these may ultimately threaten the very survival of the human race. The urgent need, therefore, is for exploring areas to minimise or contain the degradation of the habitats, plants, their behavior and their successful survival. All of these can be achieved by not resorting to or controlled, use under special case of the synthetic chemicals. The recent realisation of Biological control of plant pathogens has helped in its development as a potential strategy for checking the environmental degradation. The use of biological control strategies offers several advantages over the chemical control since it is economical, self perpetuating and usually free from residual side effects. Biological approaches for biocontrol envisages, principles, concepts, mechanisms and current practices using predators, antagonistic microbes, rhizosphere microflora, genetic engineering etc.

In this book different aspects of such biotechnological approaches in biocontrol of plant pathogens have been discussed by a group of internationally reputed scientists in the field. The book contains 12 chapters incorporating latest in the respective fields. Some of these reviews have been written for the first time and others are the updated revisions. We surmise that this book will act as a catalyst in ushering newer ideas in providing meaningful solutions to intricate problems in Plant Disease Biocontrol Technology. We also hope that the information provided in this book will be of immense value to scholars, scientists, agriculturists, administrators and enlightened farmers, for the management of disease and pests.

We are grateful to our contributors who responded instantaneously to our request for writing comprehensive review articles as well as sticking to the time schedule. It was because of their active cooperation that we completed this uphill task in less than a year. We are conscious that in a work of this nature, it is likely that some errors might have crept (inadvertently) in the text and for these lapses we own the responsibility.

We are thankful to Miss Mandeep for critically scanning the relevant literature and helping us to update the chapters.

A word of appreciation is also due to M/s. Neelam Graphics for active cooperation in preparing the neat electronic copies of the text of the chapters.

K.G. Mukerji
B.P. Chamola
R.K. Upadhyay

LIST OF CONTRIBUTORS

1. Boyetchko, Susan M.
   Agriculture and Agri-Food Canada
   Suskatoon Research Centre
   107 Science Place, Suskatoon
   Saskatchewan, CANADA S7N OX2
   Tel.: 306-956-7619
   Fax 306-956-7247
   E-mail:Boyetchkos@em.agr.ca

2. Garibaldi, Angelo
   Dipartments di Valorizzazione e
   Protezione delle
   Risorse Agroforestali Patologia Veg(
   Universita di Torino, Via Leonardo de
   Vinci 44, I-10095
   Grugliasco, Torino, ITALY
   Tel.: (39) 11 6708540
   Fax (39) 11 6708541
   E-mail: pat.veg@fileita.it

3. Grevesse, Cathy
   Unite de Phytopathologie
   Faculte Universitaire des Sciences
   Agronomiques
   2, Passage des Deportes, B-5030
   Gembloux (Belgique)
   BELGIUM
   Tel.: 32(0) 81 622431
   Fax:32(0) 81 610621
   Telax: 59482B-FSAGX
   E-mail : phytopat@fsagx.ac.be

4. Gulliono, M. Lodovica
   Dipartments di Valorizzazione e
   Protezione delle
   Risorse Agroforestali-Patologia Vegetale
   Universita di Torino, Via Leonardo de
   Vinci 44, I-10095
   Grugliasco, Torino, ITALY
   Tel.: (39) 11 6708540
   Fax (39) 11 6708541
   E-mail: pat.veg@fileita.it

5. Gupta, Rajni
   Applied Mycology Laboratory
   Department of Botany
   University of Delhi
   Delhi-110007, INDIA
   Tel.: 91-11-7257573, 7257725/383,
   91-11-7113727, 7122595
   Fax:91-11-7257830

6. Iqbal, Muhammad
   Department of Botany
   Faculty of Science
   Hamdard University, Hamdard Nagar
   New Delhi-110062, INDIA
   Tel.: 91-11-698 4685, 698 9309/207&232
   Fax: 91-11-698 8874
   E-mail:root@hamduni.ren.nic.in

7. Jijakli, M. Haissum
   Unité de Phytopathologie
   Faculté Universitaire des Sciences
   Agronomiques
   2, Passage des Deportes, B-5030
   Gembloux (Belgique)
   BELGIUM
   Tel.: 32(0) 81 622431
   Fax:32(0) 81 610621
   Telax: 59482B-FSAGX
   E-mail : jijakh:h@fsagx.ac.be

8. Kaur, Surinder
   Applied Mycology Laboratory
   Department of Botany
   University of Delhi
   Delhi-110007, INDIA
   Tel.: 91-11-7257573, 7257725/383,
   91-11-7187496, 7198809
   Fax:91-11-7257830

9. Khillam, Ken
Department of Plant and Soil Science
University of Aberdeen
Cruickshank Building
Aberdeen AB 24 3UU
Tel.: 44(0) 1224-274122
Fax: 44(0) 1224-273731
Email:agric@aberdeen.ac.uk

10. Knox, Oliver G.G.
Department of Plant and Soil Science
University of Aberdeen
Cruickshank Building
Aberdeen AB 24 3UU
Tel.: 44(0) 1224-274122
Fax: 44(0) 1224-273731
Email:agric@aberdeen.ac.uk

11. Leifert, Carlo
Department of Plant and Soil Science
University of Aberdeen
Cruickshank Building
Aberdeen AB 24 3UU
Tel.: 44(0) 1224-274122
Fax: 44(0) 1224-273731
Email:agric@aberdeen.ac.uk

12. Lepoivre, Philippe
Unite de Phytopathologie
Faculte Universitaire des Sciences Agronomiques
2, Passage des Deportes, B-5030
Gembloux (Belgique)
BELGIUM
Tel.: 32(0) 81 622431
Fax:32(0) 81 610621
Telax: 59482B-FSAGX
E-mail : phytopat@fsagx.ac.be

13. Migheli, Quirico
Dipartmento di Valorizzazione e Protezione delle Risorse Agroforestali-Patologia Vegetale
Universita Degbi Studi Torino, Via
Leonardo da Vinci 44, I-10095
Grugliasco, Torino, ITALY
Tel.: (39) 11 6708540
Fax (39) 11 6708541
E-mail: pat.veg@fileita.it

14. Mughal, M. Haseeb
Department of Botany
Faculty of Science
Hamdard University, Hamdard Nagar
New Delhi-110062, INDIA
Tel.: 91-11-698 4685, 698 9309/207&232
Fax: 91-11-698 8874
E-mail:root@hamduni.ren.nic.in

15. Mukerji, K.G.
Applied Mycology Laboratory
Department of Botany
University of Delhi
Delhi-110007, INDIA
Tel.: 91-11-7257573, 7257725/383, 91-11-7257502
Fax:91-11-7257830

16. Singh, Geeta
Division of Microbiology
Indian Agricultural Research Institute
New Delhi-110012, INDIA
Tel.: 0091-11-578 7649, 573 1270
Fax: 0091-11-575 1719, 576 6420

17. Srivastava, P.S.
Centre for Biotechnology
Faculty of Science
Hamdard University, Hamdard Nagar
New Delhi-110062, INDIA
Tel.: 91-11-698 4685, 698 9309/207&232
Fax: 91-11-698 8874
E-mail:root@hamduni.ren.nic.in

18. Sumeet
Applied Mycology Laboratory
Department of Botany
University of Delhi
Delhi-110007, INDIA
Tel.: 91-11-7257573, 7257725/383, 91-11-5122969
Fax:91-11-7257830

19. Tilak, K.B.V.R.
Division of Microbiology
Indian Agricultural Research Institute
New Delhi-110012, INDIA
Tel.: 0091-11-578 7649, 573 1270
Fax: 0091-11-575 1719, 576 6420

20. White, Duncan
Department of Agriculture
University of Aberdeen
Mac Robert Building
Aberdeen AB 24 5UA
Tel.: 44(0) 1224-274122
Fax: 44(0) 1224-273731
Email:d.white@aberdeen.ac.uk

CONTENTS

HOST PARASITE SPECIFICITY AND PATHOGENESIS

Rajni Gupta and K.G. Mukerji

Applied Mycology Laboratory
Department of Botany
University of Delhi
Delhi-110007, INDIA

1. INTRODUCTION

The word infection is derived from Latin word, "infectus" which means to mix or corrupt. The term refers to the relationship between the host and the parasite, and the competition for supremacy that takes place between them. If the host has upper hand, due to increased host-resistance, it remains healthy and the parasite is either driven away or assumes a relationship with the host. On the contrary, if the host loses the competition, disease develops. Thus diseases are the outcome of competition between host and parasite. Disease is generally considered as any change from the general state of good health. But this does not present true scientific nature of disease, since general state of good health may also be changed by a factor, other than a parasite. Thus disease is a malfunctioning process caused due to continuous irritation by the parasite.

Pathogenicity refers to the ability of a parasite to gain entry into host's tissue and bring about a change resulting in a change of health and thus disease. The word is derived from the Latin word "pathos" meaning "suffering". The term pathogen has same root and refers to an organism able to cause disease i.e. possessing pathogenecity. The symbiotic relationship between host and parasite is called parasitism. Parasites vary in their pathogenecity. Pathogens such as *Alternaria* sp., *Rhizoctonia* sp. and *Sclerotium* sp. can be considered as more hardy because, in addition to attacking tender tissues, they can also exploit comparatively mature tissues. All these pathogens have a very wide host range. They are necrotrophs, they cause extensive damage to the host tissue by employing enzymes and secondary toxic metabolites as their major weapons. Advancement towards semibiotrophs and biotrophs has probably been guided by : (a) lesser dependence of the pathogens on toxins and enzymes, (b) more involvement of phytohormones, (c) decreased deleterious effects of the parasite and increased dependence of the parasite on living host cells, (d) decreased host range, and (e) more and more synchronization of the physiological processes of host and parasite.

Biotechnological Approaches in Biocontrol of Plant Pathogens
Edited by Mukerji *et al.*, Kluwer Academic / Plenum Publishers, New York, 1999

Vascular wilts commonly induced by species of *Fusarium, Verticillium, Ceratocystis* and *Cephalosporium*, occupy a unique position among plant diseases since during the critical stages of pathogenesis, the fungus is confined within non-living xylem elements of the host. Wilt fungi are basically necrotrophs, capable of producing wall degrading enzymes and toxins. They induce water stress in plants mainly by impairing the water transport through the xylem.

Downy mildews, comprise of nine genera, are another group of obligatory biotrophic parasites. Their mode of parasitism is typified by minimal tissue damage, altered host permeability and probably by change in translocation patterns and hormone levels. On a susceptible host, they may establish a balanced relatioship in which the host remains asymptomatic while supporting profuse fungal growth and sporulation. They may cause local or systemic infection.

Powdery mildews attack all green parts of the host plant. Their infection is usually ectotrophic. A superficial mycelium draws the nutrients through the haustoria formed within the epidermal cells (Bhandari and Mukerji, 1993). These fungi have developed a remarkable ability to divert the nutrients towards the mycelium, away from the normal sinks of the host plant. The gene for gene type of relationship has been shown to occur in powdery mildews also (Mukerji *et al.*, 1998).

Rusts are among one of the most important diseases of plants. They are induced by biotrophic fungi. Host specificity is so pronounced that certain rust fungi require a second host to complete a few specific stages of their life cycle, and the reasons behind such behaviour are far from being understood.

Phytophthora sp., a semibitrophic fungi, induce a wide range of diseases on a number of host plants. They draw nutrients using a specialized organ from the living host cell, but are capable of surviving host cell death at least for a few hours. These fungi can be grown easily on synthetic and semisynthetic culture media and cause large scale and early necrosis of host cells.

2. VASCULAR WILT

Verticillium albo atrum Reinke and Berthold, the casual agent of *Verticillium* wilt, is a soil borne plant pathogen with a wide host range (Isaac and Lloyd, 1959). *Verticillium albo atrum* isolates recovered from different hosts, especially from alfalfa consistently showed differences in pathogenicity profiles in inoculation studies (Corell *et al.*, 1988). According to investigations, *V. albo atrum* consists of two groups: alfalfa isolates and non alfalfa isolates which are respectively, virulent and avirulent on alfalfa. In northern Europe, the fungus has been reported as a major threat in alfalfa (*Medicago sativa*) production (Krietlow, 1962). In North America, it has been found in alfalfa only since 1976, but has subsequently dispersed widely in many alfalfa growing areas and now limits alfalfa production (Arny and Grau, 1985). *Verticillium* wilt is also a serious problem of potato in North America. *Verticillium albo atrum* and *V. dahliae* are recognized to be the predominant species causing the disease (Platt, 1986). Pathogenecity tests have shown that *V. albo atrum* is a more severe pathogen of potato than *V. dahliae* in northeastern United States.

Fungal vascular wilts occupy a unique position among plant diseases. During the critical stages of pathogenesis, the fungal pathogen is confined within the non-living xylem elements of the host and living host tissue during the initiation and development of the wild syndrome, has promoted numerous studies into the mechanism of induction of water stress.

Duniway (1971) reported that the major cause of water stress in wilt infected plants is impaired water transport within the xylem. The extensive xylem colonization clearly visible in many vascular diseases led early workers to postulate that blockage of vessels by fungal mycelium was responsible for water stress and wilting (Smith, 1899). The amount of mycelium present is insufficient to produce the characteristic wilt syndrome and that the degree of symptom development is not invariably related to the level of hyphal colonization (Mussel and Green, 1970). Ludwig (1952) had earlier calculated that to reduce the flow rate by the determined 83%, 10 hyphae per vessel throughout the entire vascular system would be required. McHardy *et al.* (1976) correlated the growth of *Verticillium dahliae* towards vein ends in *Chrysanthemum* leaves and the relative water content and appearance of wilt symptoms in those lobes. Formation of tyloses and gels are common responses to both vascular infection and wounding in a wide range of plant species. In the vascular wilts, tylosis and gelosis are now generally regarded as resistance mechanisms rather than important factors in pathogenesis. The decreased importance placed on vascular occlusion by tyloses or gels as a mechanism of water stress is reflected in the paucity of studies that have examined the temporal relationship between such vascular occlusion and wilting.

Vascular embolism is rarely considered as a cause of vascular dysfunction by plant pathologists, although Zimmerman and McDonough (1978) considered it to be the major cause of failure of water conduction. Water within the xylem is normally under tension and thus, if a vessel wall is penetrated or degraded air may be drawn in and an embolism result.

2.1. Role of Toxins in the Induction of Water Stress

Toxins have a good role in the pathogenesis of wilts. The low molecular weight toxins that have received most attention are lycomarasmin, a dipeptide produced in ageing cultures of *F. oxysporum* f. sp. *lycopersici* and fusaric acid, produced during active growth of a range of *Fusarium* formae specialies. Gaumann (1951) has described the mode of action and effects of lycomarasmin on tomato cuttings. Selman and Pegg (1954) have considered that such low molecular weight toxins must have a role in pathogenesis. High concentration of the toxin must be applied to cuttings to produce toxicity symptoms but despite this, the toxin has not been isolated from infected plants. Mode of action of low molecular weight toxin is increasing cell permeability, in the early stages of wilt pathogenesis. Extracellular higher molecular weight substances, particularly polysaccharides and glycopeptides are produced *in vitro* by a variety of wilt inducing pathogens, both bacterial and fungal. Xylem sap are more likely to act as 'vascular toxins' causing blockage of the vascular system in much the same way as for vessel wall degradation products.

Depending on their molecular size, they may accumulate in/on pit membranes inhibiting longitudinal and lateral water flow (Dimond, 1967) or be transported to the leaves and there disrupt water transfer from veins and accumulate on cellular membranes causing marginal and interveinal chlorosis (Talboys, 1978). In some wilt diseases, such as those infected with *V. albo atrum* (Talboys, 1975), sunflower infected with *V. dahliae* (Robb *et al.*, 1975) and pea infected with *F. oxysporum* f. sp. *pisi* (Bishop and Cooper, 1983), there is an absence of leaf wilt as such; more characteristic symptoms involve a gradual acropetal chlorosis and dessication of the leaf laminae.

The study of genetic diversity helps to clarify host specialization in fungal pathosystems. Vegetative compatibility analysis was used to group 16 alfalfa isolates of *Verticillium* together, whereas 13 isolates from other hosts were vegetatively incompatible with alfalfa isolates (Corell *et al.*, 1988). Christen *et al.* (1983) suggest that these isolates consist of a genetically homogenous clonal population distributed worldwide. Direct DNA examinations of mitochondrial DNA or of genomic DNA by restriction fragment length polymorphism differentiated *V. albo atrum* from alfalfa and other isolates of *V. albo atrum* led to the suggestion of a higher taxonomic status for this group. A recent approach, random amplified polymorphic DNA based on the polymerase chain reaction, consists of amplifying portions of total DNA with oligonucleotide primers of arbitrary sequences (Williams *et al.*, 1990). The advantage of RAPD analysis is that polymorphisms are revealed by small differences between isolates such as single nucleotide changes at primary binding sites or by insertion or deletion within the amplified region. The possibilities of DNA screening for polymorphisms are enormous when different primers are used. Barasubiye *et al.* (1995) assessed the genetic diversity of several *V. albo atrum* isolates from alfalfa and potato using RAPD markers. Based on analysis of RAPD markers, genetic diversity was observed within a collection of 35 isolates of *V. albo atrum*. A significant association was noted between host of origin and DNA amplification patterns, suggesting that the *V. albo atrum* isolates formed two distinct entities one alfalfa group and one potato group. On alfalfa, the alfalfa isolates were more virulent than the potato isolates, and only the potato isolates were virulent on potato. The optimum temperature for growth on PDA was 20°C for potato isolates and 20-25°C for alfalfa isolates. At 30°C no alfalfa isolate was killed although this temperature was lethal for 75% of the potato isolates. Carder and Barbara (1991) used the RFLP technique and reached a conclusion that a new taxonomic status for *V. albo atrum* isolates recovered from alfalfa should be considered. However, the genetic identity of isolates within the potato group supports the hypothesis that at least some isolates are of clonal origin.

3. *PHYTOPHTHORA*

3.1. Host Specificity and Pathogenesis

Phytophthora sp. have features unique for pathogenic fungi, such as formation of sporangia on branched sporangiophores; the release of motile zoospores that provide a tremendous infection potential, cell walls consisting mainly of glucans, the inability to synthesize sterols which are necessary for reproduction. This genus included 43 species causing a wide range of diseases on an ever increasing list of plants, including major food crops, forest trees, subtropical and tropical fruit trees, ornamental plants and tropical export crops. *Phytophthora cinnamomi* alone cause many diseases on nearly 1000 host plants (Zentmyer, 1983). Numerous diseases caused by *Phythophthora* sp. are of minor economic and nutritional importance and as a consequence have received little attention.

Some species of *Phytophthora* produce oospores by self induction, while others require the presence of the opposite mating types known as A^1 and A^2 for sexual reproduction (Ko, 1988). Several species of the cross inducing (heterothallic) *Phytophthora* have been investigated for their possible origins (Ko, 1980; Zentmyer, 1988). The following criteria or indicators have been used individually or in combination for locating the possible center of origin of fungi (Ko, 1980).

1) Presence of the fungus in a remote area without human disturbance.
2) The center of origin of the host plants which usually is also the center of source of host resistance to the pathogen.
3) The existence of substantial variation among isolates of the species in the origin.
4) The occurrence of a male to famale ratio, or A^1 to A^2 mating type ratio in the area. Little information of the destructive leaf blight of taro (*Colocacia esculenta*).

Ko (1979) reported that all 114 isolates of *P. colocasiae* obtained from different islands in Hawaii were the A^1 type and that therefore, it was unlikely to be its center of origin. Ann *et al.* (1986) also did not consider Taiwan to be a likely center of origin of the species because all 799 isolates of *P. colocasiae* obtained from there were similar in colony appearance and behaved as the A^2 mating type. Asia was considered to be the origin of *P. colocasiae* because it is the center of origin of many wild and cultivated varieties of taro and the fungus is widespread in that region (Ko, 1979). Since taro is commonly grown and its leaf blight caused by *P. colocasiae* is widespread on this island.

3.2. Pathogen Dispersal and Persistence

The genus *Phytophthora* includes air- and soil-borne species. Sporangia are transported by air or by water. They may germinate directly or in a suitable medium such as drops of dew or rain and release zoospores. It appears that zoospores are present just below a surface film. Zoospores may be attracted by roots or root exudates (Zentmyer, 1961), as well as stems of plants and their exudates (Nandris *et al.*, 1979). Oospores, chalmydospores, sporangia and infested plant debris secure persistence of the pathogens in the absence of host (Weste, 1983). The resistance of plants to attack by *Phytophthora* sp. is not constant but depends on environmental factors, such as temperature and humidity affecting the pathogen, as well as the prediposition of the host. Temperature is one of the most important environmental factor affecting sexual reproduction in *Phytophthora* (Elliot, 1983). For most of the *Phytophthora* sp. tested, the maximum temperature permitting oospore formation is always lower than that permitting vegetative growth.

Honour and Tsao (1974) reported that the temperature ranges allowing vegetative growth and oospore production of *Phytophthora parasitica* were 6-39°C and 18-30°C. Vegetative growth of *Phytophthora palmivora* occurred between 28-32.5°C (Brasier, 1969, Satchuthananthavale, 1963). While *P. cinnamomi* produce oospores (Zentmyer *et al.*, 1979) sexual reproduction in both self inducing "homothallic" and cross inducing "heterothallic" species of *Phytophthora* has been shown to be controlled by sex hormones (Ko, 1978, 1980). A^1 mating type produces a^1 hormone capable of inducing oospore formation by setting of A^2 but not A^1 mating type, while A^2 mating type produces a^2 hormone capable of inducing oospore formation by setting of A^1 but not A^2 mating type. Zhang *et al.* (1994) reported that *Phytophthora colocasiae* was widespread on Hainan Island. Both A^1 and A^2 types were equally widespread on the island. Chern and Ko (1994) reported that A^1 and A^2 isolates of *P. parasitica* produced an average of 96200 oospores per culture block in paired cultures and grew 6.8 and 7.2 mm/day respectively at 24°C. Oogonia were produced by A^2 or A^1 isolate only during the 24 hour exposure to a^1 or a^2 hormone. At 34°C oospores developed in oogonia were either abortive or distorted. Metalaxyl inhibits vegetative growth as well as sexual and asexual reproduction of many isolates of *P. infestans* (Cohen and Coffey, 1986).

3.3. Host Penetration

The resistance of plants to attack by *Phytophthora* sp. is not constant but depends on environmental factors, such as temperature and humidity affecting the pathogen, as well as the predisposition of the host. Temperature is one of the most important environmental factors affecting sexual reproduction in *Phytophthora*. Yu *et al.* (1981) paired *Phytophthora colocasiae* $(A)^2$ as the hormone producer with *P. parasitica* A^1 as the hormone receptor on the opposite sides of a polycarbonate membrane after incubation at 25° or 30°C for 24 hours for hormone production, and further incubated at the hormone receptor at 25°C for 3 days for oospore formation. Their results showed that hormone production was strongly inhibited at 30°C but not at 25°C. However, in addition to hormone production, other processes including hormone activity, receptor formation, receptor activity and hormone reception also occur during the pairing at high temperature. Water stress increases the susceptibility of safflower plants to root rot caused by *P. cryptogea* (Duniway, 1977) and salinity increases the susceptibility of alfalfa to *P. megasperma* f. sp. *medicaginis* (Kuan and Erwin, 1980), and reduced aeration of roots may decrease phytoalexin accumulation (Murch and Paxton, 1979). Penetration through stomata, however, also occurs (Pristou and Gallegly, 1954), infection hyphae may only pierce the cuticle and the outer layer of the periclinal epidermal wall and grow within the host cell wall and parallel to leaf surface prior to completing penetration (Coffey and Wilson, 1983). The excretion of polygalacturonase galactanases and pectinesterases by *P. infestans* growing in culture or in the host have been demonstrated (Cole, 1970). These enzymes normally cause separation and death of host cells.

Phytophthora parasitica var. *nicotianae* penetrates roots of tobacco both intra and inter-cellularly (Hanchey and Wheeler, 1971). At the point of entry into the host, the hypha of *P. megasperma* f. sp. *glycinea* is often closed with a septum like structure, separating the empty appresorium from the cytoplasm containing infection hypha; this is interpreted as a sign for completion or success of penetration (Stossel *et al.*, 1980). Hyphal plugs near the point of host penetration also occur in *P. cinnamomi* infecting Eucalyptus roots (Tippett *et al.*, 1976). This fungus does not form appresoria but produces hyphal swellings apparently triggered by hindrance of penetration. In alfalfa roots of a susceptible cultivar, there is more intracellular growth of *P. megasperma* f. sp. *medicaginis* than in roots of a resistant cultivar (Miller and Maxwell, 1984).

Resistance is the common phenomenon in nature and susceptibility is the rare exception demanding highly attuned metabolism of host and parasite. Susceptibility is brought about by a pathogen derived suppressor that blocks the resistance response, such as cellular browning and rishtin accumulation induced by the pathogen's elicitor. Partially purified water soluble glucan from a compatible race of *P. infestans* suppresses rishtin accumulation more effectively than the suppressor from an incompatible race of the pathogen (Doke *et al.*, 1979). From compatible races of *P. infestans* water soluble glucan is isolated that protects protoplast against hypersensitivity induced by hyphal wall components of *P. infestans* than do the corresponding glucans isolated from the incompatible pathogen (Doke and Tomiyama, 1980). Hyphal wall components of *P. infestans* induce protoplasmic aggregation and lysis of potato protoplasts. The phenomenon of host pathogen recognition is complicated further by the fact that, according to structural criteria, the population of cells involved in the host response does not react uniformly but some cells may respond susceptibly while other apparently do not (Stossel *et al.*, 1981). Host cells already in contact with the invader may influence neighbouring cells not yet in contact with fungal constituents through the release of signals. The phenomenon of recognition is also complicated by the occurrence of tolerance, a special form of susceptibility whereby the host plant survives infection by a compatible pathogen and does not develop severe symptoms (Buzzell and Anderson, 1982). Keen (1975) reported about the evidences for a race specific elicitor of low molecular weight in incompatible races of the pathogen. Subsequently, a direct elicitor-receptor hypothesis was proposed, sugggesting host resistane genes coding for sites that recognize specific molecules of incompatible pathogen races. The data indicate high molecular weight glycoproteins located at the surface of the pathogen and recognized by resistant host plants (Keen and Legrand, 1980). Race specific glucamannan play an important physiological role in the interaction with the host, rather than the non-specific ß-glucan elicitor (Sharp *et al.*, 1984). Similar to the suppressor in the *P. infestans* potato systems, *P. magasperma* f. sp. *glycinea* also have some supressor (Ziegler and Pontzen, 1982). This race specific suppressor is an extracellular invertase, a glycoprotein containing mainly mannose residues. It reduces glyceollin accumulation in soyabean hypocotyls challenged with the pathogen derived non-specific glucan elicitor (Ziegler and Pontzen, 1982).

Phytoalexin production is one of the conferral mechanisms of natural disease resistance in plants. Phytoalexins have been demonstrated in at least 100 plant species representing 21 families. Phytoalexin production has been associated with the expression of both general and specific disease resistance (Keen and Littlefield, 1979). Phytoalexin biosynthesis is induced by a wide range of stimuli, including abiotic elicitors such as ultraviolet irradiation, organic solvents, detergents, heavy metal ions etc. and biotic elicitors such as plant cell wall constituents and fungus derived elicitors. Phytoalexin synthesis may be induced non specifically for abiotic elicitors, in general, do not discriminate between plant species, plant families. Specific recognition of the multitude of non pathogens by host receptor sites is unthinkable. It is thus believed that non host resistance depends on non specific defence reactions (Heath, 1981). Activation of genes coding for enzymes of phytoalexin biosynthesis, however, can be fairly specific: eicosapentaenoic and arachidoic acid, two biotic elicitors from *P. infestans*, induce phytoalexin accumulation in potato (Bostock *et al.*, 1982).

3.4. Phytoalexin

Phytoalexin has been associated with the expression of both general and specific disease resistance (Table-1, Keen and Brugger, 1977). Recognition of pathogen invasion and resulting defense responses primarily occur at the level of single plant cells. In the initial step of these events, plant cells perceive either plant derived (endogenous) or pathogen derived (exogenous) signals. These signal compounds are collectively referred to as elicitors (Fig. 1). Elicitors are widely used in simple experimental systems to study the molecular mechanisms of defense responses (Ebel and Cosio, 1994; Kombrink and Somssich, 1995). The exogenus elicitors of plant defense responses differ widely in the chemical nature and include proteins, oligosaccharides, glycoproteins and lipids (Fig. 2).

Woody plants respond to wounds and potential pathogens by walling off and replacing injured or infected tissue (Shigo and Tippett, 1981). Injuries in bark caused by pathogens are compartmentalized by necrophylactic periderms which form suberized, non-conductive tissue (Biggs *et al.*, 1984). Lateral meristematic activity replaces injured bark, and may enclose injured areas thus protecting exposed wood from decay fungi and pathogens (Shigo and Tippett, 1981) Anatomical changes associated with resistance to the pathogen have been reported in Australian native flora infected with *Phytophthora cinnamomi* (Cahill *et al.*, 1989). Post infection biochemical and physiological changes are also involved in resistance of Australian native plants to *P. cinnamomi* (Cahill *et al.*, 1993). Resistance to this pathogen is expressed as reduced colonization of the host and has been described as field resistance (McCredic *et al.*, 1985). A more accurate term is rate reducing resistance, which has been used to describe race non-specific resistance to *P. megasperma*. Phosphonates effectively control root rot diseases caused by several *Phytophthora* sp. (Guest and Grant, 1991). At lower plant concentrations phosphonates are thought to modify the pathogen resulting in the elicitation of host defence responses, while at higher concentrations in the plant, fungal growth is in addition directly inhibited by the chemical (Grant *et al.*, 1990). Smith *et al.* (1997) first time reported the host responses of a member of Australian flora of *P. cinnamomi*, treated with the phosphonate. Many phytopathogenic fungi produce specialized infection structures termed appresoria to achieve adhesion to the cuticle in preparation for subsequent host invasion. Appresorium formation is common in *Phytophthora* species which attack aerial parts of their hosts (Pristou and Gallegly, 1954). In *Phytophthora* adhesion of germ tubes to host surface is mediated by adhesions present at the interface (Hohl, 1991). In *Phytophthora*, IgG binding 65K Da glycoproteins are instrumental for adhesion of *Phytophthora megasperma* f. sp. *glycinea* to host mesophyll cells (Ding *et al.*, 1994), and a protein and glycoproteins apparently involved in attaching zoospores of *Phytophthora cinnamomi* to the host root surface have molecular weight of over 200 and 300K Da, respectively (Gubler and Hardhnam, 1988). Bircher and Hohl (1997) investigated the exogenous cues inducing appresorium formation in the oomycete *P. palmivora*. They revealed that this process is influenced by contact mediated signals as adhesion to a solid surface its topography of hydrophobocity. The ability of a pathogen to perceive inducive physical signals for infection structure formation requires contact between the fungus and a host surface or an artificial substrate (Nicholson, 1996). The mechanism of adhesion of fungal germlings to the plant cuticle or artificial surfaces have not been well studied. Adhesion in *Phytophthora* and other fungi appears to be mediated by surface glycoproteins (Hallenstein and Hohl, 1995).

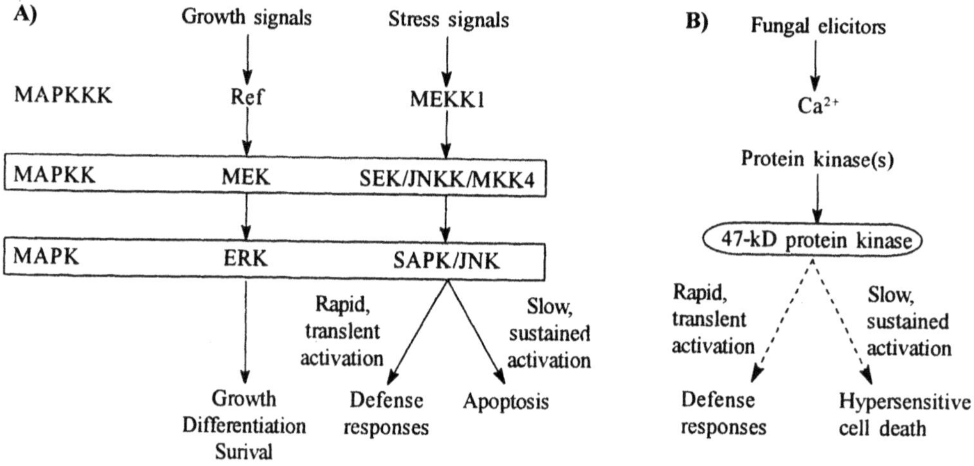

Figure 1. Comparison of models for the roles of protein kinases in signal transduction leading to defense responses and physiological cell death in plant and mammalian cells. (A) Model for the roles of mammalian MAP kinase cascades in response to extracellular signals. ERK, extracellular signal-regulated kinase; JNK, c-Jun N-terminal kinase; JNKK, JNK kinase; MAPK, mitogen-activated protein kinase; MAPKK, MAPK kinase; MAPKKK, MAPK kinase; MEK, MAPK and ERK kinase; MEKK1, MEK kinase 1; MKK4, MAPK kinase 4; SAPK, stress-activated protein kinase; SEK, SAPK and ERK kinase. (B) Proposed model for elicitor-responsive activation of the 47-kD protein kinase in tobacco cells. (Source : Suzuki and Shinshi, 1996).

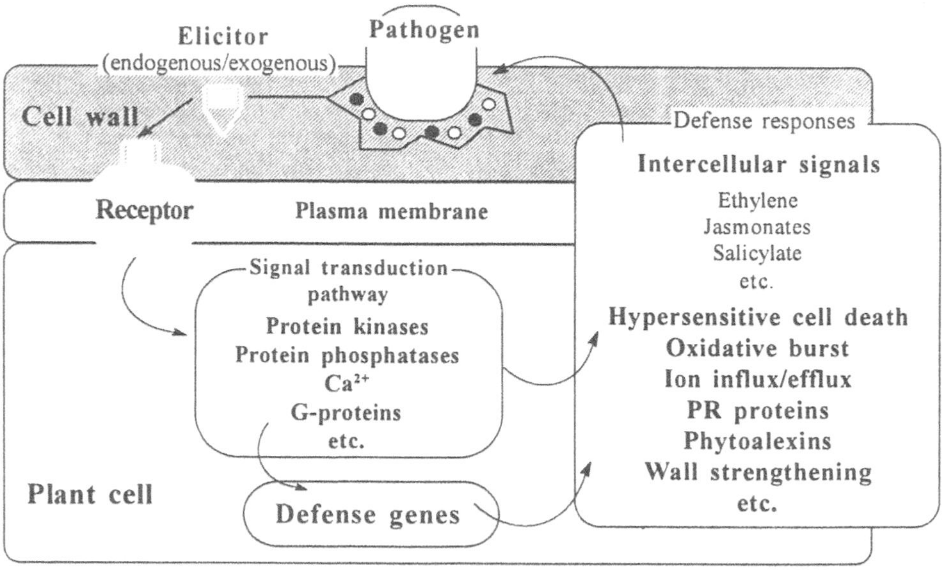

Figure 2. Schematic representation of mechanisms involved in defense responses in plant cells against pathogen infection. (Source : Suzuki and Shinshi, 1996).

Table 1. A Summary of Some of the Data Linking Phytoalexins to the Expression of Disease Resistance in Plants

1. Compatible pathogens are frequently tolerant or chemically inactive phytoalexins of their host plant but closely related nonpathogens are not.
2. Pretreatment of normally susceptible plants with agents such as UV light that elicit phytoalexin production render the plants resistance to subsequent inoculation.
3. Alterations in the environment that deleteriously affect phytoalexin production by the host with lesser or no effects on the parasite have been shown to decrease resistance.
4. Metabolic inhibitors of various kinds and preinoculation heat treatments frequently block phytoalexin production and resistance; both resistance and phytoalexin production are regained by heat-treated plants after about 3 days.
5. In the studied gene-for-gene system, substantial phytoalexin production specifically occurs only with incompatible combinations of host and parasite.
6. Resistance alleles leading to rapid cessation of parasite growth give more rapid accumulation of phytoalexins in the infection court than alleles observed to restrict pathogen development more slowly.
7. Careful studies have established that highly toxic concentrations of phytoalexins are formed at precisely the time and cellular location in which development of an incompatible pathogen race ceases.
8. Coinoculation of compatible and incompatible races of a pathogen generally results in restricted development of both, concomitant with the accumulation of high levels of phytoalexins.
9. Application of purified phytoalexins to the infection site of a normally compatible plant pathogen results in incompatibility and removal of phytoalexins results in increased compatibility.
10. Limited data suggest that specific elicitors obtained from various races of pathogens in gene-for-gene systems give the same specificity for phytoalexin production on incompatible and compatible plant genotypes as the living pathogens.

Plant pathologists can exploit HR/phytoalexin system for disease control by attempting to modulate the informational and expressive systems as shown in Table-2.

Table 2. Possible Rationales for Effecting Disease Control by Manipulating the HR/Phytoalexin System

1. Manage the major genes better.
2. Search more diligently for "strong" resistance genes or combinations of genes.
3. Launch a major effort to more fully understand the biochemistry of recognition and expression of the HR.
4. Attempt to find usable "sensitizers" or potentiators of the phytoalexin response.
5. Breed plants for the proper blend of phytoalexins.
6. Breed plants for finite constitutive levels of phytoalexins.
7. Attempt to breed plants that superproduce phytoalexins in response to biotic elicitors.
8. Attempt to apply phytoalexins to plants to effect disease control.
9. Apply phytoalexin elicitors to plants.

(Source : Bell and Stipanovic, 1978).

An additional role of surface glycoproteins in appresorium induction, namely in the perception of inducive physical signals of smooth contact substrates. Pronase E strongly reduced the frequency of germlings adhering to smooth polystrene and completely inhibited appresorium formation of adhering germlings without inhibiting germination and germ tube growth of the pathogen. The protease most probably removed proteinaceous surface components which are involved in adhesion and are essential for appresorium induction such as proteins, glycoproteins or lipoproteins. Tunicamycin, a putative inhibitor of protein glycosylation also strongly reduced the frequency of germlings adhering to smooth polystrene. This treatment prevented adhering germlings from forming appresoria while germ tube growth was only slightly affected (Elbein, 1987). Appresorium formation was inhibited by addition of IgG, which carries a putative glucose binding site (Ding *et al.*, 1994) and after addition of WGA or lectins binding to mannose or glucose residues. The inhibitory influence of WGA is most probably due to its toxic effect leading to limited germ tube growth (Hohl and Balsiger, 1986). Additionally, α-mannosidase or α-glucosidase but not β-glucosidase partly inhibited appresorium formation by germlings adhering to this substrate.

Kennedy and Duncan (1995) compared their morphology, physiology, electrophoretic banding pattern of soluble proteins and pathogenecity with that of additional papillate, *Phytophthora* sp. from raspberry namely *P. cactorum, P. citricola* and *P. syringae*. Four *P. idaei* isolates produced identical banding patterns on a vertical slab gel. The seven *P. cactorum* isolates were similar to each other but differed in a few bands from *P. idaei*. Bands scored from three replicate gels showed 84% similarity between the two groups. *Phytophthora cladenstina* also had many bands in common with the first two species, and similarly ratings were 63 and 56% respectively. *Phytophthora citricola* and *P. syringae* were distinct from each other and from the other three species. *Phytophthora cladenstina* had few bands in common with others, similarly ratings were 11 and 0% respectively. *Phytophthora iranica* and *P. citricola* also gave banding patterns which were distinct from each other and from the other species. Protein banding patterns have been used by other workers as an aid to the classification of *Phytophthora* sp. Gill and Powell (1968) found that patterns produced by undissociated proteins were different from each species. A similar result was obtained by Bielenin *et al.* (1988) using SDS-dissociated protein banding patterns. Molecular evidences regarding the relationship among the Group I species i.e. *P. cladenstina, P. iranica* and *P. pseudotsygae* included one isolate of the papillate *Phytophthora* sp. and several of *P. cactorum* in a molecular study of *P. fragariae* (Stammler *et al.*, 1993). Patterns of hybridization of total DNA with nuclear and mitochondrial probes were very similar for the two species and different from those of other species. Differences in morphology, physiology and DNA patterns were matched also by differences in pathogenicity. The papillate isolates were moderately pathogenic to raspberry but not pathogenic to any other host examined. Their pathogenicity to raspberry was not exacerbated by flooding in contrast to *P. cactorum, P. citricola* and *P. syringae*. The same isolates did not attack apple seedlings in contrast to apple and raspberry isolates of this species.

4. RUST

4.1. Host Specificity

Rusts are one of the particularly important diseases in agriculture. They occur on an extremely wide range of plant species and are distributed over a broad geographical range. They are therefore major factors in reducing agricultural productivities, especially those of gramineous and leguminous crops around the world. Rust fungi are distinguishable by their host specificity as for other biotrophic fungi. Specialization in rust diseases could be classified into three groups. One is the non host response at the level within species or formae speciales, and the other two are incompatible and compatible interactions in flax, bean, barley and sunflower (McIntosh and Waston, 1982).

Physiological races are taxonomically subordinated to special forms into which species are broken down by analogy with the higher plants. Every physiological race of the rust fungi, as testified by facts accumulated in a great number of works, by the capacity or potentiality to infect any variety or even many varieties that compose a certain species of the host plant; however, infectivity changes in dependence on the kind of host or the diversity of the host. Moreover, there are physiological races which are capable, to varying degrees, of parasitizing also on many species closely related to the main host plant. There are also rust fungi characterized by a wide polyphagy, which are capable at one or another stage of their development of parasitizing many species of host plants.

A number of studies related to plant rust fungus interaction have been carried out with uredospores which typically penetrate through stomata. Uredospores germinate by adhering germ tubes on hydrophobic surfaces of leaves and stems. No significant differences within plant species are usually seen in this process. Germ tubes stop the growth on reaching stomata, recognize the shape of stomatal opening and start appresorium formation (Wynn, 1976), or differentiate appresoria responding to volatile substances and phenolics in leaves (Grambow and Grambow, 1978). Tani and Yamamoto (1978) demonstrated that uredospores of four rust fungi which are pathogenic on gramineous plants other than oat, produced appresoria over stomata of oat leaves equally as well as those of compatible and incompatible races of *Puccinia coronata* var. *avenae*, while eight rust fungi non pathogenic on gramineous plants rarely produced appresoria.

The most distinguishable phenomenon between non host and host resistance is seen at the stage of first haustorium differentiation. In host plant tissues, the fungi produce at least one haustorium irrespective of infection types (Naito *et al.*, 1970), while they produce essentially no haustorium in non host plant (Heath, 1977). Fungal development in the post haustorial stage seems to be controlled by the expression of resistant genes and therefore varies extensively depending on the gene combination between host cultivars and rust races (Bhandari and Mukerji, 1993; Rohringer *et al.*, 1979). Mendgen (1975) reported in bean leaves the guard cells of stomata are covered during the penetration process with a layer showing peroxidase activity when inoculated with *U. phaseoli* var. *typica*. Montarbini and Elstner (1977) observed production of ethylene in bean leaves 10-18 hour after inoculation with *U. phaseoli* at the time of fungus penetration through the stomata. The amount of ethylene produced was larger in the 'early resistant' variety than in the late resistant or susceptible varieties. Ethylene is regarded as the plant hormone responsible for the induction of resistance in certain diseases.

4.2. Molecular Basis of Resistance

The molecular basis of resistance induction have been proposed in rust diseases by several workers. Hoppe *et al.* (1980) isolated glucans from germ tube cell walls of *U. phaseoli* which elicited resistance in a susceptible bean cultivar and accumulation of bean phytoalexins. Mayama (1983) demonstrated that oat phytoalexins accumulated in oat leaves treated with cell wall components of germinated uredospores of *P. coronata* var. *avenae*. Martins *et al.* (1983) reported an elicitor activity in the filtrates of water washes of autoclaved and non-autoclaved uredospores of *Hemileia vastatrix*. Elicitors released from the pathogens bind to the receptor site on plant cell membranes and the information of binding is transferred towards the plant nuclei by a second messenger (Keen, 1981).

Changes in enzyme activities in rust infected tissues have been considerably studied in relation to respiration, glycolysis and photosynthesis (Chakravorty and Scott, 1982). Peroxidases and phenylalanine ammonia lyase (PAL) have been considered to play certain roles in the resistance of various diseases. Heath (1979a) characterized the component of electron opaque deposits produced in french bean during non host response to *U. phaseoli* var. *vignae* as a material mainly composed of silicon, rather than phenolics. Both silicon deposit and lignification have been implicated as mechanical barriers for the invasion of rust fungi. Silicon deposit has also been demonstrated in electron opaque regions of haustorium encasement, host cell walls and necrotic cytoplasm in cowpea responding to an incompatible race of *U. phaseoli* var. *vignae* (Heath, 1981b). Deverall (1977) detected an antifungal resistance to *P. graminis* var. *tritici* and identified it as dihydroxymethoxy benzoxazolinone. This compound is produced from an immediate precursor by releasing glucose molecule. Phytoalexins are also regarded as compounds responsible for both general and specific resistance. Bean leaves responding with hypersensitive necrosis to *U. appendiculatus* accumulated phaseollin (Bailey and Ingham, 1971). Phaseollidin, phaseollin and phaseollinisoflavan were detected in bean leaves inoculated with an incompatible race of *U. phaseoli* (Elnagly and Heitefurs, 1976). Coniferyl alcohol and coniferyl aldehyde were isolated from flax infected by *Melampsora lini*; rapidity of accumulation paralleled with the degree of fungal growth restriction (Keen and Littlefield, 1979). Phytoalexins named avenalumins I, II and III have been shown to accumulate more rapidly and more abundantly in parallel with the degree of resistance (Mayama *et al.*, 1982). An increase in the synthesis of RNA species occurs in association with the rust infection of flax (Chakravorty and Shaw, 1971) and oat (Tani *et al.*, 1971).

4.3. Pathogenesis

Like the resistant mechanisms, the mechanism of susceptibility may involve two phases: suppressive phase of resistance and cooperative phase of metabolism essential for the completion of fungal development. Heath (1982) proposed the significance of 'basic compatibility' in which the suppression of non host resistance has been emphasized in the establishment of host pathogen specificity. Extracts of bean rust infected leaves of french bean increased the frequency of haustorium formation of *U. phaseoli* var. *vignae* in french bean, and of *U. phaseoli* var. *typica* in cowpea (Heath, 1980). Exudates from infection structures of *U. phaseoli* var. *typica* also increased the haustorium formation of *U. phaseoli* var. *vignae* in french bean (Heath, 1981a). Formation of glycoprotein is also noted with high frequency at the initiation of substomatal vesicle formation in the compatible interactions, but at relatively low frequency in the incompatible interactions. Another interesting phenomenon associated with the pathogenesis of rust fungi is the presence of common antigens between the host and parasite. Doubly *et al.* (1960) first reported the relationship in antigens between flax and *Melampsora lini*.

Chakravorty and Scott (1982) have emphasized that obligate parasites, such as rust fungi have evolved sophisticated mechanisms to control the host plants primary metabolism. In oat leaves inoculated with compatible races of *P. coronata* var. *avenea*, the treatment of leaves with blasticidin S inhibited the incorporatin of C^{14} leucine into protein, but did not affect the vegetative growth and uredial differentiation of the fungus (Tani and Yamamoto, 1979). The treatment with blasticidin S also had no effect on the susceptibility of wheat leaves to a compatible race of *P. graminis* var. *tritici* (Kim *et al.*, 1977). Infection of actinomycin D or cycloheximide into french bean leaves had little to no effect on the mycelial growth of a compatible race of *U. phaseoli* var. *typica* (Heath, 1979b).

Leaf rust caused by *Melampsora* species is the most important growth limiting disease afflicting willows in New Zealand. In Canada, England and West European countries, willows have been grown for renewable energy (McCracken *et al.*, 1992). Foliar rust caused by *Melampsora* species has been a major problem, causing premature defoliation and markedly reduced yields. Spires and Hopcraft (1996) reported the morphology and host range of *Melampsora* species attacking willows in New Zealand, which depend upon the attenuate host, apical thickening of aeciospore walls, size, thickening, location of teliospores, morphological features of urediniospores and paraphyses.

5. *PYTHIUM*

Soil borne species of *Pythium* have been shown to augment the herbicidal activity of glyphosphate by colonizing the roots of glyphosphate treated plants. This enhancement of herbicidal efficacy was termed glyphosphate synergistic interaction (GSI) by Johal and Rahe (1984). Soil borne *Pythium* sp. were found to be the first and predominant root colonizers of glyphosphate treated plants grown in different soils (Levesque *et al.*, 1993). A possible interaction of host speificity of some glyphosphate synergistic *Pythium* (GSP) was suggested from the result of an experiment by Levescue *et al.* (1992) in which *Pythium* isolate previously collected from the roots of glyphosphate treated apple and bean seedlings were glyphosphate synergists only on the plant species from which the fungus originated.

5.1. Pathogenecity

Pythium entered the germinating seeds in the soil by infecting the emerging radicle. After the initial infection, the mycelia colonized the growing embryo and cotyledons. Susceptible seeds were eventually killed and were covered with white mycelia one week after infection. Infected seeds seldom emerge from soil and those that did emerge from the soil were weak and have obvious stem necrosis immediately above the soil surface. Infected seedlings usually tople down 3-4 day after emergence and showed symptoms of stem girdling. Descalzo *et al.* (1996) reported that 15 PBI varied in their virulence on germinating bean seedlings. All isolates of *P. ultimum* and *P. irregulare* were highly virulent causing 100% and 88-98% pre emergence damping off on beans respectively. *Pythium sylvaticum* RFLP types B_2 and B_3 and Pythium 'HS' group RFLP types B_1 and B_2 were moderately virulent, causing 35-52% pre-emergence damping off. Other isolates of *P. sylvaticum* and all isolates of *P. coloratum* were moderately to weakly virulent, causing a range of pre-emergence damping off from as high as 27% to as low as 9%. The same isolates were also pathogenic to varying degrees on germinating wheat seeds. *Pythium ultimum* RFLP type B_1a was the most virulent isolate causing 95% pre emergence damping off. *Pythium ultimum* RFLP types B_1b and B_2; *P. irregulare* RFLP types B_1 and B_2 and *Pythium* HS group RFLP type B_2 caused pre emergence damping off ranging from 63% to 93%. Least virulent isolates, *P. coloratum* RFLP types B_1 and B_3; *Pythium* HS group RFLP type B_1 and *P. sylvatium* RFLP types B_4, B_5 and B_6 caused a range of pre emergence damping off ranging from 9 to 26%. Different PBI were pathogenic to varying degrees on germinating pepper seeds. All isolates of *P. ultimum, P. irregulare* and *P. coloratum* as well as *Pythium* 'HS' RFLP type B_2 and *P. sylvaticum* RFLP type B_1 caused 100% pre emergence damping off. *Pythium sylvaticum* RFLP types B_2, B_3 and B_5 caused mortalities ranging from 72 to 90%. PBI were generally less pathogenic to germinating sunflower seeds than to bean, wheat and pepper. The most pathogenic isolates were *P. ultimum* RFLP type B_2 and *P. irregulare* RFLP type B_2, causing 48% and 47% mortalities respectively. *P. ultimum* RFLP types B_1a and B_1b; *P. irregulare* RFLP type B_1; *P. sylvaticum* RFLP type B_1, B_2 and B_3 and *P. coloratum* types B_1 and B_2 caused mortalities ranging from 10 to 35%. Pathogenicity of *Pythium* species obtained from glyphosphate untreated hosts and soil were tested on germinating bean seeds. *P. aphanidermatum* caused the highest pre emergence damping off 94% followed by *P. spinosum* (66%) and *P. splendens* (27%) and *P. paroecandrum* (19%) and *P. arrhenomanes* (5%) and *P. coloratum, P. vanterpoolii, P. hypogynum* and *P. sulcatum* and *P. acanthicum* isolates did not cause mortality on germinating bean seeds.

The different species of *Pythium* varied in their pathogenecity. This is an expected result since *Pythium* in general causes diseases on diverse plant species, both monocotyledons and dicotyledons ranging from herbaceous to woody (Hendrix and Campbell, 1973). *Pythium* causes different diseases depending on the growth stage of the host. Mature plants are affected by *Pythium* at root tips and root hairs, causing a subtle overall decline of plant health over time (Mircltich, 1971). Succulent and juvenile stem and root tissues of seedlings and germinating seeds are commonly attacked below the soil line, which results in girdling of the seedling stem, and a watery decay of germinating seeds (Singh and Singh, 1984). The most prominent indication of severe *Pythium* attack on seedlings is manifested by damping off, characterized by toppling over of affected seedlings on the soil surface. *Pythium* can also be part of a disease complex on mature trees by causing synergistic interaction with other types of soil borne microorganisms as in apple replant disease (Braun, 1991) and root rot of young apple trees (Utkhede and Smith, 1991). The pathogenecity of the isolates of *Pythium* species from PWI and PBI groups on their respective reciprocal hosts showed indications of host preference when the most virulent *Pythium* isolates were considered. The most virulent isolate of *P. ultimum* in the PBI group was less virulent on beans than on wheat. PBI were generally less pathogenic on sunflower than on bean or pepper seeds. The antifungal metabolites sesquiterpene lactones present in sunflower probably have contributed to its natural resistance against *Pythium* (Spring *et al.*, 1982). Diverse *Pythium* species are capable of GSI and are generally non host specific, at least on herbaceous dicot seedlings and that these same isolates are pathogenic on germinating seeds highlights possible risks that might be associated with recurrent use of glyphosphate as a herbicide and crop desiccant.

5.2. Host Range and Specificity

Pythium periplocum Drechsler, a species with spiny oogonia, was first described as the causal agent of blossom end rot of *Citrullus vulgaris* (Drechsler, 1930). Subsequently, only sporadic reports of *P. periplocum* have been made from soil and diseased plants. It behaved as a weak plant pathogen or it was non pathogenic (Drechsler, 1939). *Pythium oligandrum* Drechsler is an aggressive parasite of other fungi, including many plant pathogens (Laing and Deacon, 1991). It has been shown to protect many crop plants against seedling disease caused by *Pythium* sp. and other fungi (Vasely, 1979). *Pythium ultimum* Trow, can germinate and infect seeds from soil borne propagules within a few hours (Nelson, 1992). *Pythium acanthicum* and *P. oligandrum* have been known as aggressive mycoparasites with a wide fungal host range (Faley and Deacon, 1986). These mycoparasites have a wide host range in comparison to other soil borne fungi such as *Syncephalis californica* (Hunter and Butler, 1975) where mycoparasitism is confined to Zygomycotina. Several species of *Pythium* were highly susceptible to *P. periplocum*. However in the case of *Phytophthora* the other oomycete groups tested, none of the species were highly susceptible to any of the mycoparasites.

Pythium periplocum is an aggressive parasite of many different fungi. Its hyphae commonly coil around host hyphae and may penetrate and grow within the host mycelium (Hockenhull *et al.*, 1992). The degree of host susceptibility was correlated with the presence of coiling and internal hyphae on the host. However, extensive coiling also has been taken as an indication of resistance (Berry *et al.*, 1993). It is possible, therefore, that some hosts recorded here as resistant would be listed as susceptible by another system. Berry *et al.* (1993) reported that *P. aphanidermatum* can antagonize *P. oligandrum* by coiling and penetrating the mycoparasite hyphae. The production of abundant oogonia and oospores by *P. periplocum* as well as by *P. oligandrum* and *P. acanthicum* when cultured with certain fungi. Berry *et al.* (1993) observed the differential production of *P. oligandrum* oospores and porpagules according to the host tested. Mycoparasite may obtain sterols essential for sexual reproduction from the host (Deacon and Henry, 1978). Host fungi provide non specialized nutrients that are in low supply in the parasite such as thiamine or certain amino acids. Finally, the observed increase in oospore formation may be due to the production of volatile chemicals or hormone like substances during the interaction between host and parasite.

In some host parasite interactions studied, the same host reaction was maintained during the entire period of observation, whereas in others, the reaction became progressively more intense. Apparently a non-progressive reaction represents a host defense response, preventing colonization by the parasite. The mechanism of host resistance is unclear at this time but it may have a purely chemical basis because there was no evidence that the parasite was walled off by the host. Madsen *et al.* (1995) reported that some newly emerging germ tubes of *P. oligandrum* did not parasitize the first host hyphae that they encountered but parasitized in later encounters and the incidence of parasitism increased significantly in relation to the distance that cysts were located from individual host hyphae. Zoospores of *Pythium* and *Phytophthora* sp. are known to adhere to a substratum when they encyst, but spores that encyst away from a surface rapidly lose the ability to adhere (Donaldson and Deacon, 1992). Cysts of *Pythium* and *Phytophthora* sp. have a pre determined point of outgrowth and that zoospores orient as they encyst on a host to which they have been attracted, so that the germination site is located next to the host (Hardhman and Gubler, 1990). The zoospores of *P. oligandrum* must have oriented to some extent during encystment, even though they were not attracted to host hyphae.

6. POWDERY MILDEW

6.1. Host Range Specificity and Pathogenesis

The powdery mildew genus *Erysiphe* occurs worldwide as a parasite on a wide range of plant species. The genus is characterized by many relatively primitive, shared telomorph characteristics. These characteristics include multiple asci within an individual ascocarp and simple, myceloid ascocarp appendages. Powdery mildew of cereals is caused by *Erysiphe graminis* formae speciales attacking all the temperate cereals like barley, wheat oats and rye (Mukerji *et al.*, 1998). Disease is an important constraint on yield in many of the cooler cereal growing areas of the world. This fungus is an obligate parasite having superficial mycelium fed by haustoria formed within epidermal cells of the host plant (Mukerji and Upadhyay, 1998). It attacks all green parts of its host to form elliptical colonies which take on their powdery appearance as chains of hyaline, single celled oidia are formed from the superficial mycelium. On the basis of primitive morphological characteristics *Erysiphe* is usually considered an ancestral group within the Erysiphaceae (Braun, 1987) but may be polyphyletic. Braun (1987) subdivided *Erysiphe*, into *Goleinomyces* and *Galeopsidis*. Further division of each section into numerous species and varieties is supported by comparison of both host range limitations and by more subtle morphological differentiation among host limited types within a section (Braun, 1987). Host range limitations suggest considerable subdivision within *Erysiphe* section *Goleinomyces*. Schmitt (1955) characterized three distinctive host limited types from within *E. cichoracearum* with each type isolated from a different host family. These host limited types correspond roughly to some of the species described by Braun (1987) within the *Erysiphe* section *Goleinomyces*. Zeller (1995) reported about information available among morphologically similar species suggest that host range may be valid characteristic for taxonomic classifications of these species since many display considerable host range limitations. Molecular analysis of ribosomal DNA differentiation among host limited forms similar to *E. cichoracearum* indicates considerable genetic distance between samples taken from different host families (Zeller and Levy, 1995) and supports subdivision within section *Goleinomyces*. Attention has been directed to increasing the durability of crop resistance to pathogens including powdery mildew. Wolfe (1984) have urged the use of cultivar mixtures in commercial crops where different components of the mixture carry different major genes for resistance. Such mixing produces the same sort of effect of decreasing epidemic build up as the use of a multiline, but a mix is practically much simpler to synthesize. Alternatively, resistance under complex genetic control might be used to increase the complexity of the genetic and physiological barriers faced by the pathogen, and hence to decrease the likelihood of rapid adaptation of the pathogen to the host genotype.

Erysiphe graminis appears to be unique amongst the powdery mildews in that its oidia regularly produce two germ tubes whilst other species form only one (Kunoh *et al.*, 1979). The optimum temperature for growth of *Erysiphe graminis* lies around 20°C and at this temperature the first signs of germination can be detected within one hour of oidium deposition. In most cases the tip of the tube attaches to the host surface, causing a host cell response, usually within 1-2 hour of its emergence. This tube is known as germ tube and have several important functions, the first being concerned with spore deposition. Another function of the germ tube may be to take up material from the host to facilitate growth of the relatively voluminous appresorial germ tube. However, in *E. graminis* f. sp. *hordei* at least firm attachment of the primary germ tube is not crucial to normal appresorial development, which indicates that recognition of the host surface by the pathogen can occur either through casual contact with the primary tube or through the oidia itself. Yang and Ellingoeb (1972) concluded that only the wax layer stimulated the formation of mature appresoria, and that the physical structure of the layer was more important than chemical composition. If the tube emerges while the spore is still air borne, it might disrupt the aerodynamic properties of the spore causing its earlier deposition.

Triazoles are an important group of systemic fungicides which act by inhibiting sterol biosynthesis and affect not only fungal development but also plant defence reactions (Smolka and Wolf, 1986). Different triazoles may interact differently with the host plant metabolism, so that the effects on the host/pathogen interactions may be unpredictable (Burden *et al.*, 1989). Heller *et al.* (1990) studied the barley/powdery mildew interaction, where other triazoles have been demonstrated to modify the extrahaustorial membrane and trigger the deposition of polysaccharidic materials by the host cells around the haustoria, leading to their encapsulation. Treatment of leaves with triazoles had no significant effect on their oidial germination and fungal development or on the host cells.

Grape powdery mildew caused by *Uncinula necator* (Schwein) Burrill (anamorph *Oidium tuckeri*) can infect all green tissues of grapevine. It is an annually re-occurring disease in vineyards of all major viticultural areas (Bulit and Laton, 1978; Hillebrand *et al.*, 1995). Control in commercial vine yards is generally based on the use of fungicides. These include the contact fungicides, sulphur, dinocap and dichlofluanid and the systemic ergosterol biosynthesis inhibiting fungicides benconazole, myclobutanil. Kresoxim-methyl (code name: BAS 490F) is currently being developed for use in field crops, fruits and vegetables (Ammermann *et al.*, 1992). It belongs to a new class of synthetic antifungal compound derived from the natural lead molecule strobilurin A. (Clough *et al.*, 1995). Strobilurins all have a common new site of inhibition, namely they strongly inhibit mitochondrial respiration in fungi by blocking the electron transfer at the cytochrome bc, complex (Becker *et al.*, 1981). Leinhos *et al.* (1997) reported fungicidal activity of Kresoxim-methyl against *U. nectator* with this he described various developmental stages of *Uncinula necator*. The growth of *U. necator* is characterized by a distinct sequence of differentiation events following germination. These include the development of appresoria, haustoria, hyphae, hyphal appresoria, oidiophores and oidia. After germination, the oidium first initiated a germ tube, a primary appresorium and an haustorium on one side of the spore. Only after the successful formation of the first functional haustorium, development of the primary hyphae at the opposite end of the same spore begin. Thereafter the secondary hyphae emerged from the primary appresorium. In *U. necator* hyphal appresorium formation and branching which occurred alternatively along the cells of the surface hyphae. Initially, single hyphal appresoria were formed, but as the hypha aged, one or two additional appresoria generally developed from the same hyphal cell. This multiple appresoria formation has been documented (Hillebrand *et al.*, 1995). Doster and Schanthorst (1985) reported that multiple appresoria formation occurred more frequently on resistant than on susceptible cultivars of grapevine. This behaviour suggests that the first haustorium may become less functional with time and further haustoria may become less functional with time and further haustoria are initiated to secure a sufficient supply of nutrients for oidiophore formation in the neighbouring hyphal cell. Mitochondrial respsiration is the essential energy delievering process for spore germination which makes this developmental stage a preferable and highly sensitive target site in fungi. Strong inhibitor of spore germination by Kresoxim methyl was proved for *U. nectator* as well as for a range of other fungal pathogens (Gold *et al.*, 1996).

7. DOWNY MILDEW

The downy mildews comprise nine genera of obligately biotrophic fungi responsible for major diseases of many important crops. Commonly recognized genera are *Basidiophora, Bremia, Bremiella, Perenosclerospora, Perenospora, Plasmopara, Pseudoperenospora, Sclerophthora* and *Sclerospora*. Infection by downy mildew fungi may originate from either primary or secondary inoculum. Primary inoculum consists of thick walled sexual oospores surviving in plant debris in soil or carried on the seed coat. Secondary inoculum is in the form of thin walled air borne sporangia. Depending upon the genus concerned, both types of inoculum may germinate either directly via a germ tube, or indirectly to release motile zoospores (Pratt, 1978). Hence downy mildews have the potential to exploit soil, air or water borne avenues of infection. Two contrasting patterns of host colonization are seen during infection by downy mildew fungi - systemic and localized. Systemic infection leads to extensive colonization of roots, stems, leaves and usually includes invasion of the apical meristem. The age of the host plant determines that it will become systemically infected or not (Bonde, 1982). Local lesions usually arise from secondary inoculum infecting aerial parts of mature plants. Such lesions occur on leaves and develop in discrete patches. With heavy inoculum loads under favourable environmental conditions, the lesions may however coalesce to cover the whole leaf.

The ability of downy mildews to form systemic infections in a wide range of herbaceous plants is well known (Spencer, 1981) but this feature is a major factor to consider in woody perennials which are vegetatively propagated as clonal material. Williamson *et al.*, (1995) described that the infectioin of leaves of *Rubus* genotype occurred mostly by penetration of the abaxial or adaxial epidermal cell walls, but infrequently infection begin from penetration of a stoma. Downy mildews which produce zoospores tend to show stomatal penetration, whereas those forming germ tubes penetrate the epidermis.

7.1. Pathogenesis

Downy mildews are biotrophic fungi, a mode of parasitism typified by minimal tissue damage, altered host cell permeability, and changes in translocation patterns and hormone levels (Cooke and Whipps, 1980). On susceptible hosts under favourable environmental conditions, downy mildews may establish a balanced relationship in which the host remains asymptomatic while supporting profuse sporulation by the fungus. Sporangiophores of vine downy mildew *Plasmopara viticola* often appear on apparently healthy green foliar tissues without any underlying lesion (Lafon and Bulit, 1981). Similarly, lettuce seedlings infected by *Bremia lactucae* show no overt symptoms until sporulation occurs (Dickinson and Crute, 1974). Cytological examination of such tissues reveals extensive development of intercellular hyphae forming haustoria in host cells. In this phase of disease development, the host cells remain viable and there is negligible necrosis of host tissues (Kluczewski and Lucas, 1982). Ultrastructural studies suggest that initially there are few degenerative changes in host cells. The plasmalemma is invaginated by the developing haustorium and remains intact as an extrahaustorial membrane (Hickey and Coffey, 1977). It has been presumed that haustoria function in the uptake of nutrients from host cells (Gay and Manners, 1981). Andrews (1975) found that in *Bremia lactucae* infecting lettuce seedlings the label from [^{3}H] glucose supplied to the host accumulated in the fungus prior to the formation of haustoria in contrast label from [^{3}H] glucose supplied to the host accumulated in the fungus prior to the formation of haustoria in constrast label from [^{3}H] leucine only accumulated once haustoria had developed.

The establishment of a successful biotrophic relationship relies upon the maintenance of viable host cells and minimal damage to host tissues (Lewis, 1973). One might therefore expect downy mildew fungi to produce limited amounts of cell wall degrading and cytolytic enzymes, and few if any phytotoxic metabolites. A study on germinating sporangia of *Bremia lactucae* has shown that the fungus does produce a range of cell wall degrading enzymes, but in low amounts and largely in cell bound forms (Cooper, 1984). Endopolygalacturonidases, the enzyme responsible for extensive wall breakdown and possibly cell death in several necrotrophic diseases. These findings are consistent with a view that downy mildew possess enzymes necessary to facilitate growth through host tissues.

Infection by downy mildews, in common with other plant pathogens leads to increased respiration (Thornton and Cooke, 1974) and increases in the activity of several enzymes including beta-glucosidase, ribonuclease and peroxidase (Ozel *et al.*, 1980). Effects on photosynthesis have scarcely been investigated although chlorophyll levels may be unchanged during the early state of infection (Thornton and Cooke, 1974). Ordinarily, infected susceptible seedlings become dessicated or collapse 10-14 days after inoculation, but in a bacteria free system the majority remain alive and support sporulation by the fungus for up to 21 days. Infection by *P. parasitica* is also known to predispose brassica crops to post harvest spoilage by bacteria (Lund and Wyatt, 1978), and a similar phenomenon may occur with other vegetable crops, for instance, lettuces infected by *Bremia lactucae*. More precise determination based on isotopic methods are required to determine the amounts and types of substances leaked, but it seems likely that loss of control of host cell permeability is a prominent feature of infected tissues, particularly in the degenerative phase of pathogenesis. Ingram (1981) has reviewed the limited information available and concludes that hormonal disturbances in downy mildew diseases represent a virtually unexpected and rewarding area for biochemical study.

7.2. Host Specificity

Plant parasites vary in their degree of host specificity. With obligately biotrophic fungi, co-evolution led to a particularly narrow form of specialization which, at its most extreme, is expressed in a physiological race host cultivar relationship conditioned by matching genes in the host and pathogen populations. Yerkes and Shaw (1959) concluded that biological specialization is not a satisfactory criterion for distinguishing species of obligately parasitic fungi, and suggested that two or more species should only be recognized if they can be separated on distinct morphological grounds. Thus, if all collections of a particular downy mildew genus from different members of a host family from a continuous overlapping series in terms of host range and biometric data, they should be grouped as a single species. Members of the genus *Bremia* parasitic on the compositae are now considered a single aggregate species (Waterhouse, 1973). Generic revision by Shaw (1978) transferred most of the species in *Sclerospora* to the new genus *Peronosclerospora*, but distinctions at the species level remain uncertain (Williams, 1984). Bonde *et al.* (1985) has shown that *Pernosclerospora phillipinensis* and *P. sacchari* are probably cospecific, and that the Thailand form of *P. sorghi* is more closely related to the *P. phillipinesis* - *P. sacchari* complex than to the form of *P. sorghi* found in the USA, Brazil and India. Host susceptibility may depend upon the inoculum used, the age of the host, environmental conditions, and even the host tissue inoculated. McMeekin (1969) compared the host range of two isolates of *P. parasitica* one from radish, *Raphanus sativus* and the other from cabbage, *Brassica oleracea*. She used a 0-5 scale to score sporulation on the cotyledons, hypocotyls and roots of seedling test hosts. Dickinson and Greenhalgh (1977) incorporated host tissue necrosis as an additional criterion for determining compatibility and showed the isolates of *P. parasitica* from *Raphanus* and *Brassica* could infect seedlings of a range of wild crucifer weed species.

Chromosomal cytology was vital in originally establishing diploidy in the oomycetes, but was scarcely been applied to downy mildews. Michelmore and Sansome (1982) observed 7-8 chromosomes in *Bremia lactucae*, while Sansome and Sansome (1974) estimated that *Perenospora parasitica* has 18-20 chromosomes. Advances in the molecular biology of eukaryotes may eventually permit detection of genomic differences between species and even perhaps physiological races. Rasheed *et al.* (1978) detected differences in soluble proteins and peroxidase isozymes between two pathotypes of *Sclerospora graminicola*, while Bonde *et al.* (1984) analyzed 12 enzymes by starch gel electrophoresis to differentiate between species of *Peronosclerospora* found on maize in different parts of the world.

Collections of a particular downy mildew species from different hosts often differ in their pathogenecity towards a range of test hosts. The existence of such variability within downy mildew species can also be inferred from instances where previously effective host resistance has broken down due to the appearance of new pathotypes. An isolate of *Perenospora manshurica* was recorded in the USA on Soyabeans carrying the Rpm resistance gene, thought to be effective against all known isolates (Lim *et al.*, 1984). Host specialization in parasitic fungi is often considered at two levels by first recognising distinct from species adapted usually to different host genera, and then physiological races differentiated on cultivars within a single host species. The genus *Bremia* is currently considered to contain only two species *B. lactucae* on the compositae, and *B. graminicola* on the grass *Arthraxon* (Waterhouse, 1973) Ling and Tai (1945) quoted 10 species of *Bremia* and proposed four from species of *B. lactucae* for the downy mildew of *Lactuca chinensis, Sonchus oleraceus, Taraxacum mangolicum* and *Crepis japorucens* respectively. Savulescu (1962) maintains *B. lactucae* as a wide species but still separate it from forms found on the host genera *Centuarea, Lapsona, Sonchus* and *Senecio* as individual species. Dickinson and Greenhalgh (1977) compared the host ranges of a *Brassica* isolate and a *Raphanus* isolate of *P. parasitica* while each isolate had a distinct host range and grew best on its host of origin, both displayed a wide spectrum of reaction types on the other test crucifiers. Kluczewski and Lucas (1983) suggested the host specialization within the genus *Brassica*. They found that an isolate from cauliflower (*B. oleracea*) was mainly restricted to *B. oleracea* hosts, while an oilseed rape (*B. napus*) isolate was capable of sporulation on a significant proportion of the *B. oleracea* hosts tested. Jafar (1963) found that *P. parasitica* collections from *Matthiola incana* and *M. bicornis* were able to cross infect, but were non pathogenic on *B. oleracea, B. napus* and *Raphanus sativus* amongst other cruciferae.

Pathogenic variation has been widely reported in the downy mildews of tropical cereals (Williams, 1984). *Sclerospora graminicola* causes a serious disease of pearl millet, but is also known to infect several other grasses, including sorghum and occasionally maize. There are two recognized strains of *Perenosclerospora sorghi* (Payak, 1975); the sorghum strain infects sorghum and maize producing oospores in both hosts, while the maize strain is non pathogenic on sorghum and does not produce oospores in maize (Bonde, 1982). The situation is made more complex however by observed differences between isolates depending upon geographical origin. For instance, a maize strain from Rajasthan, Northern India, which also infects the grass *Heteropogon*, has been proposed as a separate species *P. heteropogoni*, primarily on oospore and sporongial morphology (Sirdhana *et al.*, 1980). Variation in sorghum strain has been reported in the USA with two original pathotypes and third in Texas (Bonde, 1982).

Genomic interactions takes place leading to the gene for gene system of race cultivar specificity. Specificity at the race cultivar level is determined by interactions between gene products from the host and pathogen which either elicit or suppress host defense responses. In most cases, the resistant interaction is characterized by the early and rapid death of penetrated host cells. Often the invading pathogen is contained within one or the first few cells penetrated, although this varies depending upon the particular host isolate combination. Wehjte *et al.*, (1979) claim that resistance of sunflower to *Plasmopara halstedii* conditioned by a single gene is expressed as a necrotic reaction in root epidermal cells beneath encysting zoospores, prior to fungal penetration. Differences in the resistance of Pea cultivars to *Perenospora viciae* were not apparent until four days after inoculation and then became evident as a slower pathogen rate leading to reduced sporulation (Dickinson and Singh, 1982). A similar phenomenon has been reported in soyabean inoculated with *P. manshurica* (Riggle, 1977). Skinner and Stuteville (1985) propose a gene for gene relationship for this host pathogen system, with five *P. trifoliorum* resistance (PtR) genes distributed between alfalfa P_5 and P_6 matched by five *P. trifoliorum* avirulence (PtA) genes present in the three pathogen isolates. Difficulties with germinating *P. trifoliorum* oospores have so far percluded studies on the inheritance of virulence in the pathogen. Breeding for resistance to other *Perenospora* species has also suggested the existence of a race structure in the pathogen population, for instance, *P. parasitica* on brassicas (Natti *et al.*, 1967), *P. viciae* on peas (Heydendroff and Hoffman, 1978), *P. farinosa* on spinach (Jones and Dainello, 1982).

7.3. Host Recognition

Pre infection stages includes events leading to penetration of host : (1) triggering of germination of spores, or activation of other types of inocula; (2) orientation of pathogen growth (or swimming of motile spores) to a suitable infection site; (3) adhesion to the host surface; (4) production of an appresorium or equivalent organ, and (5) orientation of the penetration peg so that the pathogen grows into the host (Deacon, 1996). These stages are followed only after reaching an specific host (Read *et al.*, 1992).

8. CONCLUSIONS

Plants have the ability to respond to pathogen invasion by specific defense reactions. Most of the pathogen have a wide host range. They cause extensive damage to the tissue by employing enzymes and secondary toxic metabolites as their major weapons. This paper discusses the host parasite specificity and pathogenesis of various fungi namely, *Verticillium albo atrum, Phytophthora, Pythium*, Rust fungi, Downy mildews and Powdery mildews. Several possible mechanism for localised resistance have been proposed. These include phytoalexin production and alterations to plant cell walls.

REFERENCES

Ammermann, E., Lorenz, G., Schelberger, K., Wenderoth, B., Sauter, H. and Rentzea, C. 1992, BAS 490F - a broad spectrum fungicide with a new mode of action, in : Proceedings of the British Crop Protection Conference - Pests and Diseases 1, pp. 403-410, British Crop protection Council Publications, Farnham.

Andrews, J.H. 1975, Distribution of label from 3H-glucose and 3H-leucine in lettuce cotyledons during the early stages of infection with *Bremia lactucae*, Can. J. Bot. 53: 1103-1115.

Ann, P.J., Kao, C.W. and Ko, W.H. 1986, Mating type distribution of *Phytophthora colocasiae* in Taiwan, Mycopatholigia 93: 193-194.

Arny, D.C. and Grau, C.R. 1985, Importance of *Verticillium* wilt of alfalfa in North America, Canadian Journal of Plant Pathology 7: 187-190.

Bailey, J.A., and Ingham, J.L. 1971, Phaseollin accumulation in bean *Phaseolus vulgaris* in response to infection by tobacco necrosis virus and the rust *Uromyces appendiculatus*, Physiol. Plant Pathol. 1: 451-456.

Barasubiye, T., Parent, J.G., Hamelin, R.C,, Laberage, S., Richard, C. and Distaler, D. 1995, Discrimination between alfalfa and potato isolates of *Verticillium albo atrum* using RAPD markers.

Becker, W.F., Von Jagow, G., Anke, T. and Steglich, W. 1981, Ondemansin, strobilurin A strobilurin B and myxothiazol: New inhibitors of the bc_1 segment of the respiratory chain with an E-B-methoxyacrylate system as common structural element. FEBS Letters, 132: 329-333.

Bell, A.A., and R.D. Stipanovic. 1978, Biochemistry of disease and pest resistance in cotton, Mycopathologia, 65: 91-106.

Berry, L.A., Jones, E.E. and Deacon, J.W. 1993, Interaction of the mycoparasite *Pythium oligandrum* with other *Pythium* species, Biological Science and Technology, 3: 247-260.

Bhandari, N.N. and Mukerji, K.G. 1993, The Haustorium, Research Studies Press Ltd., England, pp. 308.

Bielenin, A., Jeffers, S.N., Wilcox, W.F. and Jones, A.L. 1988, Separation by protein electrophoresis of six species of *Phytophthora* associated with deciduous fruit crops, Phytopathology 78: 1402-1408.

Bircher, U. and Hohl, H.R. 1997. Surface glycoproteins associated with appressorium formation and adhesion in *Phytophthora palmivora*, Mycol. Res, 101(7): 769-775.

Bishop, C. D. and Cooper, R.M. 1983, An ultrastructural study of vascular colonisation by fungal with pathogens II, Invasion of resistant cultivars, Physiol. Plant Pathol, 24: 277.

Bonde, M.R. 1982, Epidemiology of downy mildew disease of maize, sorghum and peart millet, Trop. Pest Manag, 28: 49-60.

Bonde, M.R., Peterson, G.L. and Duck, N.B. 1985, Effects of temperature on sporulation, conidial germination and infection of maize by *Peronosclerospora sorghi* from different geographical areas, Phytopathology, 75: 122-126.

Bonde, M.R., Peterson, G.L., Dowler, W.M. and May, B. 1984b, Isozyme analysis to differentiate species of *Peronosclerospora* causing downy mildews of maize, Phytopathology, 74: 1278-1283.

Bostock, R.M., Laive, R.A. and Kuc, J.A. 1982, Factors affecting the elicitation of seqquiterpenoid phytoalexin accumulation by eicosapentaenoic and arachidonic acids in potato, Plant Physiol, 70: 1417-1424.
Brasier, C.M. 1969, The effect of light and temperature on reproduction in vitro in two tropical species of *Phytophthora*, Trans. Brit. Mycol. Soc., 52: 105-113.
Braun, P.G. 1991, The combination of *Cylindrocarpon lucidum* and *Pythium irregulare* as a possible cause of apple replant disease in Nova Scotia, Canadian Journal of Plant Pathology, 13: 291-297.
Braun, U. 1987, A monograph of the Erysiphales Powdery mildews Beih Nova Hedwigia, 89: 1-700.
Bulit, J. and Lafon, R. 1978, Powdery mildew of the vine, in: the Powdery mildews, D. Spencer ed., pp. 525-548. Academic Press: London: New York: San Francisco.
Burden, R.S., Cooke, D.T. and Carter, G.A. 1989, Inhibitors of sterol biosynthesis and growth in plants and fungi, Phytochemistry, 28: 1791-1804.
Buzzell, R.I. and Anderson, T.R. 1982, Plant loss response of soyabean cultivars to *Phytophthora megasperma* f. sp. glycinea under field conditions, Plant Disease, 66: 1146-1148.
Cahill, D., Legge, N., Grant, B. and Weste, G. 1989, Cellular and histological changes induced by *Phytophthora cinnamomi* in a group of plant species ranging from fully susceptible to fully resistant, Phytopathology, 79: 417-424.
Cahill, D.M., Bennett, I.J. and McComb, J.A. 1993, Mechanisms of resistance to *Phytophthora cinnamomi* in clonal, micropropagated *Eucalyptus marginata*, Plant Pathology, 42: 865-872.
Carder, J.H. and Barbara, J.D. 1991, Molecular variation and restriction fragment length polymorphisms RELPs within and between six species of *Verticillium*, Mycological Research, 95: 935-942.
Chakravorty, A.K. and Scott, K.J. 1982, Biochemistry of host rust interactions, in the Rust Fungi, K.J. Scott and A.K. Chakraborty eds., Academic Press, London, pp. 180-222.
Chakravorty, A.K. and Shaw, M. 1971, Changes in the transcription pattern of flax cotyledons after inoculation with flax rust, Biochem. J., 123: 551-557.
Chern, L.L. and Ko, W.H. 1994, Effects of high temperature on physiological processes leading to sexual reproduction in *Phytophthora* parasitica, Mycologia, 863: 416-420.
Christensen, A.A., Peaden, R.N., Harris, G.P. and Heale, J.B. 1983, Virulence of North American and European isolates of *V. albo atrum* on alfalfa cultivars, Phytopathology, 73: 1051-1054.
Clough, J.M. Anthony, V.M., De Fraine, P.J., Fraser, T.E.M., Godgrey, C.R.A., Godwin, J.R. and Youle, D. 1995, The synthesis of fungicidal B-methoxyacrylates, in: Proceedings of the Eighth International Congress of Pesticide Chemistry, N.N. Ragsdale, P.c. Kearney and J.R. Plimmer eds., pp. 59-73. ACS: Washington, DC.
Coffey, M.D. and Wilson, U.E. 1983, An ultrastructural study of the late blight fungus *Phytophthora infestans* and its interaction with the foliage of two potato cultivar possessing different levels of general resistance, Can. J. Bot., 61: 2669-2685.
Cohen, Y. and Coffey, M.D. 1986, Systemic fungicides and the control of oomycetes, Ann. Rev. Phytopathol, 24: 311-338.
Cole, A.L.J. 1970, Pectic enzyme activity from *Phytophthora* infestans, Phytochemistry, 9: 337-340.
Cooke, R.C. and Whipps, J.M. 1980, The evolution of modes of nutrition in fungi parasitic on terrestrial plants, Biol. Rev., 53: 341-362.
Cooper, R.M. 1984, The role of cell wall degrading enzymes in infection and damage, in: Plant Diseases: Infection, Damage and Loss, R.K.S. Wood and G.J. Jellis, eds., Blackwell Scientific Publications, Oxford.
Corell, J.C., Gordon, T.R. and McCain, A.H. 1988, Vegetative compatibility and pathogenecity of *Verticillium albo atrum*, Phytopathology, 78: 1017-1021.
Deacon, J.W. 1996, Ecological implications of recognition events in the pre-infection stages of root pathogens, New Phytol, 133: 135-145.
Deacon, J.W. and Henry, C.M. 1978, Mycoparasitism by *Pythium oligandrum* and *P. acanthicum*, Soil Biology and Biochemistry, 10: 409-415.
Descalzo, R.C., Punja, Z.K., Levesque, A.c. and Rahe, R.E. 1996, Identification and role of *Pythium* species as glyphosphate synergists on beam *Phaseolus vulgaris* grown in different soils, Mycological Research, 100: 971-978.
Deverall, B.J., Keogu, R.C. and Maleod, S. 1977, Responses of soyabean to infection by and to germination fluids from urediniospores of *Phakopsora pachyschizi*, Trans. Br. Mycol. Soc., 69: 411-415.
Dimond, A.E. 1967, Physiology of wilt disease, in: Molecular Constituents in Plant Parasite Interactions, C.J. Mirocha and I. Vritani eds., pp. 100, American Phytopathoogical Society, St. Paul, Minn.

Dickinson, C.H. and Crute, I.R. 1974, The Influence of seedlings age and development on the infection of lettuce by *Bremia lactucae*, Ann. Appl. Biol., 76: 49-61.
Dickinson, C.H. and Greenhalgh, J.R. 1977, Host range and taxonomy of *Peronospora* on crucifiers, Trans. Br. Mycol. Soc., 69: 111-116.
Dickinson, C.H. and Singh, H. 1982, Colonisation and sporulation of *Peronospora viciae* on cultivars of *Pisum sativum*, Plant Pathol., 31: 333-341.
Ding, H., Balsiger, S., Guggenbuhl, C. and Hohl, H.R. 1994, A putative IgG binding 65 KDa adhesin involved in adhesion and infection of soybeans by *Phytophthora megasperma* f. sp. *glycinea*, Physiological and Molecular Plant Patholgogy, 44: 363-378.
Doke, N. and Tokiyama, K. 1980, Suppression of the hypersensitive response of potato tuber protoplasts to hyphal wall components by water soluble glucans isolated from *Phytophthora infestans*, Physiol. Plant Pathol., 16: 177-186.
Doke, N., Garas, N.A. and Kuc, J. 1979, Partial characterization and aspects of mode of action of a hypersensitivity inhibiting factor HIF isolated from *Phytophthora* infestans, Physiol. Plant Pathol., 15: 127-140.
Donaldson, S.P. and Deacon, J.W. 1992, Role of calcium in adhesion and germination of zoospore cyst of *Pythium*, A model to explain infection of host plants, Journal of General Microbiology, 138: 2051-2059.
Doster, M.A. and Schnathorst, W.C. 1985, Effects of leaf maturity and cultivar resistance on development of the powdery mildew fungus on grapevines, Photopathology, 75: 318-321.
Doubly, J.A., Flor, H.H. and Clagett, C.D. 1960, Relation of antigens of *Melampsora lini* and *Lineun usitalissimum* to resistance and susceptibility, Science, 131: 229.
Drechsler, C. 1930, Some new species of *Pythium*, Journal of the Washington Academy of Sciences, 20: 398-418.
Drechsler, C. 1939, Several species of *Pythium* causing blossom and rot of watermelons, Phytopathology, 29: 391-422.
Duniway, J.M. 1971, Resistance to water movement in tomato plants infected with *Fusarium*, Nature, 230-252.
Duniway, J.M. 1977, Predisposing effect of water stress on the severity of *Phytophthora* root rot in Safflower, Phytopathology, 67: 884-889.
Ebel, J. and Cosio, E.G. 1994, Elicitors of plant defense responses, Int. Rev. Cytol., 148: 1-36.
Elaaghy, M.A. and Heitefuss, R. 1976, Permeability changes and production of anti-fungal compounds in *Phaseolus vulgaris* infected with *Uromyces phasealli* II, Role of Phytoalexins, Physiol. Plant Pathol., 8: 269-277.
Elbein, A.d. 1987, Inhibitors of the biosynthesis and processing of N-linked oligosaccharide chains, Annual Review of Biochemistry, 56: 497-534.
Elliott, C.G. 1983, Physiology of sexual reproduction in *Phytophthora*, in: Phytophthora, D.C. Erwin, S. Bartnicki-Garcia and P.H. Tsao eds., APS Press, St. Paul, Minnesota, pp. 71-80.
Faley, M.f. and Deacon, J.W. 1986, Susceptibility of *Pythium* sp. and other fungi to antagonism by the mycoparasite *Pythium oligandrum*, Soil Biology and Biochemistry, 18: 91-95.
Gaumann, Z., 1958, The mechanism of fusaric acid injury, Phytopathology, 48: 670.
Gay, J.L. and Manners, J.M. 1981, Transport of host assimilates to the Pathogen, in: Effects of Disease on the Physiology of the Growing Plant, P.G. Ayres eds., Cambridge University Press, Cambridge, pp. 85-100.
Gill, H.S. and Powell, D. 1968, The use of polyacrylamide gel disc electrophoresis in delimiting three species of Phytophthora, Phytopathologische Zeitschrift, 63: 23-29.
Gold, R.E., Ammermann, E., Kohle, H., Leinhos, G.E.M., Lorenz, G., Speakman, J.B., Starkurnau, M. and Sauter, H. 1996, The synthetic strobilurin, BAS 490F: profile of a modern fungicide, in: Modern fungicides and antifungal compounds H. Lyr, P.E. Russell and H.D. Sisler eds., Intercept. Andover, U.K., pp. 79-92.
Grambow, H.J. and Grambow, G.E. 1978, The involvement of epicuticular and cell wall phenols of the host plant in the in vitro development of *Puccinia graminis* f. sp. *tritici*, Z. Pflanzenphysiol, 90: 1-9.
Grant, B.R., Danstan, R.H., Griffith, J.M., Niere, J.O. and Smillie, R.H. 1990, The mechanism of phosphonic acid action in *Phytophthora*, Australian Plant Pathology, 19: 115-121.
Gubler, F. and Hardham, A.R. 1988, Secretion of adhesive material during encystment of *Phytophthora cinnamomi* zoospores, Characterized by immunogold labelling and with monoclonal antibodies to components of peripheral vesicles, Journal of Cell Science, 90: 225-235.
Guest, D.I. and Grant, B.R. 1991, The complex action of phosphonates as antifungal agents, Biological Reviews of the Cambridge Philosophical Society, 66: 159-187.
Hallenstein, L. and Hohl, H.R. 1995, The influence of IgG mannose and proteases and glycosidas on

adhesion to and infection of potato leaves by *Phytophthora infestans*, in: *Phytophthora* infestans 150 L.J. Dowley, E. Bannon, L.R. Cooke, T. Keane and E. O'Sullivan eds., Boale Press Ltd: Dablin Ireland, pp. 64-70.
Hanchey, P. and Wheeler, H. 1971, Pathological changes in ultrastructure: tobacco root infected with *Phytophthora parasitica* var. nicotianae, Phytopathology, 61: 33-39.
Hardhman, A.R. and Gubler, E. 1990, Polarity of attachment of zoospores of a root pathogen and pre-alignment of the emerging germ tubes, Cell Biology, International Reports, 14: 947-956.
Heath, M.C. 1977, A comparative study of non-host interactions with rust fungi, Physiol. Plant Pathol., 10: 73-88.
Heath, M.C. 1979a, Partial characterization of the electron opaque deposits formed in the non-host plant, french bean, after cowpea rust infection, Physiol. Plant Pathol., 15: 141-148.
Heath, M.C. 1979b, Effects of heat shock, actinomycin D. cycloheximide and blasticidins on non host interactions with rust fungi, Physiol. Plant Pathol., 15: 211-218.
Heath, M.C. 1980, Effect of infection by compatible species or injection of tissue extracts on the susceptibility of non host plants to rust fungi, Phytopathology, 70: 356-360.
Heath, M.C. 1981, A generalized concept of host-parasite specificity, Phytopathology, 71: 1121-1123.
Heath, M.C. 1981a, The suppression of the development of silicon containing deposits in french bean leaves by exudates of the bean rust fungus and extracts from bean rust infected tissue, Physiol. Plant. Pathol., 18: 149-155.
Heath, M.C. 1981b, Insoluble silicon in necrotic cowpea cells following infection with an incompatible isolate of the cowpea rust fungus, Physiol. Plant Pathol., 19: 273-276.
Heath, M.C. 1982, The absence of active defense mechanism in compatible host pathogen intractions, in: Active Defence Mechanisms in Plants, R.K.S. Wood ed., Plenum Press, London, pp. 143-156.
Heller, A., Grossman, F., Frenzel, B. and Hippe, S. 1990, A cytological study of the development of *Erysiphe graminis* in its host barley, as influenced by the two fungicides ethirimol and propiconazole, Canadian Journal of Botany, 68: 2618-2628.
Hendrix, F.F. Jr. and Campbell, W.A. 1973, *Pythium* as plant pathogens, Annual Review of Phytopathology, 11: 77-98.
Heydendorff R.C. Von and Hoffmann, G.M. 1978, Physiological specialization of *Peronospora* pisi, Syd. Z. Pflanzenkr Pflanzenschutz, 85: 561-569.
Hickey, E.L. and Coffey, M.D. 1977, A fine structural study of the pea downy mildew fungus *Peronospora pisi* in its host *Pisum sativum*, Can. J. Bot., 55: 2845-2858.
Hillebrand, W., Lovenz, D. and Louis, F. 1995, Rebschutz. Fachverlag Fraund:Mainz. Germany.
Hockenhull, J., Jensen, D.f. and Yudiarti, T. 1992, The use of *P. periplocum* to control damping off of cucumber seedlings caused by *P. aphanidermatum*, in: Biological Control of Plant Diseases, E.S. Tjaimos, G.C. Papaviza and R.J. Cook ed., New York: Plenum Press, pp. 203-206.
Hohl, H.R. 1991, Surface related host pathogen interactions in *Phytophthora*, in: *Phytophthora* J.A. Lucas, R.C. Shattock, D.S. Shaw and L.R. Cooke, eds., Cambridge University Press, Cambridge, U.K, pp. 70-89.
Hohl, H.R. and Balsiger, S. 1986, Probing surfaces of soybean protoplasts and germ tubes of the soybean pathogen *Phytophthora megasperma* f. sp. *glycinae* with lectins, Botanica Hebetica, 2: 289-297.
Honour, R.C. and Tsao, P.H. 1974, Production of oospores of *Phytophthora parasitica* in liquid medium. Mycologia, 66: 1031-1038.
Hoppe, H.H., Humne, B. and Heitefuss, R. 1980, Elicitor induced accumulation of phytoalexiris in healthy and rust infected leaves of *Phaseolus vulgaris*, Phytopathol. Z., 97: 85-88.
Hunter, W.E. and Butler, E.E. 1975, *Syncephalis californica*, a mycoparasite inducing giant hyphal swellings in species of Mucorales, Mycologia, 67: 863-872.
Ingram, D.S. 1981, Physiology and biochemistry of host parasite interaction, in: the Downy mildews, D.M. Spencer ed., Academic Press, London, pp. 143-163.
Issac, I. and Lloyd, A.T.E. 1959, Wilt of lucerine caused by species of *Verticillium* II, Seasonal cycle disease: range of pathogenecity: host parasite relations: effects of seed dressing, Annals of Applied Biology, 47: 673-684.
Jafar, H. 1963, Studies on downy mildew *Perenospora mattiolae* Roumeguerre Gaumann on stock *Matthiola incana* R. Br. N. Z., J. Agric. Res., 6: 70-82.
Johal, G.S. and Rahe, J.E. 1984, Effect of soilborne plant pathogenic fungi on the herbicidal action of glyphosphate on bean seedlings, Phytopathology, 74: 950-955.
Jones, R.K. and Dianello, F.G. 1982, Occurrence of race 3 of *Perenospora effusa* on spinach in *Texas* and identification of sources of resistance, Plant Dis, 66: 1078-1079.
Keen, N.T. 1975, Specific elicitors of plant phytoalexin production: determinants of race specificity in pathogens, Science, 187: 74-75.

Keen, N.T. 1981, Specific recognition in gene for gene host parasite system, Adv. Plant Pathol, 1: 35-82.

Keen, N.T. and Bruegger, B. 1977, Phytoalexins and chemicals that elicit their production in plants, in: Host Plant Resistance to Pests, P. Hedin ed., Am. Chem. Soc. Symp. Ser., 62, pp. 1-26.

Keen, N.T. and Legrand, M. 1980, Surface glycoproteins: evidence that they may function as the race specific phytoalexin elicitors of *Phytophthora megasperma* f. sp. *glycinea*, Physiol. Plant Pathol., 17: 175-192.

Keen, N.T. and Littlefield, L.J. 1979, The possible association of phytoalexins with resistance gene expression in flax to *Melampsora lini*, Physiol. Plant Pathol., 14: 265-280.

Kennedy, D.M. and Duncan, J.M. 1995, A papillate *Phytophthora* species with specificity to Rubs, Mycol. Res., 99(1): 57-68.

Kim, W.K., Rohringer, R. and Samborski, D.J. 1977, Effect of blasticidin S, ethionine and polyoxin D on stem rust development and host cell necrosis in wheat near isogenic for gene Sr 6, Can. J. Bot., 55: 568-573.

Kluczewski, S.M. and Lucas, J.A. 1982, Development and physiology of infection by the downy mildew fungus *Perenospora parasitica* (Pers. ex. Fr.) Fr. in susceptible and resistant Brassica species, Plant Pathol., 31: 373-389.

Kluczewski, S.M. and Lucas, J.A. 1983, Host infection and oospore formation by *Perenospora parasitica* in agricultural and horticultural Brassica species, Trans. Br. Mycol. Soc., 81: 591-596.

Ko, W.H. 1978, Heterothallic *Phytophthora*: evidence for hormonal regulation of sexual reproduction, J. Gen. Microbiol., 107: 15-18.

Ko, W.H. 1979, Mating type distribution of *Phytophthora colocasiae* on the island of Hawaii, Mycologia, 71: 434-437.

Ko, W.H. 1980, Hormonal regulation of sexual reproduction in Phytophthora, J. Gen. Microbiol., 116: 459-463.

Ko, W.H. 1980, Sexuality, evolution and origin of *Phytophthora*, Pl. Protect. Bull. Taipei, 22: 141-151.

Ko, W.H. 1988, Hormonal heterothallism and homothallism in *Phytophthora*, Annual. Rev. Phytopathol, 26: 57-73.

Kombrink, E. and Somssich, I.E. 1995, Defense responses of plant to pathogens, Adv. Bot. Res., 21: 1-34.

Krietlow, K.W. 1962, *Verticillium* wilt of alfalfa, A destructive disease in Britain and Europe yet observed in the United States, USDA-ARS Technical Bulletin, 34-20, pp. 15.

Kuan, T.L. and Erwin, D.C. 1980, Predisposition effect of water saturation of soil on *Phytophthora* root rot of alfalfa, Phytopathology, 70: 981-986.

Kunoh, H., Aist, J.R. and Israel, H.W. 1979, Primary germ tubes and host cell penetrations from appresoria of *Erysiphe graminis hordei*, Ann. Phytopath. Soc., Japan, 45: 326-332.

Lafon, R. and Bulit, J. 1981, Downy mildew of the Vine, in: the Downy Mildews, D.M. Spenser ed., Academic Press, London, pp. 606-614.

Laing, S.A.K. and Deacon, J.W. 1991, Video microscopical comparison of mycoparasitism by *Pythium oligandrum, P. nunn* and anunnamed *Pythium* species, Myco. Res., 95: 469-479.

Leinhos, G.E., Gold, R.E., Duggelin, M. and Guiggenheim, Richard 1997, Development and morphology of *Uncinula* necator following treatment with the fungicides Kresoxim - methyl and pencanazole, Mycological. Research, 1019: 1033-1046.

Levesque, C.A., Beckenbach, K., Baillie, D.A. and Rahe, J.E. 1993, Pathogenecity of DNA restriction fragment length polymorphism of isolates of *Pythium* sp. from glyphosphate treated seedlings, Mycological Research, 97: 307-312.

Levesque, C.a., Rahe, J.E. and Eaves, D.M. 1992, The effect of soil heat treatment and microflora on the efficacy of glyphosphate in seedlings, Weed Research, 32: 363-373.

Lewis, D.H. 1973, Concepts in fungal nutrition and the origin of biotrophy, Biol. Rev., 48: 261-278.

Lim, S.M., Bernard, R.L., Rickell, C.D. and Gray, L.E. 1984, New physiological race of *Peronospora manshurica* virulent to the gene Rpm in Soybeans, Plant Dis., 68: 71-72.

Ling, L. and Tai, M.C. 1945, On the specialization of *Bremia lactucae* on composital, Trans. Br. Mycol. Soc., 28: 16-25.

Lucas, J.A., Dolan, T.e. and Coffey, M.D. 1985, Non-transmissibility to regenerants from protected tobacco explants of induced resistance to *Perenospora hyoscyani*, Phytopathology, 75: 1222-1225.

Ludwig, R.A. 1952, Studies on the physiology of hydromycotic witting in the tomato plant, Mac Donald College Technical Bulletin, 20.

Lund, B.M. and Wyatt, G.M. 1978, Post harvest spoilage of cauliflower by downy mildew, *Peronospora parasitica*, Plant Pathol., 27: 143-144.

Madsen, A.M., Robinson, H.L. and Deacon, J.W. 1995, Behaviour of zoospore cysts of the mycoparasite *Pythium oligandrum* in relation to their potential for biocontrol of plant-pathogens, Mycol. Res., 9912: 1417-1424.

Martins, E.M.F., Guzzo, S.D., Alba, A.P.C., Roveratti, D.S., Miglioli, D.M.M. and Moraes, W.B.C. 1983, Differences between the biochemical behaviour of filtrates of water washes of autoclaved and non autoclaved urediniospores of Hemileia vastatrix, Abstr. 4th International Congress of Plant Pathology, p. 187.

Mayama, S. 1983, The role of avenalumin in the resistance of oats to crown rust, Mem. Fac. Agr. Kagawa Univ. No. 42, 1-64.

Mayama, S.J., Matsuura, H., II Da and Tani, T. 1982, The role of avenalumin in the resistance to oat to crown rust, *Puccinia coronata*, f. sp. *avenae*, Physiol. Plant Pathol., 20: 189-199.

Mc Hardy, W.E., Busch, L.V. and Hall, R. 1976, *Verticillium* wilt of *Chrysanthemum*: Quantiative relationship between increased stomatal resistance and local vascular dysfunction preceding wilt, Can. J. Bot., 54: 1023.

McCracken, A.R. and Dawson, M. 1992, Clonal response in *Salix* to *Melampsora* rusts in short rotation coppice plantation, European Journal of Forest Pathology, 22: 19-28.

McCredic, T.A., Dixon, K.W. and Sivasithamparam, K. 1985, Variability in the resistance of Banksia L.f. species to *Phytophthora cinnamomi* Rands, Australian Journal of Botany, 33: 629-637.

McIntosh, R.A. and Watson, I.A. 1982, Genetics of host pathogen interactions in rusts, in: Rust Fungi, K.J. Scott and A.K. Chakravorty ed., Academic Press London, 122-149.

McMeekin, D. 1969, Other hosts for *Perenospora parasitica* from cabbage and radish, Phytopathology, 59: 693-696.

Mendgen, K. 1975, Ultrastructural demonstration of different peroxidase activities during the bean rust infection process, Physiol. Plant. Pathol., 6: 275-282.

Michelnore, R.W. and Sausome, E.R. 1982, Cytological studies of heterothallism and secondary homothallism in *Bremia lactucae*, Trans. Br. Mycol. Soc., 79: 291-297.

Miller, S.A. and Maxwell, D.P. 1984, Light microscope observations of susceptible, host-resistant, and non host-resistant interactions of alfalfa with *Phytophthora megasperma*, Can. J. Bot., 62: 109-116.

Mircetich, S.M. 1971, The role of *Pythium* in feeder roots of diseased trees and orchard soils in peach tree decline, Phytopathology, 61: 357-360.

Montarbini, P. and Elstner, E.F. 1977, Ethylene evolution by rust infected detached bean leaves susceptible and hypersensitive to *Uromyces phaseoli* (Pers.), Wint. Planta., 135: 301-306.

Mukerji, K.G., Upadhyay, R.K. and Rawat, R. 1998, Biology of Cereal Powdery mildews and management options, in: IPM System in Agriculture, Vol. 3. Careals (R.K. Upadhyay, K.G. Mukerji, and R.L. Rajak eds.), Aditya Book Pvt. Ltd., New Delhi, India, pp. 61-74.

Mukerji, K.G. and Upadhyay, R.K. 1998, Powdery mildews and their control, in: IPM System in Agriculture, Vol. VII, Key Pests and Diseases (R.K. Upadhyay, K.G. Mukerji and O.P. Dubey eds.), Aditya Books Pvt. Ltd., New Delhi, India.

Murch, R.S. and Paxton, J.D. 1979, Rhizosphere anaerobiosis and glyceollin accumulation in Soyabean, Phytopathol. Z., 96: 91-94.

Mussel, H.W. and Green, R.J. 1970, Host colonisation and polygalacturonase production by two tracheomyotic fungi, Phytopathology, 60: 192.

Naito, N., Tani, T. and Araki, T. 1970, Relationship between parasite development and infection type in Oat, wheat and barley inoculated with *Puccinia coronata*, Trans. Mycol. Soc. Jap., 11: 16-22.

Nandris, D., Vallavielle, C. De and Bouvier, J. 1979, Studies on some interactions between potatoes and *Phytophthora* infestans, Physiol. Plant. Pathol., 15: 1-12.

Nelson, E.B. 1992, Bacterial metabolism of propagule germination stimulants as an important trait in the biocontrol of Pythium seed infections, in: Biological control of Plant Diseases (E.S. Tjamos, G.C. Papacizas and R.J. Cook ed.), pp. 353-357. Plenum Press: New York.

Nicholson, R. 1996, Adhesion of fungal propagules. In Histology, Ultrastructure and Molecular Cytology of Plant Microorganisms Interactions, (M. Nicale and V. Gianinazzi, Pearson ed.), Kluwer Academic Publishers, Dordrecht, The Netherlands, pp. 117-134.

Ozel, A.F., Krober, H. and Petzold, H. 1980, Peroxidase activity in the compatible and in the incompatible system with downy mildew of spinach, Phytopathol, 97: 156-168.

Payak, M.M. 1975, Epidemiology of maize downy mildews with special reference to those occurring in Asia, Trop. Agric. Res., 8: 81-91.

Platt, H.W. 1986, Varietal response and crop loss due to *Verticillium* wilt of potato caused by *Verticillum albo atrum*, Phytoprotection, 67: 123-128.

Pratt, R.G. 1978, Germination of oospores of *Sclerospora sorghi* in the presence of growing roots of host and non host plants, Phytopathology, 68: 1606-1613.

Pristou, R. and Gallegly, M.E. 1954, Leaf penetration by *Phytophthora* infestans, Phytopathology, 44: 81-86.

Rahe, J.E., Levesque, C.a. and Johal, G.S. 1990, Synergistic role of soil fungi in the herbicidal efficacy of glyphosphate, in: Microbes and Microbial products as Herbicides (R.E. Hoagland ed.), pp. 260-275. ACS Symposium Series, 439.

Rasheed, A., Shetty, H.S. and Safeculla, K.M. 1978, Existence of pathogenic race in *Scleroscopora graminicola* Schroet attaching pearl millet, Abstracts of the Third International Congress of Plant Pathology, Munich, 1978.

Read, N.D., Kellock, L.J., Knight, H. and Trewavas, A.J. 1992, Contact sensing during infection by fungal pathogens, in: Perspectives in Plant Cell Recognition (Callow, J.A. and Green, J.R. eds.), Cambridge University Press, Cambridge, 137-172.

Riggle, J.H. 1977, *Perenospora manshurica* on a non host and on resistant, susceptible, and intermediate soybeans, Can. J. Bot., 55: 153-157.

Robb, J., Busch, Z.V., Brisson, J.D. and Lu, B.C. 1975, Ultrastructure of wilt syndrome caused by *Verticillium dahliae* II, In Sunflower leaves, Can. J. Bot., 53: 2725.

Rohringer, R., Kim, W.K. and Samborski, D.J. 1979, A histological study of interactions between avirulent races of stem rust and wheat containing resistance genes Sr^5, Sr^6, Sr^8, or Sr^{22}, Can. J. Bot., 57: 324-331.

Sansome, E. and Sansome, F.W. 1974, Cytology and life history of *Peronospora parasitica* on *Capsella brusa pastoris* and of *Albugo candida* on *C. brusa pastoris* and on *Lunaria annua*, Trans. Br. Mycol. Soc., 62: 323-332.

Satchuthananthavale, V. 1963, Complimentary strains of *Phytophthora palmivora* from Ceylon rubber, Phytopathology, 53: 729.

Savulescu, D. 1962, A systematic study of the general *Bremia* and *Bremiella* Wilson, Revue Biol., 7: 43-62.

Schmitt, J.A. 1955, The host specialization of Erysiphe *Cichoracearum* from Zinia, Phlox and cucurbits, Mycologia, 47: 608-701.

Selman, I.W. and Pegg, G.F. 1957, An analysis of the growth response of young tomato plants to infection by *Verticillium albo atrum*, Appl. Biol., 45: 674.

Sharp, J.K., McNeil, M. and Albersheim, P. 1984, The Primary structure of one elicitor active and seven elicitor inactive hexa B-D-glucopyranosyl-D-glucitols isolated from the mycelial walls of *Phytophthora megasperma* f. sp. glycinea, J. Biol. Chem., 259: 1321-1336.

Shaw, C.G. 1978, *Perenosclerospora* species an other downy mildews of Gramineae, Mycologia, 10: 594-604.

Shigo, A.L. and Tippett, J.T. 1981, Compartmentalization of American elm tissues infected by *Ceratocystis ulmi*, Plant Disease, 65: 715-718.

Singh, R.S. and Singh, U.S. 1984, *Pythium* in host and soil, in: Progess in Microbial Ecology (K.G. Mukerji, V.P. Agnihotri and R.P. Singh ed.), pp. 229-271. Printhouse, India.

Siradhana, B.S., Dange, S.R.S., Rathore, R.S. and Singh, S.D. 1980, A new downy mildew on maize in Rajasthan, India, Current Science, 49: 316-317.

Smith, B.J., Shearer, B.L. and Sivasithamparam, K. 1997, Compartmentalization of Phytophthora *cinnamomi* in stems of highly susceptible Banksia brownii treated with phosphate, Mycological Research, 101(9): 1101-1107.

Smith, E.F. 1899, Wilt disease of cotton, water melon and cowpea, U.S. Dept. of Agriculture Bulletin 17.

Smolka, S. and Wolf, G. 1986, Cytological studies on the mode of action of systemic fungicides on the host pathogen complex barley - powdery mildew (*Erysiphe graminis* f. sp. hordei Marchal), Pesticide Science, 17: 249-255.

Spencer, D.M. 1981, The Downy mildews, Academic Press: London.

Skinner, D.Z. and Stuteville, D.L. 1985, Genetic of host parasite interaction between alfalfa and *Perenospora trifoliorum*, Phytopathology, 75: 717-721.

Spires, A.G. and Hopcroft, D.H. 1996, Morphological and host range studies on Melampsora rust's attacking *Salix* species in New Zealand, Mycological Research, 100(10): 1163-1175.

Spring, O., Albert, K. and Hager, A. 1982, Three biologically active Heliangolides from *Helianthus annuus* Phytochemistry, 21: 2551-2553.

Stammler, G. Seemuller, E. and Duncan, J.M. 1993, Analysis of RFLPs in nuclear and mitochondrial DNA and the taxonomy of Phytophthora fragariae, Mycol. Res., 97: 150-156.

Stossel, P., Lazarovitis, G. and Ward, E.W.B. 1981, Electron microscope study of race specific and age related resistant and susceptible reactions of soyabeans to *Phytophthora megasperma* var. sojae,

Phytopathology, 71: 617-623.
Stossel, P., Lazarovitis, G. and Ward, E.W.B. 1980, Penetration and growth of compatible and incompatible race of *Phytophthora megasperma* var. *sojae* in hypocotyl tissue differing in age, Can. J. Bot., 58: 2594-2601.
Suzuki, K. and Shinshi, H. 1996, Protein Kinases in Elicitor Signal Transduction in Plant cells, J. Plant Res., 109: 253-263.
Talboys, P.W. 1975, Simulation of hop *Verticillium* with syndrom by applications of ethrel, Ann. Appl. Biol., 81: 126.
Talboys, P.W. 1978, Dysfunction of the water system, in: Plant Disease, J.G. Horfall and E.B. cowling eds., Vol. 3, Academic Perss, New York, pp. 141.
Tani, T. and Yamamoto, M. 1978, Nucleic acid and protein synthesis in association with the crown rust resistance of oat leaves, Physiol. Plant Pathol., 12: 215-223.
Tani, T., Yoshikawa, M. and Naito, N. 1971, Changes in ^{32}P ribonucleic acids in oat leaves associated with susceptible and resistance reactions to *Puccinia coronata*, Ann. Phytopathol. Soc. Jap., 37: 43-51.
Tani, T. and Yamamoto, H. 1979, RNA and protein synthesis and enzyme changes during infection, in: Specificity in Plant Host parasite interactions, J.M. Daly and I. Uritani eds., Japan Sci. Soc. Press, Tokyo, Univ. Park Press, Baltimore, pp. 273-287.
Thornton, J.D. and Cooke, R.C. 1974, Changes in respiration, chlorophyll content and soluble carbohydrates of detached cabbage cotyledons following infection with *Perenospora parasitica* (Pers. ex. Fr.), Fr., Physiol. Plant Pathol., 4: 117-125.
Tippett, J.T., Holland, A.A., Marks, G.C. and Obrien, T.P. 1976, Penetration of *Phytophthora cinnamomi* into disease tolerant and susceptible eucalyptus, Arch. Microbiol, 108: 231-242.
Utkhede, R.S. and Smith, E.M. 1991, *Phytophthora* and *Pythium* species associated with root rot of young apple trees and their control, Soil Biology and Biochemistry, 23: 1059-1063.
Vasely, D. 1979, Use of *Pythium oligandrum* to protect emerging sugar beet, in: Soil borne Plant Pathogens, B. Schippens and W. Gams ed., Academic Press, London, pp. 593-595.
Waterhouse, G.M. 1973, Perenosporales, in: The Fungi, Vol. IVb, G.C. Ainsworth, F.K. Sparrow and A. Sussman eds., Academic Press, London, pp. 165-185.
Weste, G. 1983, Population dynamics and survival of *Phytoph* thora, in: *Phytophthora*, Its Biology, Taxonomy, Ecology and Pathology, D.C. Erwin, S. Bartnicki Garcia and P.H. Tsqo eds., American Phytopathological Society St. Paul MI, pp. 237-257.
Wehtje, G., Littlefield, L.J. and Zimmer, D.E. 1979, Ultrastructure of compatible and incompatible reactions of Sunflower to *Masmopara halstedii*, Can. J. Bot., 57: 315-323.
Williams, R.J. 1984, Downy mildews of tropical cereals, Adv. Plant Pathol., 2: 1-103.
Williams, J.G.K., Kubelik, A.r., Livak, K.J., Rafalski, J.A. and Tingey, S.V. 1990, DNA polymorphisms amplified by arbitrary primers are useful as genetic markers, Nucleic acid Research, 18: 6531-6535.
Williamson, B., Breese, W.A. and Shattock, R.C. 1995, A histological study of downy mildews *Perenospora rubi* infection of leaves - flowers and developing fruits of Tummelberry and other *Rubus* sp., Mycological Research, 99(11): 1311-1316.
Wolfe, M.S. 1984, Trying to understand and control powdery mildew, Plant Pathol. 33 : 451-466.
Wynn, W.K. 1976, Appressorium formation over stomata by the bean rust fungus: response to a surface contact stimulus, Phytopathology, 66: 136-146.
Yang, S.L. and Ellingloe, A.H. 1972, Criticle layer as a determining factor for the formation of mature appresoria of *Erysiphe gramines* on wheat and barley, Phytopathology, 62: 708-714.
Yerkes, W.D. and Shaw, C.G. 1959, Taxonomy of the *Perenospora* species on Cruciferae and Chenopodiaceae, Phytopathology, 49: 499-507.
Yu, J.Y., Chang, H.S. and Ko, W.H. 1981, Factors affecting the induction of sexual reproduction in *Phytophthora parasitica* by *P. colocasiae*, J. Gen Microbiol., 123: 249-252.
Zeller, K.A. 1995, Phylogenetic relatedness within the genus *Erysiphe* estimated with morphological characteristics, Mycologia, 87(4): 525-531.
Zeller, K.A. and Levy, M. 1995, Intra species differentiation in the powdery mildew *Erysiphe Chicoracearum* determined with rDNA, RFLPs. Molec. Ecol. 4 in Press.
Zentmyer, G.A. 1961, Chemotaxis of zoospores for root exudates, Science, 133: 1595-1596.
Zentmyer, G.A. 1983, The world of *Phytophthora*, in: *Phytophtora* : Its Biology, Taxonomy Ecology and Pathology, D.C. Erwin, S. Bartnicki-Garcia and P.H. Tsao Eds, American Phytological Society, St. Paul. MI, pp. 1-7.
Zentmyer, G.A. 1988, Origin and distribution of four species of *Phytophthora*, Trans. Brit. Mycol. Soc., 91: 367-378.
Zentmyer, G.A., Klure, L.J. and Pond, E.C. 1979, The influence of temperature and nutrition on

formation of sexual structures by *Phytophthora cinnamomi*, Mycologia, 71: 55-67.
Zhang, K.M., Zheng, F.C. and Zi, Y.D. 1994, Isolates of *Phytophthora colocasiae* from Hawian Island in China: evidence suggesting an Asian origian of this species, Mycologia, 86(1): 108-112.
Ziegler, E. and Pontzen, R. 1982, Specific inhibition of glucan-elicited glyceollin accumulation in soybeans by an extracellular mannan-glycoprotein of *Phytophthora megasperma* f. sp. *glycinea*, Physiol. Plant. Pathol., 20: 321-331.
Zimmerman, M.H. and Mc Donough, J. 1978, Dysfunction of the flow of food, in: Plant Disease, J.G. Horsfall and E.B. Cowling eds., Academic Press, New York, Vol. 3, p. 117.

YEAST SPECIES FOR BIOCONTROL OF APPLE POSTHARVEST DISEASES : AN ENCOURAGING CASE OF STUDY FOR PRACTICAL USE

M. Haïssam Jijakli, Philippe Lepoivre and Cathy Grevesse

Unité de Phytopathologie, Faculté Universitaire des Sciences Agronomiques, 2, Passage des Déportés, B-5030 Gembloux, Belgium

1. INTRODUCTION

Since early 1970's, postharvest diseases of apple annually cause losses of 15-25 % despite modern storage facilities including controlled atmosphere (CA) or Ultra Low Oxygen (ULO) facilities (Bondoux, 1992). Factors that favour microbial growth, such as physiological senescence of fruits, mechanical injuries, as well as physiological disorders due to undesirable storage conditions can promote and explain these postharvest decays.

In Belgium and France (Bondoux, 1992) most losses are attributable to *Penicillium expansum* LK, *Botrytis cinerea* Pers. and *Gloeosporides* sp. (including *Cryptosoporiopsis malicorticis* (Cordl.) Nannf., *Colletotrichum gloeosporioides* (Penz.) and *Trichoseptoria fructigena* Maubl.). In USA and UK, *Botrytis* and *Penicillium* are the most important agents of postharvest diseases (Rosenberger, 1991). During storage, wounds are the first sites of infection for the initiation of diseases caused by the grey mould (*B. cinerea*) or the blue mould (*P. expansum*) agents. Both fungal agents infect fruits by deposit of airborne or waterborne conidia on wounds during harvesting and handling before storage. Infections of the *Gloeosporides* sp. occur in the field but quiescent latent and escape notice at harvest (Bondoux, 1992).

Control measures are principally based on the protection of fruits from pre- and post-harvest infection with pre- and postharvest fungicide treatments. However, in the context of consumer reluctance to chemical residues in food and public concern for environmental safety, there is an increasing demand to develop alternative methods to control diseases. That demand becomes a critical need with respect to the possible deregistragion of effective and widely used fungicides (Wellings, 1996) and the development of fungicide-resistant strains of postharvest pathogens (Franclet, 1994). Several novel approaches are emerging as possible

Biotechnological Approaches in Biocontrol of Plant Pathogens
Edited by Mukerji *et al.*, Kluwer Academic / Plenum Publishers, New York, 1999

alternatives to synthetic fungicides, including induction of natural defence mechanisms of harvested products, application of natural biocides, genetic resistance, and biological control agents (BCA's) (Wilson *et al.*, 1994).

The first phases of development of a biological control product can be considered as being very similar to the development of a chemical pesticide: early screening of a reliable activity, high efficacy in realistic conditions and adequate formulation to meet the expectation of the growers. However, because biological control agents (BCA) used for the control of plant diseases are primarily living organisms, their mass production, the study of their mode of action and their safety requirements are likely to be somewhat different than those for chemical pesticides. Public policy must translate that specificity in homologation procedures based on a case by case study.

Although the relevance of BCA appears limited until now, biological control of postharvest diseases are often presented as a specially promising area as testifying by already commercialised agents. In this respect, the paper exemplifies the potential and limits of such biocontrol strategy by presenting the case study of two yeast's, *Pichia anomala* (strain K) and *Debaryomyces hansenii* (strain O) exhibiting high and reliable protective activity against *Penicillium* sp. and *Botrytis cinerea* in postharvested apples.

2. POSTHARVEST DISEASES OF FRUITS : A PREDILECTION FIELD OF BIOLOGICAL CONTROL

Biological control is generating a great enthusiasm to play a role in sustainable agriculture although the relevance of BCA's in plant pathology appears limited until now. If everybody recognises the existence of natural phenomenon's of microbial antagonism, the question is to know how to manipulate naturally occurring antagonistic micro-organisms to achieve a reliable and effective strategy of disease control meeting the requirements of the market. In this respect, postharvest biological control could be considered as specially promising on a practical point of view because (1) the application sites are limited to the fruits, (2) the environmental conditions are defined and stable in storage rooms, and (3) the harvested commodities are of high value.

Literature presents numerous examples of biocontrol of fruit diseases (Table 1). Furthermore, biocontrol products such as Biosave™ (*Pseudomonas syringae* van Hall, Esc-11) and Aspire™ (*Candida oleophila* Montrocher, I-182) are already commercialised by Ecogen Inc. (Longhorn, PA) and Ecoscience Corp. (Worcester, MA), respectively and are used among other on postharvest apples against wound diseases.

Before becoming an economically feasible alternative to chemical control, BCA's have to satisfy different requirements related to biological, technological and toxicological properties. We will review the requirements that an "ideal antagonist" must meet through our research dealing with the biological protection of postharvest apples. An "ideal antagonist" should have the following characteristics: effective at low concentrations in several postharvest host pathogen combinations; able to survive under adverse environmental conditions such as low temperatures and controlled atmospheres adopted prevailing in storage facilities ; amenable to inexpensive production and formulation with a long shelf life; easy to dispense; compatible with commercial handling practices; genetically stable ; non pathogenic for the consumer and for the host commodity.

Table 1. Examples of BCA's used to control diseases on apples.

Disease	Antagonist	Nature of BCA and application period	Mode of action	Reference
B. cinerea	*Trichoderma pseudokoningii*	Fungus, blossoming to harvest	Parasitism and antibiosis	Tronsmo and Ystaas (1980)
	Trichoderma harzianum	Fungus, blossoming to harvest	Parasitism	Tronsmo and Ystaas (1980)
	Metschnikowia pulcherrima	Yeast, storage	Competition	Piano *et al.*, (1997)
	Cryptococcus humicola, Filobasidium floriforme, Rhodosporidium toruloides	Yeast's, storage	Competition	Filonow *et al.*, (1996)
B. cinerea and *P. expansum*	*Pseudomonas cepacia*	Bacteria, storage	Antibiosis (pyrrolnitrine) and competition	Janisiewicz and Roitman (1988); Chalutz and Droby (1993)
	Acremonium breve	Yeast, storage	Competition and induction of host resistance	Janisiewicz (1987, 1988)
	Cryptococcus laurentii	Yeast, storage	Competition	Roberts (1991); Wilson *et al.*, (1993); Chalutz and Droby (1993)
	Pichia guilliermondii	Yeast, storage	Competition, parasitism	Wilson and Wisniewski (1989); Chalutz and Droby (1993)
	Trichosporon sp. and *Candida* sp.	Yeast, storage	Competition	Aloi *et al.*, (1991)
B. cinerea, P. expansum; and *Mucor* sp.	*Pseudomonas syringae* (Biosave ä)*	Bacteria, storage	Induction of host resistance	Ecogen Inc. (Longhorn, PA)
	Candida oleophila (Aspire ä)*	Yeast, storage	Competition	Ecoscience Corp. (Worcester, MA)
Venturia inaequalis	*Athelia bombacina*	Fungus, blossoming to harvest	Competition and antibiosis	Janisiewicz (1991)

* : commercial name of the BCA

3. ISOLATION AND SELECTION OF BIOCONTROL AGENTS

3.1. Isolation and Characterisation of Epiphytic Micro-organisms

Theoretically, the isolation procedure of potential antagonists depends on the characteristics of the infection by the pathogen. To control post-harvest diseases, investigators usually isolated naturally occurring micro-organisms from fruits just before harvesting or during storage (Aloi *et al.*, 1991; Gullino *et al.*, 1991b; Janisiewicz, 1991). Nevertheless, an absolute relationship between efficacy and origin of isolation doesn't exist. Actually, micro-organisms exhibiting antagonistic properties against *B. cinerea* and *P. expansum* have been isolated from soil, leaves of apple tree or from fruits or leaves of other plants (Janisiewicz, 1988; Janisiewicz and Roitman, 1988; Wilson and Wisniewski, 1989). In this respect, an elegant and fast method of antagonist isolation has been adopted by Wilson *et al.*, (1993). They applied rinsing waters from tomatoes and apples directly on wounds inoculated with the pathogen (*B. cinerea*) and isolated antagonistic micro-organisms from wounds which did not exhibit any symptom.

We isolated micro-organisms from rinsing waters of Golden Delicious at harvest and after different periods of storage (Jijakli, 1996). We pointed up the coexistence of filamentous fungi, yeast's and bacteria on surface of apples whatever the storage period. Fungi belonging to the genera *Cladosporium*, *Penicillium*, *Aureobasidium*, *Alternaria*, *Mucor*, *Fusarium*, *Cephlosporium*, *Stemphylium*, *Trichoderma*, *Epicoccum*, *Pyrenochaeta*, *Pithomyces* and

Stigmella were principally observed. After two months of storage, epiphytic populations of bacteria and yeast's reached a maximum density of 5 x 10^7 cfu/ml and 3 x 10^6 cfu/ml respectively. In our study, the dominant population in term of density level was attributed to bacteria for each isolation period, while Janiciewicz (1996) observed that yeast's were the dominant population on Golden Delicious in term of different number of species.

Until now, microbial populations and their ecology on fruit surface are poorly known (Leibinger *et al.*, 1997). A better knowledge of characteristics of epiphytic micro-organism should lead to a more rational isolation scheme of antagonists well fit to their habitat and competitive with regard to other naturally occurring micro-organisms. This gap could be explained by the lack of reliable methods to correctly take a census of micro-organism populations on the plant surface. For example, the plating method under-estimated yeast populations from rinsing tissues of plants (Fokkema, 1991), although it is still the sole technique used in the scope of postharvested fruits.

3.2. Selection of Efficient BCA's

Most of the investigators (Aloi *et al.*, 1991; Gullino *et al.*, 1991a; Janisiewicz, 1987, 1988; Janisiewicz and Roitman, 1988; Wilson and Wisniewski, 1989) evaluated the efficacy of micro-organism strains on artificially wounded apples since *in vitro* antagonistic properties of a strain does not always lead to *in vivo* protection activity (Elsheriff and Grossmann, 1994).

Among 329 epiphytic micro-organisms (yeast's and bacteria), we selected two yeast's, *Pichia anomala* (strain K) (Hansen) Kurtzman and *Debaryomyces hansenii* (strain O) (Zopf) Lodder and Kreger-van Rij for their high and reliable biocontrol activity against infection by *B. cinerea* or *Penicillium* sp. on wounded Golden Delicious (Jijakli and Lepoivre, 1993). Treatment of wounded sites with 50 µl of yeast suspension (107 cfu/ml) was sufficient to inhibit rot development induced by 50 µl (106 spores/ml) of either *B. cinerea* or *Pencillium* sp. at 5°C and 25°C.

Numerous yeast strains exhibiting antagonism against *Botrytis* and/or *Penicillium* have been reported in the literature : *Acremonium breve* W. Gams (Janisiewicz, 1988), *Candida sake* (Saito and Ota) van Uden and Buckley, *C. tenuis* Diddens and Lodder (Wilson and Wisniewski; 1989), *C. oleophila* (Mercier and Wilson, 1994), *Candida* sp. (McLaughlin *et al.*, 1990), *C. guilliermondii* (Castellani) Langeron and Guerra , and *Kloechera apiculata* Janke (McLaughlin and Wilson, 1992), *Candida* sp. and *Trichosporon* sp. (Aloi *et al.*, 1991, Gullino *et al.*, 1991a), *Sporobolomyces roseus* Kluicer et van Niel (Janisiewicz *et al.*, 1994). The high frequency of yeast among the antagonistic agents reported could be related to the fact that yeast's are tolerant to extreme environmental conditions of storage rooms (temperature close to 0 °C, high relative humidity) and adapted to apple characteristics (high sugar concentration, high osmotic pressure and low pH) (Janicisiewicz, 1991). However, bacteria such as *Bacillus subtilis* (Ehrenberg) Cohn (Pusey and Wilson, 1984; Sholberg *et al.*, 1995), *Pseudomonas cepacia* Palleroni et Holmes (Janisiewicz and Roitman, 1988) or *P. syringae* van Hall (Janisiewicz and Bors, 1995) were also reported as effective BCA's on apples against both pathogens.

3.3. Major Parameters Controlling the Level of Protection

The study of parameters affecting the level of protective activity under sub-optimal conditions contributes to the selection of the most efficient BCA's. This study determines also the conditions giving a high and reproducible protective activity against postharvest diseases.

Our work (Jijakli *et al.*, 1993b) demonstrated that antagonistic activity of different strains of yeast's, isolated from apple surface were firstly dependent on the incubation time before inoculation of *B. cinerea* or *Penicillium* sp. (Table 2). Protection level increased with time between application of the antagonist and inoculation of the pathogen. The most efficient strains, *P. anomala* (strain K) and *D. hansenii* (strain O) reduced significantly the diameter of decay lesion, even when inoculation of the pathogen and application of the yeast were performed simultaneously.

In other respect, there was a quantitative relationship between spore concentration of *B. cinerea* and the amount of antagonist required for disease control (Table 3). The application of 108 cfu/ml of *P. anomala* (strain K) was necessary to completely inhibit lesions caused by *B. cinerea* upon inoculation with 50 µl of 106 spores/ml whereas 108 cfu/ml of *D. hansenii* (strain O) protected fruits against inoculation with 50 µl of 105 spores/ml of *B. cinerea*. The same parameters influencing the level of protection were already identified by other authors (Gullino *et al.*, 1991b ; Janisiewicz *et al.*, 1994 ; McLaughlin *et al.* ,1990 ; Mercier and Wilson, 1995 ; Roberts, 1991).

The temperature of fruit incubation and the humidity at the wound site were also identified as factors controlling the protective level (Gullino *et al.*, 1991a ; Mercier and Wilson, 1995). When apple wounds were artificially dried, Mercier and Wilson (1995) observed a decreased of both population level of *C. oleophila* and protection level against *B. cinerea*.

Table 2. Lesion development (mm) on wounded Golden Delicious apples treated with 50 µl of antagonistic yeast suspension (about 107 cfu/ml); and then inoculated with 50 µl of pathogen suspension (106 spores/ml) after different incubation times of the antagonist (Jijakli *et al.*, 1993a).

	B. cinerea[b]				*Penicillium* sp.[b]			
Incubation times	0 h	12 h	24 h	48 h	0 h	12 h	24 h	48 h
2.13[c]	21.6[a]	9.4[d]	6.9[d]	0.5[d]	20.2	15.6[d]	6.9[d]	5.4[d]
1.58	24.9	12.2[d]	11.2[d]	9.1[d]	16.4[d]	15.6[d]	14.2	9.0
9C5	24.7	19.5	7.0[d]	3.1[d]	20.2	22.6	19.1	10.4
5F2	27.0	13.9	7.1[d]	19.7[d]	18.5	16.7[d]	15.5	4.1[d]
K	10.7[d]	8.6[d]	0.0[d]	0.1[d]	10.6[d]	12.2[d]	3.7[d]	0.0[d]
O	15.2[d]	5.7[d]	4.7[d]	3.0[d]	19.2	18.1	3.1[d]	2.0[d]
9A4	22.0	19.5	14.1	5.4[d]	25.4	22.6	19.2	14.2
Control[e]	33.8	25.1	31.9	29.4	23.6	23.4	22.6	19.0

a: Data represent average lesion diameter (mm) measured 5 days after pathogen inoculation.

b: Pathogen

c: Antagonistic strains

d: Means of lesion diameters of the antagonist-treated apples are significantly different (p=0.001) from the control mean (in the same column) according to Dunnett's procedure.

e : Untreated apples inoculated with the pathogen only.

Data shown for 1 of the 2 trials (Data of separate trials were not pooled because variances differed significantly)

Table 3. Lesion development (mm) on wounded Golden Delicious apples inoculated with various spores concentrations of *B. cinerea* or *Penicillium* sp. 24 h after treatment of different concentration of *P. anomala* (strain K) or *D. hansenii* (strain O) (Jijakli *et al.*, 1993a).

Yeast concentration (cfu/ml)	*B. cinerea* spores concentration (spores/ml)				*Penicillium* sp. spores concentration (spores/ml)			
	10^7	10^6	10^5	10^4	10^7	10^6	10^5	10^4
P. anomala								
10^8	3.5[ad]	0.0[d]	0.0[d]	0.0[d]	4.6[d]	0.0[d]	1.1[d]	0.0[d]
10^7	4.7	0.7[d]	1.4[d]	1.6[d]	9.4[d]	2.6[d]	0.0[d]	3.7[d]
10^6	17.0	5.2[d]	10.2[d]	0.0[d]	17.2[d]	18.0	14.1	20.2
10^5	24.4	11.2[d]	5.9[d]	5.0[d]	18.1[d]	12.9[d]	15.4	4.9[d]
control[b]	24.5	28.1	25.0	22.1	21.0	21.4	21.6	19.1
D. hansenii								
10^8	9.5[ad]	2.9[d]	0.0[d]	0.0[d]	4.7[d]	6.2[d]	8.5[d]	4.6[d]
10^7	14.2	2.6[d]	3.4[d]	0.0[d]	15.5	4.0[d]	2.1[d]	0.0[d]
10^6	17.1	18.6[d]	9.1[d]	6.7[d]	11.0	15.2	10.1[d]	4.7[d]
10^5	13.2	7.9[d]	2.6[d]	0.9[d]	14.9	16.1	8.4[d]	9.2[d]
control[b]	24.5	28.1	25.6	22.1	21.0	21.4	21.6	19.1

a: Data are the average lesion diameter (mm) measured 5 days after pathogen inoculation.

b: Untreated apples inoculated with the pathogen only.

c: Means of lesion diameters of the antagonist-treated apples are significantly different to the control mean (in the same column) according to Dunnett's procedures ($p=0.001$)

Data shown for 1 of the 2 trials (Data of separate trials were not pooled because variances differed significantly)

4. MASS PRODUCTION OF BCA's

A screening of the potential efficient antagonists based on their capacity to be produced and dried in mass must follow the study of the protective properties. The capacity to produce in mass and to dry the micro-organisms has to be evaluated. The cost production of a micro-organism includes the culture media and the energy needed for production and drying process. Appropriate adjuvants relative to the drying process (protectants, carriers,...) must be added to increase the viability of BCA's during the different processes.

P. anomala (strain K) and *D. hansenii* (strain O) were tested for their technological properties. They were produced in fermentor (media 863) by CWBI (Centre Wallon de Bio-Industrie, Gembloux, Belgium) and fermentation products were dried by lyophilisation. Both antagonistic strains can be produced in fermentor and dried while maintaining their antagonistic activity (Jijakli *et al.*, 1993b).

The technologies of mass production are too often neglected and published data in this area remains spare because of industrial secrets. A closer collaboration between plant pathologists and teams working on mass production and formulation of micro-organisms would greatly stimulate the crossing of BCA's from laboratory to practical use. In this respect, useful insights could be expected from research's on BCA's against insect pests and weeds (Guillon, 1993).

5. FORMULATION OF BCA's

A formulation has to be developed to ensure a reasonably long shelf life of the BCA's and to facilitate their preparation and their application with the standard equipment of the producer. A good formulation will protect the antagonist from adverse conditions, increase the survey and/or enhance the efficiency of micro-organisms. A formulation which reduced antagonist concentration without affecting the protective level, will improve the economical feasibility of the product.

Nutrients are frequently reported as adjuvants which stimulate or stabilise the protective level of antagonistic strains in postharvest diseases (Janisiewizc, 1994; Janisiewizc *et al.*, 1992). We selected 15 carbohydrates and 16 nitrogenous compounds as potential adjuvants of yeast formulation (Jijakli *et al.*, 1993a). Only one sugar analogous (2-deoxy-D-glucose or 2-gluc) showed a protective effect against *B. cinerea* when applied alone and increased the level of protection from about 60 % to 90 % when added to *P. anomala* (strain K) or *D. hansenii* (strain O) suspension (105 cfu/ml). The sole application of 2-gluc inhibited also the development of *B. cinerea* on bean (Jejelowo *et al.*, 1988). This analogous of glucose reduced *in vitro* spore germination and hyphal growth of *B. cinerea* and *P. expansum* (Janisiewizc 1994; Jejelowo *et al.*, 1988; Jijakli *et al.*, 1993a) and could act as a competitive inhibitor of glucose metabolism (Janisiewizc, 1994). 2-gluc could be a proper additive if toxicological requirements of any additive of formulation are fulfilled.

On the other hand, none of the other nutrients (L-asparagin, L-proline, galactose, mannitol, riblose and sorbitol), selected for their *in vitro* and/or *in vivo* antagonist stimulation or pathogen inhibition either by Janisiewicz *et al.*, (1992) or Harper *et al.*, (1981) enhanced the protective activity of *P. anomala* (strain K) or *D. hansenii* (strain O) in our experiments. These different results show that an effect observed in a specific plant-antagonist-pathogen combination is not automatically transposable to an other system when the nutrient affects specifically the antagonistic agents. In opposite, when the nutrient has a specific effect on basic metabolism of the pathogen with no subsequent inhibition of the antagonist, we can expect an easier transposition in different systems.

The application of calcium chloride (2 % w/v) in mixture with *P. anomala* (strain K) or *D. hansenii* (strain O) enhanced the protective level against *B. cinerea* and *Penicillium* sp. McLaughin *et al.*, (1990) and Gullino *et al.*, (1991b) observed similar results when calcium is applied together with an antagonistic strain. The influence of calcium on increasing the resistance of host tissues is often reported (Conway, 1991 ; Messiaen, 1994), while its action on micro-organisms is still poorly studied. Nevertheless, the synergistic action of a combined treatment calcium-antagonist leads McLaughlin *et al.*, (1990) to suggest the secretion of new antifungal metabolites by the BCA's.

6. MODE OF ACTION

6.1. Difficulties and Importance of the Study of Modes of Action

As we move towards the application of biological control, new research problems are emerging among which the mechanisms of action of BCA's were too often considered only as an academic concern. The understanding of the modes of action of biocontrol agents is a prerequisite to (1) developing rational selection procedures yielding a second generation of more effective antagonistic microbial strains, (2) carrying out appropriate production and formulation enhancing antagonistic efficacy (3) providing a quality control procedure, and (4) fulfilling some requirements of the toxicological and registration procedure for commercial use.

Unfortunately, knowledge's on the mode of action of many antagonists of postharvest diseases are still limited. The comprehension of the mechanisms of action are hampered by the complex interactions between host-pathogen-antagonist. Moreover, the mechanisms studied *in vitro* in order to simplify these complex interactions, do not necessarily reflect *in situ* reality. Nevertheless, in the absence of antibiotics production by most antagonistic yeast's, it appears that the mode of action of yeast's could comprise one or several of the following processes : nutrient or site competition, direct interactions between the biocontrol agent and the pathogen, and induced host resistance (Wilson and Wisnieswski, 1994).

6.2. Nutrient Competition as Part of Mode of Action of *P. anomala* (strain K) and *D. hansenii* (strain O) Against *B. cinerea*

Experimental evidence of the implication of nutrient competition in the antagonistic relationship is still missing although several studies suggested that competition for nutrients might play a role in the antagonistic activity. Some authors (Chalutz *et al.*, 1991; Roberts, 1991; Wisniewski *et al.*, 1989) highlighted the ability of antagonists to rapidly multiply at the wounded sites of fruits but without relating this aptitude to their protective activity. Results of other worker (Droby *et al.*, 1989; Wisniewski *et al.*, 1991) showed that the addition of nutrients can restore both the germination of the pathogen and its development on fruits in presence of the antagonist.

We also investigated the ability of *P. anomala* (strain K) and *D. hansenii* (strain O) to colonise the wounds in relation with their protective activity against *B. cinerea* on apples and with regard to the *in situ* conidial germination of *B. cinerea* (Jijakli *et al.*, 1993a). Populations of *P. anomala* (strains K) and *D. hansenii* (strain O) in wounds increased at 25°C to reach a maximum density (approximately 1 log unit over the initial density) after 12 hours of incubation (Figure 1A and B) similarly to the protection level against *B. cinerea* which also reached a maximum after 12 hours of yeast incubation.

On the other hand, *in situ* spore germination of *B. cinerea* was markedly reduced on wounded sites treated with strain K or strain O, even when pathogen and yeast were applied simultaneously with no subsequent protection (Figure 1C). This suggested that other factor(s) than inhibition of spore germination may be involved in biocontrol effectiveness.

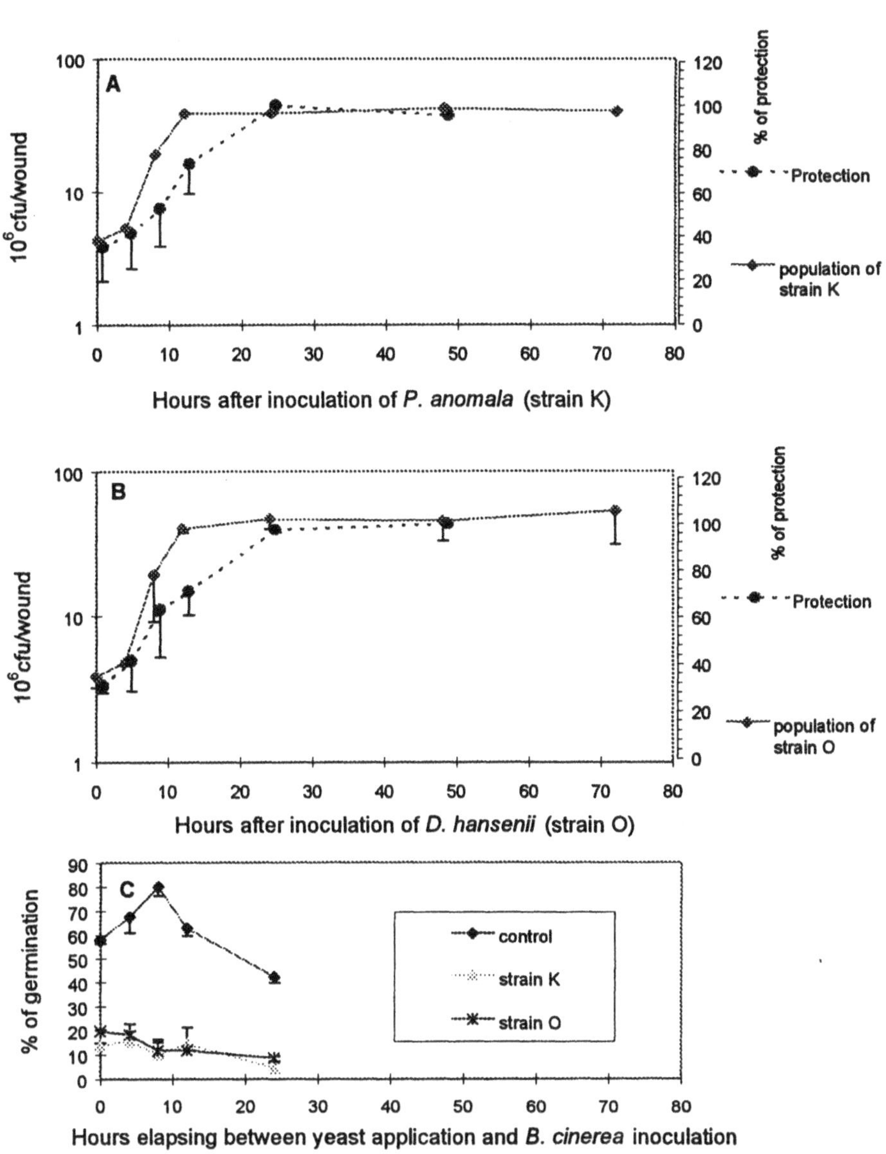

Hours elapsing between yeast application and *B. cinerea* inoculation

Figure 1. Effect of population densities of *P. anomala* (A) and *D. hansenii* (B) on level of protective activity against *B. cinerea*. Data from population densities represent mean colony from three trials (one wound site/trial). Each wound was triplicate-plated. Data from protection level represent the mean % of protection (as compared to the control which was not treated with yeast before inoculation of *B. cinerea*) from two trials (6 wounds/trial). Vertical bars represent standard error of the mean. (Jijakli *et al.*, 1993a).

Effect of *P. anomala* and *D. hansenii* on spore germination (C). Data represent the mean % and associated standard error (vertical bars) of spore germination from one of the two trials (1 replicate per trial). Twenty fields were observed per replicate and a spore was considered as germinated when the germinating tube was longer than the spore.

6.3. ß-1,3-glucanases as Part of Mechanism of Action *P. anomala* (strain K) Against *B. cinerea*

Production of hydrolytic enzymes which degrade cell walls of phytopathogens has been regularly reported as a mechanism of suppression of soil borne pathogens by some biocontrol agent such as *Trichoderma* genus (Bélanger *et al.*, 1995; Benhamou and Chet, 1996; Elad, 1996; Lorito *et al.*, 1994) but lytic enzymes have rarely been reported in biological control of postharvest diseases (Wisnieswski *et al.*, 1991).

We investigated the possible role of ß-1,3-glucanases and chitinases in the antagonistic properties of *P. anomala* (strain K) against *B. cinerea* on apples (Jijakli and Lepoivre, 1998). While chitinase activities were not detected from culture filtrates of strain K after various growth conditions (incubation period, source of carbon), endo- (EC 3.2.1-39) and exo-ß-1,3-glucanase (EC 3.2.1-58) activities were detected in the culture filtrates of strain K (Figure 2). Higher specific activities for both enzymes were obtained in media containing 2g/l of *B. cinerea* cell wall preparation (CWP) as sole carbon source as compared to media with glucose or laminarin (2 g/l). Endo- and/or exo-ß-1,3-glucanases activity from *T. harziamum* Rifai, *P. guilliermondii* Wickerham or *Serratia marescens* was also higher in media supplemented with fungal cell walls than in media containing laminarin (Elad *et al.*, 1982, Ordentlich *et al.*, 1988, Wisnieswski *et al.*, 1991). Exoglc1, an exo-ß-1,3-glucanase purified until homogeneity from *P. anomala* (strain K) culture filtrates, showed a stronger inhibitory effect on germinative tube growth than on conidial germination of *B. cinerea* (Figure 3). Moreover, the enzyme caused morphological changes such as leakage of cytoplasm and cell swelling on *B. cinerea*. Hydrolytic enzymes produced by other antagonists such as *Stachybotrys elegans* Barron or *Schizophyllum commune* Fries : Fries (Chiu and Tzean, 1995; Tweddell *et al.*, 1995) caused similar effects on hyphal growth of *Rhizoctonia solani* (Künh) and *Fusarium monoliforme* Wollenweber.

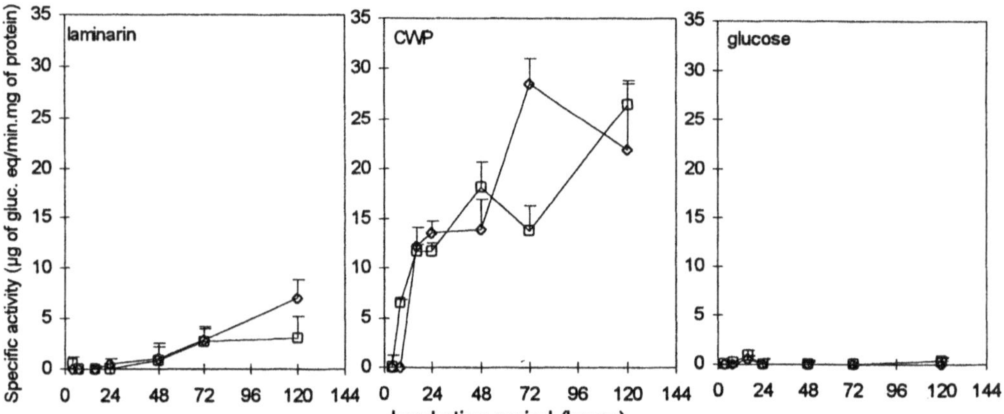

Figure 2. Time courses of endo-ß-1,3-glucanase(◇) and exo-ß-1,3-glucanase(□) activity from strain K supernatant of YNB medium containing *B. cinerea* cell wall fragments (CWP), laminarin or glucose as sole carbon source (2mg/ml). One specific unit of ß-1,3-glucanase (endo-or exo-) is defined as the amount of enzyme causing the release of 1µg of glucose equivalent per miligram of protein per min. Assays were performed in triplicate and the entire experiment was repeate three times. Values presented are averages of the 3 experiments. The error bars represent standard deviations of the means of the 3 experiments (Jijakli and Lepoivre, 1998).

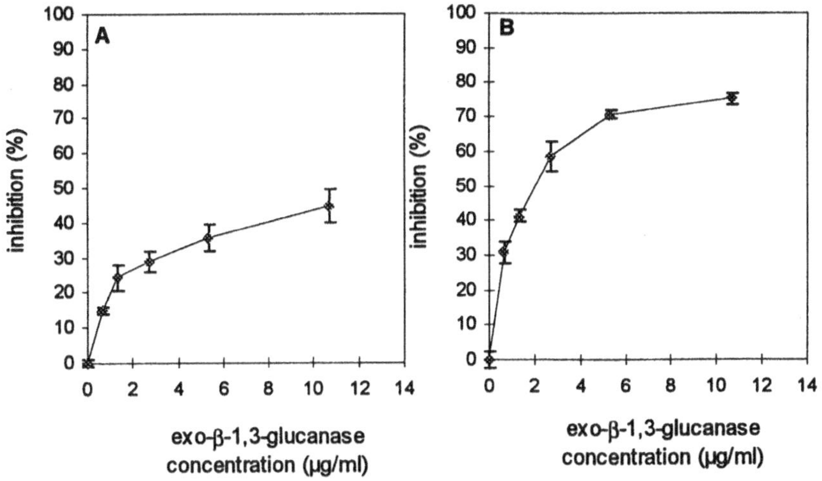

Figure 3. Effect of purified exo-ß-1,3-glucanase from *P. anomala* on spore germination (A) and germ tube elongation (B) of *B. cinerea*.
The percentage of conidia germinating was determined as the percentage of the first 100 spores found on a microscope slide. Length of 20 germ tubes was measured and averaged. For each enzyme concentration, the mean percent of inhibition and the standard deviation (error bars) were calculated. The data collected from two separate experiments with four replicates per treatment were pooled (Jijakli and Lepoivre, 1998).

In other respect, water used to rinse apple wounds treated with strain K and with CWP were assayed for exo-glucanase activity (Table 4). Exo-ß-1,3-glucanase activity was detected on apples treated with strain K. That activity appears to be related to exoglc1 on the basis of its electrophoretic mobility in native gel. Moreover, the addition of CWP to suspension of *P. anomala* stimulated both *in situ* exo-ß-1,3-glucanase activity and protection level against the pathogen (Table 4). To our knowledge, this is the first demonstration of *in vivo* glucanases production by yeast's.

Table 4. *In vivo* detection of exo-ß-1,3-glucanase activity from *P. anomala* strain K and protection from this strain against *B. cinerea* on apples (Jijakli and Lepoivre, 1998)

Treatment[x]	exo-ß-1,3-glucanase activity (μg eq. gluc./wound)[y]	Decay lesion (mm) after *B. cinerea* inoculation[z]
Control	3.4 ± 0.47 a	35.5 ± 8.43 a
CWP	2.1 ± 1.58 a	40.3 ± 4.65 a
Strain K	10.8 ± 0.52 b	15.6 ± 4.31 b
Strain K + CWP	32.7 ± 5.51 c	6.9 ± 3.65 c

x :Wounded sites of apples were treated with 50 μl of distilled water (Control and CWP) or 50 μl of strain K suspension (105cfu/ml) (Strain K). After 24 hours, 50 μl of CWP were applied at the wound sites (CWP and strain K + CWP). Treatments were arranged in a total randomized design.

y : 24 h after treaments, exo-ß-1,3-glucanase activities were measured. Means of enzymatic activity (± standard deviation) were expressed as μg equivalent of glucose/wound. Treatments were arranged in a total randomized design (three replicates per treatment). The experiment was conducted three separate times. After analysis of variance (ANOVA), data from the three experiments were pooled. Resulting means with a common letter do not differ significantly at P = 0.01(Fisher's LSD).

z : 24 h after treaments, wounded sites were inoculated with *B. cinerea* suspension (106 spores/ml). Treatments were arranged in a total randomized design. The experiment was conduted in two replicates searately. After analysis of variance (ANOVA), data from the two experiments were pooled. Resulting means were separated by the Ficher's LSD at P = 0.01.

The possible production of exoglc1 on apples and its effect on *B. cinerea in vitro* strengthen the hypothesis that exo-ß-1,3-glucanase contributes to the biocontrol of *B.cinerea* by *P.anomala* strain K without demonstrating its action. The following criteria may be used to determine if a particular compound is directly involved in biological control of fungal pathogens : (1) the purified compound shows fungicidal or antimicrobial properties, (2) the compound may be detected *in situ*, when producing strains are present, (3) the biocontrol ability of mutants defective in compound of interest is reduced in the laboratory and in practical conditions, (4) the complementation of the mutant with DNA sequence restoring the synthesis of the compound of wild strain restores also biocontrol ability. Because of an easier way to engineer prokaryotes, all of these criteria have been already used to determine the involvement of metabolites produced by antagonistic bacteria (Glick and Bashan, 1997; Wilson and Wisnieswski, 1994). In opposite, the criteria involving molecular tools were not yet reported to support the implication of a compound produced by antagonistic yeast's. Moreover, their possible polyploidy complicates the genetic investigations. However, in contrast to most other organisms, integrative recombination of transforming DNA in yeast's proceeds exclusively via homologous recombination and that property constitutes the basis of our further investigations on the mechanisms of biocontrol of *P. anomala* (strain K).

We isolated two exo-ß-1,3-glucanases encoding genes from a *P. anomala* (strain K) genomic library (Grevesse *et al.*, 1998). These genes named *PAEXG*1 and *PAEXG*2 were found to share significant similarities at the deduced amino acid level with exo-ß-1,3-glucanases from other fungi. Evidence that *PAEXG*2 coded for the purified exoglc1 was given by the sequencing of the N-terminal region of exoglc1. Because possible allelic effects render genetic studies more complicated, *P. anomala* (strain K) isolated from an apple in its diploid form, was induced to sporulate and ascospores were isolated by micromanipulation. Segregation of *PAEXG*1 and *PAEXG*2 in 10 haploid segregants was studied by Southern blots in relation with their *in vitro* exo-ß-1,3-glucanase activity production and their *in vivo* protective activity against *B. cinerea*. All segregants showed an exo-ß-1,3-glucanase activity production equivalent to the production of the diploid strain (or even higher) and retained some significant biocontrol activity at a lower or equivalent level in comparison with the diploid strain. No relation was found between these properties and the segregation of *PAEXG*1 or *PAEXG*2 showing that either *P. anomala* (strain K) is homozygous at both loci and/or other genetic factors (genes or regulatory elements) are active in the protective effect. The implication of *PAEXG*1 and *PAEXG*2 in the antagonism will be studied *in vivo* by their disruption through integrative transformation in the genome of the haploid material. This disruption strategy will give the experimental evidence for the hypothetical action of exo-ß-1,3-glucanase in the protective properties of *P. anomala* (strain K).

7. INTEGRATION OF BIOLOGICAL CONTROL TO OTHER TREATMENTS AGAINST DISEASES ON APPLES

The complete substitution of pesticides by biological control does not constitute a realistic goal. Biological control must be considered like a new potential strategy to be integrated to a panel of other methods. In this respect, the compatibility of BCA's treatments must be firstly evaluated with regard to their integration to the succession of fruit conditioning operations before storage.

A first strategy to widen the spectrum of antagonistic action consists in mixing several BCA's (Falconi and Megden, 1994; Janisiewicz, 1996; Leibinger *et al.*, 1997). A higher protection of postharvest apples against *P. expansum* was observed with the application of a mixture of *S. roseus* and *P. syringae* in comparison with the effects of their separate application (Janisiewicz and Bors, 1995). The development of five different wound diseases on pears was totally controlled by the treatment of yeast combination [*Cryptoccocus laurentii* (Kufferath) Skinner, *C. infirmo-miniatus* (Okunuki) Phaff and Fell and *Rhodotorula glutinis* (Fresenius) Harrison] (Chand-Goyal and Spotts, 1996)

The combination of chemical treatments with BCA's was more efficient than the application of the sole antagonistic micro-organsism. The success of biocontrol integration to chemical treatments will depends on the selection of the best combinations of fungicide-wax-antagonist (Pusey, 1986). The compatibility of a strain of *P. guilliermondii* with a commercial wax and a fungicide (thiabendazole) was evaluated against *P. digitatum* on postharvest citrus (Droby *et al.*, 1993). Other investigators observed the same phenomenon with other bio-chemical treatments (McGuire, 1994 ; Chand-Goyal and Spotts, 1996). In some cases, the combination of biological and chemical applications allowed the reduction of chemical concentration (Droby *et al.*, 1993).

The integration of the biological agents to other physical measures (thermotherapy, gamma and UV irradiation, film-forming polymers treatments) could be an other solution to widen the spectrum of activity of BCA's (Wilson *et al.*, 1994).

The grower attitude will depend on the capacity of non chemical strategy to inhibit both latent and wound infections usually targeted by traditional fungicide treatments. The control of *Gloeosporium* infections developing postharvest apple rots from orchard latent infections, has to be assumed on postharvest apples in addition to the protection of wound pathogens (*Botrytis* and *Penicillium*) (Jijakli and Lepoivre, 1995). In this context, heat treatments with water (45° C for 10 min) appeared to be efficient against *Gloeosporides* lenticel infections (*C. gloeosporioides*, *C. malicorticis* and *T. fructigena*) by inactivation of spores or hyphae located on skin and on external part of fruit flesh (Bondoux, 1992 ; Eckert, 1975). However, thermotherapy treatment with water bath can enhance the sensitivity of apples to wound pathogens such as *Alternaria* sp., *B cinerea* or *P. exansum* (Edney and Burchill, 1967).

In order to integrate biological control to thermotherapy, we applied three different treatments separately or in combination on postharvest Golden Delicious : (1) dipping the apples in water at 45°C for 10 min, (2) dipping the fruits for 2 min in a water suspension of two antagonistic yeast's, *Pichia anomala* (strain K) and *Candida sake* (strain O) (107 cfu/ml each), or (3) dipping the fruits in an emulsion of a 2% film-forming antitranspirant (2 % Nu-film-P or NFP, 96 % of poly-1-p-Menthen, Miller laboratory) for 2 min. Thermotherapy alone reduced the incidence of *Gloeosporides* lenticel infections from 54.4 % (untreated apples) to 4.6 %, but enhanced sensitivity of the apples to *Penicillium* spp (Table-5). The higher sensitivity to *Penicillium* infections could be explained by lenticel damage (Edney and Burchill, 1967) or partial sterilization of fruit surface enhancing the subsequent *Penicillium* contamination. Control of this pathogen was restored by dipping the fruits in yeast suspension in NFP emulsion after the heat treatment. The quality parameters (weight, size, skin color, firmness, acidity and refractometric indice) of the fruits were not affected by any of the treatments (Jijakli *et al.*, 1993a). During a second year of trials, combination of heat treatment and yeast application could entirely control infections caused by *B. cinerea* and *Gloeosporium* rot and the percentage of apples rotted by *Penicillium* spp. was reduced from 18.2 % (untreated fruits) to 3.8 % without affecting any quality parameters (Table-6), thus confirming the efficacy of such integrated treatments against major fungal diseases on postharvest apples. Nevertheless, the evaluation of the practical feasibility of an integrated approach combining biological control and thermotherapy must be further studied with a less intensive scheme of chemical application.

Table 5. Separate and combined effects of thermotherapy, biological control and antitranspirant application on post-harvest diseases of apples in 1993 (Jijakli *et al.*, 1993b)

	Control	45°C[b]	45°C +NFP[c]	45°C K+O[d]	45°C+ K+O+ NFP	K+O	NFP	NFP+ K+O
Total infection	72.4a	55.2	61.3	50.0	40.2	56.2	57.5	62.5
Gloeosporides	54.4	4.6	4.6	2.3	1.5	37.7	33.8	41.5
Penicillium spp.	8.2	20.0	32.3	19.2	11.5	3.1	10.0	6.1
Alternaria spp.	4.8	13.1	10.0	10.8	11.5	4.6	5.4	7.7
Fusarium spp.	3.4	7.7	5.4	3.1	6.9	0.8	0.8	2.3
Cylindrocarpon spp.	3.4	3.8	3.8	5.4	6.9	3.8	1.5	3.8
Rhizopus spp.	0.0	1.5	0.0	0.0	0.0	0.0	0.0	0.0
Non determined	4.1	8.5	11.5	9.2	9.2	10.0	10.0	6.9

a = mean percentage of infection from 130 apples, b = apples were dipped in heat water (45°C) for 10 min,
c = apples were dipped in Nu-Film-P (2%), d = apples were dipped in strain K et O suspension (107 cfu of each strain/ml).

Table 6. Separate and combined effects of thermotherapy, biological control and antitranspirant application on post-harvest diseases of apples in 1994

	Control	45°C+K+O +NFP[b]
Total infection	24.2 a	4.6
Gloeosporides	1.5	0.0
Penicillium spp.	18.2	3.8
Botrytis cinerea	6.1	0.0
Fusarium spp.	0.0	0.0
Alternaria spp.	0.0	0.0
Non determined	0.8	0.8

a = mean percentage of infection from 130 apples, b = apples were successively dipped in heat water (45°C) for 10 min and in a suspension of yeast's in a 2 % NFP emulsion, c = dry heat treatment carried out at 38° C for 24 h, d = apples were successively heated in a steam room and dipped in a suspension of yeast's in a 2 % NFP emulsion.

8. HOMOLOGATION

Biological agents used for the control of plant diseases are primarily living organisms. The impact on the environment and therefore, safety requirements are likely to be somewhat different than those for chemical pesticides.

OECD, with particular inputs from the Directorate of Sciences and Technology and Environment, published a booklet which concerned the release of micro-organisms in the environment. Most countries, including the European Union, have incorporated the main conclusions of the OECD reports (OECD, 1995) in their national regulation.

In this respect, the European Union Council Directive of 15 July, 1991 (91/414/EEC) concerns the marketing of plant protection products. According to this directive, any products used in the biological control of plant diseases must be considered as plant protection products and have therefore to fulfil different specific requirements. The technical dossier of registration of the BCA must supply the necessary information's for evaluating the foreseeable risks, whether immediate or delayed, which the substance may entail for humans and the environment. These requirements mainly concern the identity of the organism, its biological properties, the description of the analytical methods available for its identification, its toxicological impact (pathogenicity and infectivity), and the ecotoxicological properties. The associated costs for the registration of plant protection products containing biological agents are directly related to the level of requirements and rapidly become so high that the development of products for the biological control of plant disease could be discouraging for companies and leads into temptation of circumventing these regulations. Public policy could thus promote biological alternative to chemical control by authorising realistic registration procedures based on a case by case study. If the active substance belongs to a group of micro-organisms well known for their non pathogenicity, different costly studies such as long term toxicity should be not required for the evaluation of the dossier. While the responsibility of the registration usually falls upon the company developing a commercial product, some requirements also concern the criteria of the screening tests set up for the selection of the agents. In this respect, we must emphasise possible problems encountered in the registration of BCA's whose mode of action include antibiosis or that belong to groups of micro-organisms comprising some potential pathogenic species or strains.

Large scale introduction of biocontrol agents also requires specific evaluation of their environmental impact. A requisite to any study related to the persistence of artificially introduced micro-organisms in the environment is the ability to track them. In this purpose, molecular fingerprinting of *P.anomala* (strain K) was attempted by means of random amplification of DNA (RAPD). Polymorphism was observed in the RAPD profiles of different strains of *P.anomala* and between different species of yeast. The nucleotide sequence of a strain K-specific amplified product was used to design specific primers which allow the unequivocal recognition of the antagonist. That method primarily used for identification of the BCA and monitoring its population on the fruit surface, could also be used to perform studies of ecophysiology of the BCA, with the aim of understanding and predicting its antagonistic activity as affected by environmental conditions of storage of the fruits.

9. CONCLUSIONS

Many challenges must be met before biocontrol of postharvest diseases can be successfully used on a commercial basis (Figure 4). Biological control is often generating a great enthusiasm although the still limited relevance of BCA today. The postharvest environment may be one in which the best chance to develop successful biological control exists because many aspects of it can be controlled. Moreover, consumer demand for alternatives to postharvest chemical treatments constitutes a major and special impetus in the search for biocontrol agents. Already commercially available products (such Aspire™ and Biosave™) demonstrate the realism of the approach.

However, encouragement's from environmental agencies and central government as well as grower education remain crucial in determining the economic climate within biological control will operate.

It remains doubtful that the first generation of biocontrol agents will easily provide by themselves the efficacy and the consistency associated with conventional fungicides. However, biocontrol agents already selected in different laboratories could constitute a significant part of integrated systems including physical treatments or lower doses of chemicals to provide adequate control of postharvest diseases.

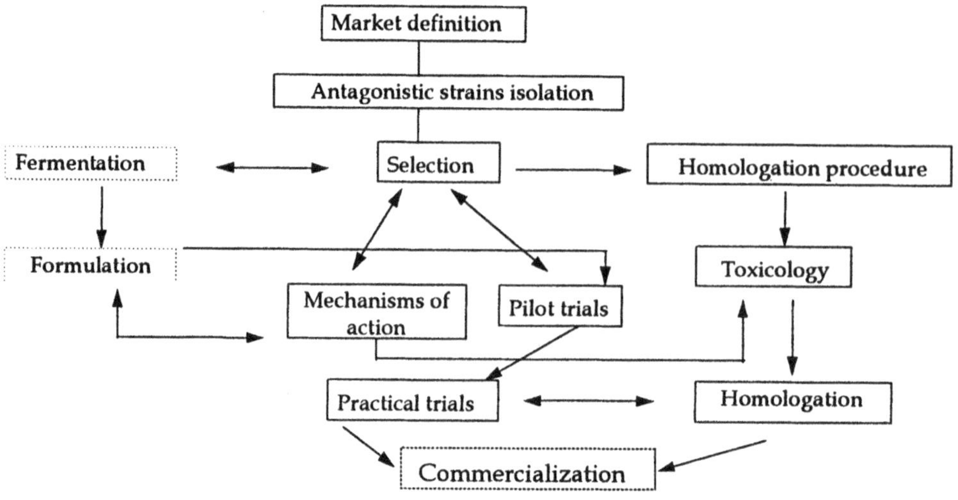

Figure 4. Steps leading to the practical use of BCA's

In the long term, basic information's on the genetically determined factors that control survival, colonisation, effectiveness in the field and storage and properties of mass production are required to overcome the random process of selection.

ACKNOWLEDGEMENTS

This work has been as a part of the "CAMAR" project "Biological control of rot diseases of fruit and vegetable crops" funded by EC.

REFERENCES

Aloi, C., Benzi, D., Testoni, A. and Gullino, M. L. 1991, Biological control of postharvest diseases of apple : experimental results, in : ANPP-Third International Conference on Plant Diseases, Tome II. Ministère de l'Agriculture et Ministère de la Recherche et de la Technologie, Bordeaux, France pp. 695-702.

Belanger, R. R., Dufour, N., Caron, J. and Benhamou, N. 1995, Chronological events associated with the antagonistic propreties of *Trichoderma harzianum* against *Botrytis cinerea* : indirect evidence for sequential role of antibiosis and parasitism, Biocontrol Science and Technology 5: 41-53.

Benhamou, N. and Chet, I. 1996, Parasitism of sclerotia of *Sclerotium rolfsii* by *Trichoderma harzianum* : ultrastructural and cytochemical aspects of the interaction, Phytopathology 86: 405-416.

Bondoux, P., (ed.), 1992, Maladies de conservation des fruits à pépins, pommes et poires, INRA et PHM revue horticole, Paris, France.

Chalutz, E. and Droby, S. 1993, Mechanisms of antagonisms of yeasts against postharvest diseases, in : Biological Control of Fruit and Foliar Disease, P. Lepoivre, ed., Proceedings of an EC Workshop, Gembloux, Belgium pp. 67-76.

Chalutz, E., Droby, S., Cohen, L., Weiss, B., Barkai-Golan, R., Daus, A. and Fushs, Y. 1991, Biological control of *Botrytis*, *Rhizopus* and *Alternaria* rots of tomato fruit by *Pichia guillermondii*, in : Proccedings of an International Workshop on Biological Control of Postharvest Diseases of Fruits and Vegetables, USA-USDA Publ. ARS-92, Shepherdstown, West Virginia, USA pp. 71-85.

Chand-Goyal, T. and Spotts, R. A. 1996, Control of postharvest pear diseases using natural saprophytic yeast colonies and their combination with a low dosage of thiabendazole, Postharvest Biology and Technology 7: 51-64.

Chiu, S. C. and Tzean, S.S. 1995, Glucanolytic enzyme production by *Schizophyllum commune* Fr. during mycoparasitisme, Physiological and Molecular Plant Pathology 46: 83-94.
Conway, W. S. 1991, Postharvest calcium treatment of apple fruit to provide broad-spectrum protection against postharvest pathogens, Plant Diseases 75: 620-622.
Droby, S., Chalutz, E., Wilson, C. L. and Wisniewski, M. 1989, Characterization of the biocontrol activity of *Debarryomyces hansenii* in the control of *Penicillium digitatum* on grapefruit, Can. J. of Microbio. 35: 308-313.
Droby, S., Hofstein, R., Wilson, C. L., Wisniewski, M, Fridlender, B., Cohen, L., Weiss, B., Daus, A., Timar, D. and Chalutz, E. 1993, Pilot testing of *Pichia guilliermondii* : a biocontrol agent of postharvest diseases of Citrus fruit, Biological Control 3: 47-52.
Eckert, J.W 1975, Postharvest diseases of fresh fruits and vegetables - Etiology and control, in : Symposium Postharvest Biology and Handling of Fruits and Vegetables, N.F. Haard and D.K. Salunkhe, eds., The AVI Publishing Company Inc., Westport, Connecticut, USA pp. 81-117.
Edney, K. L. and Burchill, R. T. 1967, The use of heat to control the rotting of Cox's Orange Pippin apples by *Gloeosporium* spp., Annals of Applied Biology 59: 389-400.
Elad, Y. 1996, Mechanisms involved in the biological control of *Botrytis cinerea* incited diseases, European Journal of Plant Pathology 102: 719-732.
Elad, Y., Chet, I. and Henis, Y. 1982, Degradation of plant pathogenic fungi by *Trichoderma harzianum*, Can. J. Microbio. 28: 719-725.
Elsherif, M. and Grossmann, F. 1994, Comparative investigations on the antagonistic activity of fluorescent pseudomonas against *Gaeumannomyces graminis* var. *tritici in vitro* and *in vivo*, Microbiol. Res. 149: 371-377.
Falconi, J. and Mengden, K. 1994, Eppiphytic fungi on apples leaves and their value for control of the postharvest pathogens *Botrytis cinerea*, *Monilia fructigena* and *Penicillium expansum*, J. Pl. Dis. Protec. 1: 38-47.
Filonow, A.B., Vishniac, H. S., Anderson J. A. and Janisiewicz, W. J. 1996, Biological control of *Botrytis cinerea* in apple by yeasts from various habitats and their putative mechanisms of antagonism, Biological control 7: 212-220.
Fokkema, N. J. 1991, The phyllosphere as an ecologically neglected milieu : a plant pathologist's point of view, in : Microbial Ecology of Leaves, J. H. Andrews and S. S. Hirano, eds., Springer-Verlag, Madison, USA pp. 3-18.
Franclet, J. 1994, Enquête ANPP - Maladies de conservation des pommes, L'arboriculture fruitière 473: 24-26.
Glick, B. R. and Bashan, Y. 1997, Genetic manipulation of plant growth-promoting bacteria to enhance biocontrol of phytopathogens, Biotech. Adv. 15: 353-378.
Grevesse, C., Jijakli, M. H., Duterme, O., Colinet, D. and Lepoivre, P. 1998, Preliminary study of exo-ß-1,3-glucanase encoding genes in relation to the protective activity of *Pichia anomala* (strain K) against *Botrytis cinerea*, Proceedings of the Workshop on Molecular Approaches in Biological Control, Délemont, Suisse, (in press).
Guillon, M. 1993, Formulation of biopesticides, an industrial view of biological agents, in : Biological Control of Fruit and Foliar Disease, P. Lepoivre, ed., Proceedings of an EC Workshop, Gembloux, Belgium pp. 99-106.
Gullino, M. L., Aloi, C. and Garibaldi, A. 1991a, Integrated control of grey mould of tomato, IOBC/WPRS Bulletin : Integrated control in protected crops under mediterranean climate 14, pp. 211-215.
Gullino, M.L., Aloi, C., Benzy, D. and Garibaldi, A. 1991b, Attempts at biological control of postharvest diseases of apple, Med. Fac. Landbouw, Univ. gent 56: 195-202.
Harper, A. M., Strange, R. N. and Langcake, P. 1981, Characterization of the nutrients required by *Botrytis cinerea* to infect broad bean leaves, Physiological Plant Pathology 19: 153-167.
Janisiewicz, W. 1987, Postharvest biological control of blue mold on apples, Phythopathology 77: 481-485.
Janisiewicz, W. 1988, Biocontrol of postharvest diseases of apples with antagonist mixtures, Phytopatology 78: 194-198.
Janisiewicz, W. 1991, Biological control of postharvest fruit diseases, in : Handbook of Applied Mycology, Volume 1, D.K. Arora, B. Rai, K.G. Mukerji and G. R. Knudsen, eds., Marcel Dekker, Inc., New York, USA pp. 301-325.
Janisiewicz, W. 1994, Enhancement of biocontrol of blue mold with the nutrient analog 2-deoxy-D-glucose on apples and pears, Appl. Environ. Microbiol. pp. 2671-2676.
Janisiewicz, W. 1996, Ecological diversity, niche overlap, and coexitence of antagonists used in developing mixtures for biocontrol of postharvest diseases of apples, Phytopathology 86: 473-479.
Janisiewicz, W. and Bors, R. 1995, Development of a microbial community of bacterial and yeast antagonists to control wound-invading postharvest pathogens of fruits, Appl. Environ. Microbiol. pp. 3261-3267.
Janisiewicz, W. and Roitman, J. 1988, Biological control of blue mold and gray mold on apple and pear with

Pseudomonas cepacia, Phytopathology 78: 1697-1700.
Janisiewicz, W., Peterson, D. L. and Bors, R. 1994, Control of storage decay of apples with *Sporobolomyces roseus*, Plant Diseases 78 : 466-470.
Janisiewicz, W., Usall, J. and Bors, B. 1992, Nutritional enhancement of biocontrol of blue mold on apples, Phytopathology 82: 1364-1370.
Jejelowo, O. A., Epton, H. A. S. and Trinci, A. P. J. 1988, Effects of paramorphogens and 2-deoxy-D-glucose on development of lesions of *Botrytis fabae* on *Vicia faba*, Trans. Br. Mycol. Soc. 91: 661-669.
Jijakli, M. H. 1996, Etudes des propriétés antagonistes de deux souches de levures vis-à-vis de *Botrytis cinerea* Pers. sur pommes en conservation, Thèse de doctorat, Faculté Universitaire des Sciences Agronomiques, Gembloux, Belgium.
Jijakli, M. H. and Lepoivre, P. 1993, Biological control of postharvest *Botrytis cinerea* and *Penicillium* on apples, IOBC/WPRS Bulletin : Biological Control of Foliar and Post-Harvest Diseases 16, pp. 106-110.
Jijakli, M. H. and Lepoivre, P. 1995, Utilisation de biopesticides pour la protection des pommes en conservation, Le Fruits Belge 445: 83-88.
Jijakli, M. H. and Lepoivre, P. 1998, Characterization of an exo-ß-1,3-glucanase produced by *Pichia anomala* strain K, antagonist of *Botrytis cinerea* on apples, Phytopatholgy (in press).
Jijakli, M. H., Lepoivre, P., Tossut, P. and Thonart, P. 1993a, Biological control of *Botrytis cinerea* and *Penicillium* on postharvest apples by two antagonistic yeasts, Med. Fac. Landbouw, Univ. Gent. 58: 1349-1358.
Jijakli, M.H., Choutka, C. and Lepoivre, P. 1993b, Formulation and integrated use of two antagonistic yeasts to postharvest treatments against diseases on apples, in : Biological Control of Fruit and Foliar Disease, P. Lepoivre, ed., Proceedings of an EC Workshop, Gembloux, Belgium pp. 107-117.
Leibinger, W., Breuker, B., Hahn, M. and Mengden, K. 1997, Control of postharvest pathogens and colonization of the apple surface by antagonistic microorganisms in the field, Biological Control 87: 1103-1110.
Lorito, M., Hayes, C. K., Di Pietro, A., Woo, S. L. and Harman, G. E. 1994, Purification, characterization and synergisic activity of a glucan 1,3-b-glucosidase and an N-acetyl-b-glucosaminidase from *Trichoderma harzianum*, Phytopathology 84: 398-405.
McGuire, R. G. 1994, Application of *Candida guilliermondii* in commercial citrus coatings for biocontrol of *Penicillium digitatum* on grapefruits, Biological Control 4: 1-7.
McLaughlin, R. J. and Wilson, C.L. 1992, Biological control of postharvest diseases of grape, peach and apple with the yeasts *Kloeckera apiculata* and *Candida guilliermondii*, Plant Diseases 76: 470-473.
McLaughlin, R. J., Wisniewski, M. E., Wilson, C. L. and Chalutz, E. 1990, Effect of inoculum concentration and salt solutions on biological control of postharvest diseases of apple with *Candida* sp., Phytopathology 80: 456-461
Mercier, J. and Wilson, C. L. 1994, Colonization of apple wounds by naturally occurring microflora and introduced *Candida oleophila* and their effect on infection by *Botrytis cinerea* during storage, Biological Control 4: 138-144.
Mercier, J. and Wilson, C. L. 1995, Effect of wound moisture on the biocontrol by *Candida oleophila* of gray mold rot *(Botrytis cinerea)* of apple, Postharvest Biology and Technology 6: 9-15.
Messiaen, J. 1994, La transduction du signal chez les végétaux supérieurs en réponse à une attaque pathogène : Identification de réponses membranaires, cytosoliques et nucléaires induites par des olgogalacturonides, Thèse de doctorat, Presse Universitaire de Namur, Belgium.
OECD, 1995, Safety consideration for biotechnology : scale-up of crops plants, OECD, Paris, France.
Ordentlich, A., Elad, Y. and Chet, I. 1988, The role of chitinase of *Serratia marescens* in biocontrol of *Sclerotium rolfsii*, Phytopathology 78: 84-88.
Piano, S., Neyrotti, V., Migheli, Q. and Gullino, M.L. 1997, Biocontrol capability of *Metschnikowia pulcherrima* against *Botrytis postharvest* rot of apple, Postharvest Biology and Technology 11: 131-140.
Pusey, P. L. 1986, Compatibility of *Bacillus subtilis* for postharvest control of peach brown rot with commercial fruit waxes, Dicloran and cold-storage conditions, Plant Diseases 70: 587-590.
Pusey, P. L. and Wilson, C. L. 1984, Postharvest biological control of stone fruit brown rot by *Bacillus subtilis*, Plant Diseases 68: 753-756.
Roberts, R. G. 1991, Postharvest biological control of gray mold of apple by *Cryptococcus laurentii*, Phytopathology 80: 1051.
Rosenberger, D.A. 1991, Postharvest diseases (Blue mold- Gray mold), in : Compendium of Apple and Pear Diseases, A.L. Jones and H.S. Aldwinckle, eds., APS Press The American Phytopathological Society, USA pp. 53-58.
Sholberg, P. L., Marchi, A. and Bechard, J. 1995, Biocontrol of postharvest diseases of apple using *Bacillus* spp. isolated from stored apples, Can. J. Microbiol.41: 247-252.
Tronsmo, A. and Ystaas, J. 1980, Biological control of *Botrytis cinerea* on apple, Plant Diseases 64: 1009.
Tweddell, R. J., Jabaji-Hare, S. H. and Charest, P. M. 1995, Production of chitinases and b-1,3-glucanases by

Stachybotrys elegans, a mycoparasite of *Rhizoctonia solani*, Appl. Environ. Microbiol. pp. 489-495.

Wellings, P. W. 1996, The role of public policy in biological control : some global trends, Entomophaga 41: 435-441.

Wilson, C. L. and Wisnieswski, M. E., (eds.), 1994, Biological Control of Postharvest Diseases : Theory and Practice, CRC Press, London, UK.

Wilson, C. L. and Wisniewski, M. E. 1989, Biological control of postharvest diseases of fruits and vegetables : an emerging technology, Annual Review Phytopathology 27: 425-441.

Wilson, C. L., El Ghaouth, A., Chalutz, E., Droby, S., Stevens, C., Lu, J. Y., Khan, V. and Arul, J. 1994, Potential of induced resistance to control postharvest diseases of fruits and vegetables, Plant Diseases 78: 837-844.

Wilson, C. L., Wisniewski, M.E., Droby, S. and Chalutz, E. 1993, A selection strategy for microbial antagonists to control postharvest diseases of fruits and vegetables, Scientia Horticulturae 53: 183-189.

Wisniewski, M. E., Wilson, C. L. and Hershberger, W. S. 1989, Characterization for inhibition of *Rhyzopus stolonifer* germination and growth by *Enterobacter cloacea*, Can. J. Bot. 67: 2317-2323.

Wisniewski, M., Biles, C., Droby, S., McLaughlin, R., Wilson, C. and Chalutz, E. 1991, Mode of action of the postharvest biocontrol yeast, *Pichia guillermondii*, I, Characterization of attachment to *Botrytis cinerea*, Physiol. Mol. Pl. Path. 39: 245-258.

BIOLOGICAL CONTROL AGENTS OF CANOLA AND RAPESEED DISEASES - STATUS AND PRACTICAL APPROACHES

Susan M. Boyetchko

Agriculture and Agri-Food Canada, Saskatoon Research Centre, 107 Science Place, Saskatoon, Saskatchewan, Canada S7N 0X2

1. INTRODUCTION

Although initial exploration in biological control of plant pathogens began in the mid 1920s, it is a technology in plant pathology that has experienced more intensive investigation by researchers and industry over the last 25 to 30 years (Cook, 1993; Utkhede, 1996; Weller, 1988). "The general accepted definition of biological control, proposed by Baker and Cook (1974), is the reduction of inoculum density or disease-producing activities of a pathogen or parasite in its active or dormant state, by one or more organisms, accomplished naturally or through manipulation of the environment, host, or antagonist, or by mass introduction of one or more antagonists".

Cook (1993) listed three major strategies to consider for the development of introduced fungal and bacterial microorganisms for biological control of plant diseases: i) reduce or limit the population of the plant pathogen; ii) control or prevent infection by the pathogen; and iii) control the spread and development of disease following infection. A number of desirable characteristics in a biocontrol agent have been suggested. It must have a high and reliable performance on the target pest but have no detrimental effects on non-target organisms. It must be easy to produce and use, have an acceptable shelf-life (approximately 1-2 years), and meet acceptable environmental and toxicological safety standards. The biocontrol agent must also be highly competitive with indigenous microbes and persist in the environment for an acceptable period of time (not too short or too long). In addition, it must be highly effective under a broad range of environmental conditions (Campbell, 1986; Cook, 1993; Elad and Chet, 1995; Jacobsen and Backman, 1993; Utkhede, 1996). Not all of these criteria are easy to achieve in one biocontrol product.

Biotechnological Approaches in Biocontrol of Plant Pathogens
Edited by Mukerji *et al.*, Kluwer Academic / Plenum Publishers, New York, 1999

One must wonder why the number of biological control agents and products is limited despite the number of years that biocontrol research has been under investigation. Lack of consistent performance in the field and the narrow spectrum of bioactivity have been cited as reasons for the lack of biological products (Elad and Chet, 1995; Jacobsen and Backman, 1993). The term biopesticide, often used for biological control agents of many pests, implies that they are analogues of chemicals and exert the same level of control as synthetic fungicides (Cook, 1996). Moreover, regulation of biologicals has often been subject to the same criteria developed for chemical pesticides and our expectations for biocontrol products have often exceeded our expectations for chemicals. Furthermore, complete control of plant diseases is not always achieved with chemicals and two or more fungicidal applications may be required (Bailey, 1997). It has also been argued that funding, resources, and infrastructure dedicated to the development of biological control products are small compared to that spent on the development of chemical pesticides (Campbell, 1986; Jacobsen and Backman, 1993; Utkhede, 1996). The development of a chemical pesticide from discovery to the marketplace may take 8 to 10 years and cost as much as $80 million (Utkhede, 1996). However, it has been estimated that the development of a biopesticide product only costs $0.8 to 1.6 million and that the period from discovery to commercialization is much shorter than for chemical pesticides. It is only recently that investment in scientific expertise, infrastructure, and research into biological control by the private sector is being pursued more vigorously.

Major considerations that have prompted investment in the development of biological control products include lack of chemical fungicides for control of specific plant pathogens, de-registration of fungicides, high costs associated with discovery and registration of new fungicides, spray drift and pesticide residues in soil, water, and food, and development of pesticide resistance in plant pathogens (Elad and Chet, 1995; Jacobsen and Backman, 1993). Biological control has a role in addressing these issues. However, have we over-sold the idea of biological control and exceeded our ability to develop a commercial product without considering the level of performance and efficacy of the biocontrol agents and the technical challenges in mass production, formulation and delivery? Biocontrol agents do not necessarily eradicate disease, but reduce damage or disease symptoms to acceptable levels. Most researchers would agree that biological control should be considered one strategy for farmers to use as part of an integrated pest management program (Bailey, 1997; Jacobsen and Backman, 1993; Lumsden *et al.*, 1995).

This book chapter will focus on a few diseases of canola and rapeseed and the various fungal and bacterial microorganisms that have been investigated as potential biological control agents. In addition, strategies for the development and improvement of these biocontrol agents as biopesticide products will be discussed.

2. DISEASES OF CANOLA AND RAPESEED

Economically important diseases of canola and rapeseed include sclerotinia stem rot [*Sclerotinia sclerotiorum* (Lib.) de Bary], seedling blight and root rot (*Rhizoctonia solani* KYhn), alternaria blackspot [*Alternaria brassicae* (Berk.) Sacc., and *A. raphani* Groves and Skolko], and blackleg [*Leptosphaeria maculans* (Desm.) Ces. & De Not.] (Kharbanda and Tewari, 1996; Tewari, 1985).

Sclerotinia sclerotiorum is a pathogen with a worldwide distribution and an extremely wide host range, affecting over 400 plant species in 75 plant families (Boland and Hall, 1994; Morrall and Dueck, 1982; Purdy, 1979). The pathogen survives primarily by sclerotia that can persist in the soil for at least 3 years (Williams and Stelfox, 1980). The sclerotia in the top 2 to 3 cm of soil are reported to be capable of producing apothecia (Abawi and Grogan, 1979) and germination of sclerotia occurs under cool, moist conditions (Abawi and Grogan, 1975). Disease symptoms in canola and rapeseed begin as a soft, watery rot and developing into pale grey zonate stem lesions. Under optimal conditions, these lesions expand until they completely girdle the stem. Stems often appear bleached and shredding can occur as a result of the action of cell-wall degrading enzymes produced by the pathogen. Disease caused by *S. sclerotiorum* is mostly through carpogenic infection (i.e. germination of sclerotia to produce apothecia) that results in the release of ascospores (Andrews, 1992). These ascospores germinate and colonize flower petals and eventually infect and invade living tissue (such as stems). In some cases, the sclerotia of *S. sclerotiorum* germinate myceliogenically and infect tissue at or near the base of the plant (Purdy, 1979). Sclerotinia stem rot of canola is difficult to control because the pathogen is widespread, omnivorous, and survives over several years in the soil. Cultural and chemical practices are currently used to manage this disease.

Rhizoctonia solani is a soilborne fungal pathogen causing pre- and post-emergence damping-off and root rot of several members of the Brassicaceae (Tewari *et al.*, 1987; Verma, 1996; Yitbarek *et al.*, 1988). The pathogen survives in the soil as mycelium and microsclerotia, and saprophytically on plant debris. *Rhizoctonia solani* isolates are separated into different anastomosis groupings, with AG2-1 being the most common group associated with canola in western Canada (Gugel *et al.*, 1987; Tewari *et al.*, 1987; Yitbarek *et al.*, 1987). Group AG2-1 causes pre- and post-emergence damping off and is favored by cool temperatures (15°C) (Kaminski and Verma, 1985; Tewari *et al.*, 1987). Group AG4 is also associated with canola and attacks mainly adult plants, causing basal stem rot, and is favored by warmer temperatures (20°C) (Teo *et al.*, 1988). High soil moisture favors plant disease development, particularly for AG2-1 (Teo *et al.*, 1988). Symptoms of pre- and post-emergence damping-off include death of seeds and or seedlings prior to emergence, and constrictions and discoloration of the hypocotyl which may lead to death of the emerged seedling (Calman, 1990; Teo *et al.*, 1988; Tewari, 1985; Tewari *et al.*, 1987).

Alternaria blackspot is a significant disease of canola and rapeseed throughout the world (Conn, 1991; Tewari, 1985; 1993). The causal agent has a wide host range and occurs throughout Canada, several countries in Europe and southeast Asia. The most commonly isolated pathogens in western Canada are *A. brassicae* and *A. raphani*, but the former is the most important of the two pathogens (Tewari, 1985). Symptoms on the leaves include necrotic lesions that appear as black points and as they develop, they form concentric zones, often surrounded by chlorotic halos. This eventually leads to defoliation, early senescence, and premature ripening (Conn, 1991). The stems and siliques (pods) of canola show small brown to black spots that enlarge and may cause shattering of the siliques. Disease is favored by moist, warm weather (Kolte, 1985) and disease severity increases after heavy rainfall and with relative humidity over 80% during crop development (Chahal and Kang, 1979; Kolte, 1985). *Alternaria brassicae* is known to overwinter on infected canola stubble and on infected plant debris (Tewari, 1985).

Blackleg of canola and rapeseed, caused by *L. maculans*, occurs throughout the world in countries such as Canada, France, Germany, England, and Australia (Gugel and Petrie, 1992; Humpherson-Jones, 1986; Krüger, 1979; Lacoste *et al.*, 1969; McGee and Emmett, 1977; Seidel *et al.*, 1984). Disease symptoms include greyish to dirty white lesions on the cotyledons and leaves, often with black pycnidia at the centres of the lesions (Gugel and Petrie, 1992). Stem cankers found at the base of the stem appear similar to the leaf lesions and have characteristic black or purple borders. Black pycnidia are often evident in the stem cankers. Primary inoculum consists of pycnidia, pseudothecia and mycelium in crop residues (Hall, 1992) as well as seed infected with *L. maculans*, particularly where blackleg is not previously present (Kharbanda and Tewari, 1996). The fungus can survive in crop stubble for more than five years, but the most damage occurs from inoculum produced from 2- or 3-year old stubble (Petrie, 1986).

3. BIOCONTROL AGENTS

A number of fungal and bacterial microorganisms have been successfully developed as biocontrol products for control of plant diseases (Table 1). *Rhizoctonia, Pythium, Botrytis cinerea* Pers.: Fr, and *Fusarium* may be controlled by fungal biocontrol agents such as *Gliocladium virens* Miller & Foster (GlioGard"), *Trichoderma harzianum* Rifai (TRICHODEX"), and *Streptomyces* sp. (Mycostop"). In addition, several bacterial agents, such as *Bacillus subtilis* (Quantum 400"), *B. subtilis* (Kodiak"), and *Pseudomonas fluorescens* (Dagger G") have been registered for use against several soilborne plant pathogens. Some of these agents, along with other unregistered microbial agents, have been evaluated for biological control against canola and rapeseed pathogens. The results of these investigations on each of the target pathogens will be discussed in greater detail.

Table 1. Examples of commercially available fungal and bacterial biocontrol agents

Biocontrol Agent	Target Pathogen/ Disease	Trade Name	Company/Country
Agrobacterium radiobacter	*A. tumefaciens*	Galltrol-A"	AgBioChem, Inc., USA
Agrobacterium radiobacter	Biobactericide crown gall	Digall"	ICI, New Zealand
Bacillus subtilis	*Pythium, Rhizoctonia*	Quantum 400"	USA
Pseudomonas fluorescens	*Rhizoctonia, Pythium*	Dagger G"	Ecogen, USA
Phlebia (Peniophora) gigantea	*Fomes annosus*	Rotstop	USA
Pythium oligandrum	*Pythium* spp. *Rhizoctonia*	Polygandron	Chechnia & former USSR
Gliocladium virens	*Rhizoctonia, Pythium*	GlioGard"	W.R. Grace & Co., USA
Trichoderma harzianum	*Botrytis cinerea*	TRICHODEX"	Israel
Streptomyces griseoviridis	root rot diseases in ornamental greenhouses	Mycostop"	Kemira OY, Finland
Ampelomyces quisqualis	powdery mildew on specialty soft fruits and vegetable crops	AQ-10"	Ecogen Inc., Israel
Bacillus subtilis	seed inoculant of *Fusarium* sp. and *Rhizoctonia* in various field crops	Kodiak"	Gustafson Div. of Uniroyal Chemical Co. USA
Pseudomonas cepacia	root rot fungi and nematodes in various field crops	Blue circle inoculant	Stine Seeds Co.
Trichoderma harzianum/ polysporum	wood decay fungi	BINAB T	Binab, USA

3.1. Fungi

Several fungi have exhibited biological control activity against plant pathogenic fungi of rapeseed and canola (Table 2). These include *Trichoderma* spp., *G. virens*, *G. roseum* (Link) Bain., *Myrothecium verrucaria* (Alb. & Schw.) Ditmar, *Coniothyrium minitans* Campbell, and *Talaromyces flavus* (Klöcker) Stolk and Sampson (Adams, 1990; Baker and Dickman, 1993; Dube and Podile, 1988; Papavizas, 1985). Strategies used for control of these diseases included reducing initial inoculum, reducing secondary inoculum and spread, and controlling infection in the rhizosphere and phyllosphere.

Table 2. Examples of fungal microorgansms evaluated as potential biocontrol agents for canola and rapeseed diseases

Biocontrol Agent	Pathogen	References
Coniothyrium minitans	*Sclerotinia sclerotiorum*	Huang & Hoes, 1976; Tu, 1984; Huang & Kokko, 1987; 1988, Jones & Watson, 1969; Jones *et al.*, 1974, Whipps *et al.*, 1993; McQuilken *et al.*, 1995
Cyathus olla, C. striatus	stubble-borne diseases	Tewari & Briggs, 1995; Shinners & Tewari, 1997; Tewari *et al.*, 1997
Gliocladium catenulatum	*S. sclerotiorum*	Huang, 1978
G. roseum	*S. sclerotiorum*	Pachenari & Dix, 1980
G. roseum	*Alternaria brassicae*	Boyko, 1995
Gliocladium virens	*S. sclerotiorum*	Tu, 1980; Howell & Stipanovic, 1983; Phillips, 1986; Whipps & Budge, 1990
G. virens	*Rhizoctonia solani*	Lewis & Papavizas, 1991
Myrothecium verrucaria	*A. brassicae*	Boyko, 1995
Nectria inventa	*A. brassicae*	Tsuneda *et al.*, 1976; Tsuneda & Skoropad, 1977, 1978, 1980
Sporidesmium sclerotivorum	*S. sclerotiorum*	Adams & Ayers, 1981, 1983
T. hamatum	*R. solani*	Lewis & Papavizas, 1991
T. harzianum	*R. solani*	Akhtar, 1969; Calman, 1990
Talaromyces flavus	*S. sclerotiorum*	McLaren *et al.*, 1989

3.1.1. *Sclerotinia sclerotiorum*

For biological control of *S. sclerotiorum*, fungi that parasitize the sclerotia, the major overwintering structure, have been investigated. Several of these sclerotial mycoparasites include *C. minitans* (Huang, 1977; McLaren *et al.*, 1994; Trutmann *et al.*, 1980, 1982; Whipps and Gerlagh, 1992), *G. virens* (Phillips, 1986; Tu, 1980), *G. roseum* (McCredie and Sivasithamparam, 1985), and *Trichoderma* spp. (Dennis and Webster, 1971). A significant amount of evidence has demonstrated that *C. minitans* kills both the mycelium and sclerotia of the pathogen (Huang and Hoes, 1976). It was shown that the hyphae of *C. minitans* penetrate and parasitize the hyphae and sclerotia of *S. sclerotiorum* (Huang and Kokko, 1987, 1988; Tu, 1984) and that degradation of the mycelia and sclerotial tissue occurs as a result of the production of cell wall degrading enzymes such as ß-(1-3)-glucanase and chitinase (Jones and Watson, 1969; Jones *et al.*, 1974). Ultrastructural evidence has indicated that lysis of the melanized rind of sclerotia by chemical activity is required for the penetration by *C. minitans* hyphae (Huang and Kokko, 1987). In addition, Whipps *et al.* (1993) reported, for the first time, the isolation of *C. minitans* from petals of winter oilseed rape. Although the petals containing the mycoparasite did not contain *S. sclerotiorum*, it could not be confirmed if *C. minitans* originated from spores or if it was actively growing on the petals. In field-grown oilseed rape, McQuilken *et al.* (1995) demonstrated that *C. minitans* reduced survival and carpogenic germination of sclerotia and production of apothecia, resulting in lower inoculum potential of the pathogen. However, disease control by this mycoparasite as a soil inoculant was not achieved. It was speculated that the long-distance spread of ascospores between plots may have contributed to the inability of *C. minitans* to control the disease. It may therefore be necessary to explore the potential of using *C. minitans* or other mycoparasites as a foliar spray to reduce hyphal growth and infection of *S. sclerotiorum*.

Gliocladium virens has been evaluated as a mycoparasite of mycelia and sclerotia of *S. sclerotiorum* (Phillips, 1986; Tu, 1980; Whipps and Budge, 1990). Decay of sclerotia occurred over a wide range of soil moistures and pH (pH 5-8), but activity was reduced at temperatures below 15°C (Phillips, 1986). Whipps and Budge (1990) demonstrated that the type of substrate used to raise *G. virens* inoculum affected its ability to infect and reduce viability of sclerotia. Sand used as a substrate and spores as inoculum provided the easiest method for screening *G. virens* as a mycoparasite. Other researchers also indicated that the ability of various strains of *G. virens* to parasitize *S. sclerotiorum* varied and that strain selection will play an important role in the mycoparasite-pathogen interaction (Phillips, 1986; Tu, 1980). As a mycoparasite, *G. virens* has great potential because of its ability to grow and sporulate quickly, spread rapidly, and produce metabolites such as glioviren, an antibiotic with significant antagonistic properties (Howell and Stipanovic, 1983). Other *Gliocladium* spp. (*G. roseum* and *G. catenulatum* Gilman & Abbott) that parasitize the hyphae and sclerotia of *S. sclerotiorum* produced toxins and cell wall degrading enzymes such as ß-(1-3)-glucanase and chitinase (Huang, 1978; Pachenari and Dix, 1980).

Several *Trichoderma* spp. have been used to reduce the number of sclerotia in soil. *Trichoderma viride* and *T. harzianum* parasitized sclerotia and mycelium of *S. sclerotiorum* by penetrating into host mycelia and producing lytic enzymes such as ß-(1-3)-glucanases (Dennis and Webster, 1971; Jones and Watson, 1969; Jones *et al.*, 1974). Other fungi that have been described as parasites of *S. sclerotiorum* are *T. flavus* and *Sporidesmium sclerotivorum* Uecker, Ayers, and Adams (Adams and Ayers, 1981, 1983; McLaren *et al.*, 1989). The sclerotia of *S. sclerotiorum* were penetrated by germ tubes of *Sporidesmium sclerotivorum*, and infection hyphae spread intercellularly throughout the sclerotia and ß-glucanase activity was detected (Adams and Ayers, 1981, 1983). McLaren *et al.* (1989) described the penetration and infection of *S. sclerotiorum* sclerotia by *T. flavus* and indicated that lytic enzymes produced by the mycoparasite may be responsible for the destruction of the sclerotia. In field experiments, the initial inoculum (sclerotia) was reduced, indicating the potential of this fungus as a biocontrol agent for *S. sclerotiorum*.

3.1.2. *Rhizoctonia solani*

There have been few studies exploring the use of fungal biological control agents to control damping-off and root rot diseases in canola and rapeseed. Researchers have reported the suppression of damping-off of canola, caused by *R. solani* (AG2-1 and AG4), with binucleate *Rhizoctonia* isolates (Gugel *et al.*, 1986). In controlled environment and field experiments, the incidence of disease was significantly reduced when soil was infested with a mixture of both the binucleate and the virulent isolates AG2-1 or AG4 of *R. solani*, compared to soil infested with the virulent isolates alone. Cardosa and Echandi (1987) demonstrated that, on snap bean, binucleate *Rhizoctonia*-like fungi reduced the incidence and severity of root rot caused by *R. solani* AG4, but hyperparasitism and antibiosis did not play a role in the suppression of the disease. Although the precise mechanism was not identified, it was speculated that either competition for nutrients or host-induced resistance could play an important role.

Lewis and Papavizas (1992) demonstrated that *Trichoderma hamatum* (Bonord.) Bain and *G. virens* were effective biocontrol agents for *R. solani* AG4, causing damping-off of cotton. Formulation of these biocontrol agents in bran/germling preparations reduced the saprophytic activity of the pathogen and disease severity while cotton plant stands were similar for disease-infested soils with the biocontrol agents and non-infested soils. Akhtar (1969) also demonstrated that damping-off in mustard could be substantially reduced if soils were amended with *T. harzianum*. *Trichoderma harzianum* applied as a seed treatment or soil amendment decreased symptoms on canola seedlings by 46.4% in *R. solani* infested soils, while seedling emergence was increased by 39.2% (Calman, 1990). Extensive coiling of *T. harzianum* hyphae around *R. solani* hyphae was also observed. Although *Trichoderma* spp. and *G. virens* are good candidates as fungal biocontrol agents for damping-off and root rot of canola and rapeseed, much of this research has not progressed past the discovery phase. With the emerging technology in fermentation, formulation, and application technology, these fungal agents have high potential for development as biocontrol products.

3.1.3. *Alternaria* spp.

Nectria inventa Pethybridge has been reported to be a destructive mycoparasite of *A. brassicae* (Tsuneda and Skoropad, 1977; 1978; Tsuneda *et al.*, 1976). The relationship between *N. inventa* and *A. brassicae* resulted in ultrastructural changes, leading to eventual degeneration of the host cells. The mycoparasite preferred to attach to the septal area or the basal portion of the host germ tube and conidia were more commonly parasitized than hyphae. Enzymatic activity has been suggested as the primary mode of penetration. In susceptible *Alternaria* fungi, the cytoplasm became granulated and eventual vacuolation of host cytoplasm occurred (Tsuneda and Skoropad, 1980). *Nectria inventa* reduced vegetative growth and sporulation of *A. brassicae* on excised leaves, but parasitism on intact leaves took four days, thus preventing the mycoparasite from reducing the amount of alternaria blackspot disease (Tsuneda and Skoropad, 1978). Conidia of *N. inventa* require an exogenous source of nutrients to germinate. More extensive research to provide the biocontrol agent with these requirements is necessary to ensure effective control of the disease. This may be possible by providing *N. inventa* with the appropriate formulation to enhance conidial germination and effective mycoparasitism of *A. brassicae*.

Two additional *A. brassicae* antagonists that have been investigated are *G. roseum* and *M. verrucaria* (Boyko, 1995). *Myrothecium verrucaria* inhibited conidial and mycelial growth through the production of metabolites and their effects were concentration-dependent. Although these metabolites were not identified, other researchers reported that a number of metabolites produced by *M. verrucaria* belong to the trichothecenes (Moreau, 1979). *Gliocladium roseum* was demonstrated to be parasitic to *A. brassicae*, but it did not cause any significant reduction in alternaria blackspot disease (Boyko, 1995). Further research is required to investigate the role of *M. verrucaria* metabolites on *A. brassicae* and to determine if *G. roseum* has any potential as a biocontrol agent for the disease.

Biocontrol of seed-borne diseases of oilseed crops by pathogens such as *Alternaria brassicicola* (Schw.) Wiltshire, *A. raphani* Groves and Skolko, and *A. alternata* (Fr.) Keissl. was also explored (Turhan, 1993; Vannacci and Harman, 1987; Wu and Lu, 1984). Some of the antagonists reported were *G. virens, T. harzianum, T. koningii* Oudem., *Chaetomium globosum* Kunze ex Steud., and *M. verrucaria*. Some of the antagonists coiled around the germ tubes or hyphae of *A. brassicicola* while others induced plasmolysis or granulation of the conidia or hyphae (Wu and Lu, 1984). Fungal antagonists, such as *G. roseum* and *Myrothecium* spp. produced antibiotics that caused distortion and collapse of *A. alternata* conidia (Turhan, 1993).

3.1.4. *Leptosphaeria maculans*

Innovative research using bird's nest fungi for decomposition of crop residues, such as canola, is being conducted (Shinners and Tewari, 1997; Tewari and Briggs, 1995; Tewari *et al.*, 1997). It has been suggested that *Cyathus olla* (Batch) ex. Pers. and *C. striatus* (Huds.) ex Pers. may be used for the degradation of crop residues that are inhabited by stubble-borne diseases such as blackleg disease (*L. maculans*) and sclerotinia stem rot (*S. sclerotiorum*). Calcium oxalate crystals were associated with the hyphae of these fungi and greater amounts of crystals were produced on calcium-rich V8 juice agar, compared to PDA. It has been well documented that secretion of oxalic acid by fungi results in the sequestration of calcium from plant cell walls making access easier for cell wall degrading enzymes involved in plant tissue decomposition (Dutton *et al.*, 1993; Punja and Jenkins, 1984; Rao and Tewari, 1987). Further work on the role of these bird's nest fungi in residue decomposition and biocontrol of stubble-borne diseases in canola is underway.

3.2. Bacteria

A number of bacterial strains have been explored as potential biocontrol agents of several foliar and soilborne diseases (Table 3) (Dube and Podile, 1988; Weller, 1988). Many of the bacterial strains belong to the fluorescent pseudomonads and *Bacillus* spp. and many of the strategies have focused on control of soilborne pathogens.

Table 3. Examples of fungal microorgansms evaluated as potential biocontrol agents for canola and rapeseed diseases

Biocontrol Agent	Pathogen	References
Bacillus spp.	*Sclerotinia sclerotiorum*	LYth *et al.*, 1993
Bacillus sp.	*Leptosphaeria maculans*	Kharbanda *et al.*, 1997; Yang *et al.*, 1997a, 1997b, 1998
B. subtilis	*Rhizoctonia solani*	Fiddaman & Rossall, 1993, 1995
Burkholderia cepacia	*R. solani*	Reddy *et al.*, 1993a, 1993b, 1994; Pastula *et al.*, 1997
Pseudomonas fluorescens	*R. solani*	Dahiya & Woods, 1987; Dahiya *et al.*, 1988

3.2.1. *Sclerotinia sclerotiorum*

Very little research on bacterial agents for control of *S. sclerotiorum* has been conducted. LYth *et al.* (1993) demonstrated that strains of *Bacillus* spp. applied to soil are capable of reducing apothecium formation to 36%, when compared to untreated soil. This resulted in a significant reduction in yield losses in oilseed rape. However, it is likely that these bacteria may be more suitable at controlling the disease from myceliogenic infection rather than by reducing infection by apothecia and ascospore production. It should be borne in mind what life cycle of the pathogen has the most potential for creating disease and which strategy should be used to minimize the impact of the pathogen. Although reduction of primary inoculum (i.e. sclerotial production) is feasible with many biocontrol agents, foliar spread of the disease from neighbouring fields through ascospore dispersal and via infected weeds and other plants growing in patches in the field and along roadsides will be a concern. It may therefore be worthwhile to investigate the utility of foliar-applied bacteria that inhibit *S. sclerotiorum* through various mechanisms such as antibiosis.

3.2.2. *Rhizoctonia solani*

A strain of *Pseudomonas fluorescens* isolated from the rhizosphere of canola seedlings was found to be antagonistic to *R. solani* (Dahiya and Woods, 1987; Dahiya *et al.*, 1988). This bacterial strain produced three antibiotics (pyocyanin, pyrrolnitrin, and phenazine carboxamide) that inhibited the growth of the pathogen. The cytoplasm of *R. solani* mycelium coagulated when treated with pyrrolnitrin, but not with pyocyanin and phenazine carboxamide and it was concluded that pyrrolnitrin was more inhibitory than the other two antibiotics. Survival of canola seedlings grown in *R. solani* infested soil increased if the seeds were treated with the bacterial strain at the time of planting (Dahiya and Woods, 1987; Dahiya *et al.*, 1988). Percent survival with pyrrolnitrin treated seed was 90%, compared to 18% in the untreated control. However, seed treatment with pyocyanin and phenazine carboxamide was 65% and 48%, respectively, compared to 20% and 23% in the controls. It was suggested that the *P. fluorescens* strain is a highly effective biocontrol agent for *R. solani* damping-off of canola and that the bacterially-derived antibiotics, particularly pyrrolnitrin, are effective protectants against the disease.

Three bacterial strains, *Pseudomonas fluorescens, P. aureofaciens*, and *Burkholderia cepacia*, have demonstrated antifungal activity against *R. solani* resulting in significant increases in canola stand establishment under high disease pressure (Reddy *et al.*, 1993b, 1994). Bacterization of canola seed caused a reduction in damping-off of canola and combining the bacterial strains with Vitavax RS as a fungicidal seed treatment improved stand establishment when grown in *R. solani* infested soil. Water soluble and volatile compounds with antifungal properties were detected (Pastula *et al.*, 1997). All strains produced ammonia and low levels of indole acetic acid, while two strains produced siderophores. No phosphatase or chitinase activity was detected.

Other research has described the evaluation of a *Bacillus subtilis* strain with inhibitory properties to *R. solani* (Fiddaman and Rossall, 1993; 1995). Plant stands of oilseed rape plants bacterized with *B. subtilis* and sown in *R. solani* infested soils were similar to those plants sown in non-infested soils. The bacterial strain produced an antibiotic metabolite as well as a volatile compound(s), resulting in hyphal swelling and extensive vacuolation (Fiddaman and Rossall, 1993). This suggested more than one mode of action by the bacterial strain. Temperature and nutritional factors of the growth media had a profound effect on the production of the volatile compounds. Although a great deal of screening for bacterial biocontrol agents is conducted using *in vitro* techniques that are rapid and relatively cost-effective, whole plant screening methods would allow the selection of field-competent bacterial strains that inhibit plant pathogens by mechanisms other than antibiosis.

3.2.3. *Leptosphaeria maculans*

Bacterial suspensions and culture filtrates of a *Bacillus* sp. isolated from canola stubble reduced the germination and germ tube length of *L. maculans* pycnidiospores (Yang *et al.*, 1997a,b). The viability of the pycnidiospores was reduced but no changes in germ tube morphology were observed. Development of blackleg disease in the presence of the bacteria and the pathogen was significantly reduced. Detection of the bacterial agent using digoxigenin-labelled probes is currently underway (Yang *et al.*, 1998). These probes appear to be highly strain specific and the PKB1 strain used to control *L. maculans* could be detected from infested compost. Initial characterization of the antifungal metabolite produced by the *Bacillus* sp. revealed it to be a small lipophilic peptide with a molecular weight of of about 2,000 Da (Kharbanda *et al.*, 1997). This bacterial agent was also found to be inhibitory to *A. brassicae*, *R. solani*, and *S. sclerotiorum*. More research on this bacterial agent and the nature of its metabolite is in progress.

4. MECHANISMS

Biological control activity by fungal and bacterial agents may be attributed to a variety of mechanisms. They include parasitism and lysis, antibiosis, competition, and induced resistance. Understanding the interactions between the biological control agent and the plant pathogen may provide clues to developing strategies and facilitate practical implementation of microorganisms for plant disease management.

4.1. Parasitism and Lysis

Parasitism is a common mechanism amongst fungal biocontrol agents such as *Trichoderma* spp., *G. virens*, and *C. minitans*. It often results in the destruction of fungal pathogens through the lysis of fungal cell walls containing chitin (Baker and Dickman, 1993; Handelsman and Stabb, 1996; Jacobsen and Backman, 1993; Kloepper, 1993). Hyphae of the mycoparasite may grow alongside the host hypha or coil around it, followed by the lysis of the host wall through the production of enzymes such as chitinases, chitobiosidase, N-acetyl-ß-glucosaminidase, and glucan 1,3-ß-glucosidases (Chet *et al.*, 1990; Handelsman and Stabb, 1996; Tronsmo and Hjeljord, 1998). For some *Trichoderma* spp., the amount of inhibition of the fungal pathogen was found to be regulated by the amount of chitin in fungal cell walls (Tronsmo and Hjeljord, 1998; Ulhoa and Peberdy, 1991). Researchers have also observed that many of these enzymes may act either individually or in combination with antifungal metabolites (Di Pietro *et al.*, 1993; Lorita *et al.*, 1993; 1994). For example, low concentrations of gliotoxin and endochitinase produced by *G. virens*, caused little or no inhibition of *B. cinerea* mycelium, but when applied in combination, 95% inhibition was observed (De Pietro *et al.*, 1993).

4.2. Antibiosis

One of the most frequently cited modes of action of fungal and bacterial biocontrol agents is antibiosis (Baker and Dickman, 1993; Chet *et al.*, 1990; Handelsman and Stabb, 1996; Kloepper, 1993). Perhaps one of the reasons why this appears to be a common and popular mechanism is the bias for screening microorganisms for antibiosis. One of the disadvantages is that antibiosis *in vitro* is not always correlated with antibiosis *in vivo*. However, screening for this mode of action is attractive because it is relatively easy and inexpensive to incorporate into a biocontrol screening program. Also, more than one antibiotic may be produced by one strain of bacteria or fungi.

Several antibiotics have been identified from bacteria and fungi (Table 4). Antibiotics such as pyrrolnitrin, pyocyanin, and phenazine carboxamide produced by *P. fluorescens* were involved in suppression of *R. solani* (Dahiya and Woods, 1987; Dahiya *et al.*, 1988), while the nature of the antibiotic metabolites of *Bacillus* spp. for control of *R. solani* and *L. maculans* was not identified (Fiddaman and Rossall, 1993, 1995; Yang *et al.*, 1997a,b). Two antibiotics, zwittermicin A and kanosamine, were produced by *Bacillus cereus*, which was found to be inhibitory to oomycetous fungi (Milner *et al.*, 1996; Silo-Suh *et al.*, 1994).

Table 4. Examples of antibiotics produced by bacterial and fungal biocontrol agents

Biocontrol Agent	Antibiotic
Pseudomonas spp.	pyoluteorin
	pyrrolnitrin
	2,4-diacetylphloroglucinol
	hydrogen cyanide
	phenazine 1-carboxylate
	2-hydroxyphenazine 1-carboxylate
Bacillus cereus	kanosamine
	zwittermicin A
Gliocladium virens	gliovirin
	gliotoxin

Fungi known to produce antibiotics are *M. verrucaria*, *Gliocladium virens*, and *Trichoderma* spp. (Baker and Dickman, 1993). Although the metabolites produced by M. *verrucaria* showing antifungal activity against *A. brassicae* were not identified (Boyko, 1995), some antibiotics reported from *M. verrucaria* were verricarin, roridin, myrothecine, dehydroverrucan, and muconomycin (Cook and Baker, 1983). The role of antibiotics in *G. virens* and *Trichoderma* spp. are the most extensively studied for fungal biocontrol agents. These two fungi produce several antimicrobial compounds, including gliotoxin and viridin, that suppress a number of plant diseases. Baker and Dickman (1993) indicated that sufficient amounts of the antibiotics must be produced in order to have adequate biological control activity, and they must be produced at the appropriate time when the plant pathogen is susceptible to the antibiotic-producing agent. This may be one reason why antibiosis demonstrated *in vitro* is not always exhibited *in vivo*. Understanding the life cycle and stage of susceptibility of the plant pathogen to antibiotics will improve consistency and field performance of the biocontrol agent.

4.3. Competition

Competition as a mechanism for biocontrol agents refers to competition for infection sites and nutrients (Baker and Dickman, 1993; Chet *et al.*, 1990; Kloepper, 1993). This mechanism often involves the fungal or bacterial biocontrol agent competing for food and essential nutrients in the same infection court (either in the rhizosphere, rhizoplane, phyllosphere, or phylloplane). It was suggested that pseudomonads catabolize a variety of nutrients quickly and have fast generation times, thereby making them excellent candidates as biocontrol agents by competition for nutrients (Kloepper, 1993; Weller, 1985). Another form of competition is the microbial production of siderophores, which are low molecular weight compounds with high affinity for $Fe3+$, that mediate iron transport into the cell (Kloepper *et al.*, 1980). Specific bacterial strains, such as fluorescent pseudomonads, produce these siderophores (e.g. pseudobactin) that chelate iron that is already limited in the rhizosphere. This results in making iron unavailable to other rhizosphere-inhabiting microorganisms such as soilborne plant pathogens.

4.4. Induced Systemic Resistance

Induced systemic resistance (ISR) refers to enhanced changes in the level of plant resistance leading to reduced disease or increased tolerance of the plant to a pathogen (Baker and Dickman, 1993; Handelsman and Stabb, 1996). In theory, avirulent microorganisms (such as fluorescent pseudomonads) are pre-inoculated on the plant to induce a plants natural defense machinery, analogous to an immunization process, that finally results in the prevention of diseases caused by a variety of plant pathogens (Conn and Tewari, 1992; Kloepper *et al.*, 1992). Phytoalexin elicitation, as one mechanism in plant disease resistance, may be used as a plant protectant in biological control strategies. Brassinin, methoxybrassinin, and cyclobrassinin were the first phytoalexins reported from *Brassica* spp. (Pedras and Séguin-Swartz, 1992). In addition, brassilexin and cyclobrassinin sulphoxide have reportedly been elicited by *L. maculans*. ISR may also activate a variety of potential defense mechanisms that include higher activity of chitinases, ß-1-3-glucanases, peroxidases, and pathogenesis-related (PR) proteins (Lawton and Lamb, 1987). ISR may result in the protection of the plant from a wide spectrum of plant pathogens and the effect is systemic, unlike other mechanisms that may be localized and at the point of infection. The development of biocontrol agents whose mechanism is through ISR is very attractive since the number of plant pathogens that may be potentially controlled is increased and it may be possible to apply a mixture of ISR-mediated bacterial strains for biological control. Screening of bacterial strains for this mode of action may be somewhat more expensive, time-consuming, and labor-intensive than screening for antibiotic production.

5. PRACTICAL APPROACHES FOR IMPLEMENTATION

Research in biological control has progressed from the initial discovery and evaluation phase to the development of practical application techniques. With the emergence of numerous strains or isolates of bacterial and fungal agents, researchers have investigated methods that will lead to the practical implementation of biocontrol agents. With increased understanding into the mechanisms by which these agents control plant diseases, selection and screening criteria can be evaluated and new technology in fermentation and formulation can be evaluated.

5.1. Selection and Screening of Biocontrol Agents

The search for biological control agents for commercial application must be practical and rational. Initial investigations should consider where are the most suitable sites for isolating the fungal and bacterial agents. Campbell (1986) offered a pragmatic approach for selecting search sites and screening methods for biocontrol agents. It was suggested that culture collections are generally unsuitable because the microorganisms may have high nutritional requirements that are impractical for a commercial product and lack survival abilities in harsh environments. Practical considerations for selecting sites include areas where there is high disease pressure and where the plants show little or no disease symptoms. Disease suppressive soils offer a logical site for selecting naturally-occurring biocontrol agents for soilborne diseases. It may also be practical to isolate microorganisms from agroecosystems under various farm management practices where a crop is being grown. This may aid in choosing microorganisms that are compatible with certain farm management practices such as fertilizer and pesticide applications and tillage practices.

It has been argued that screening methods that focused on antibiosis *in vitro* may not be suitable and should be avoided (Campbell, 1986). Screening for this mode of action is easy and inexpensive and permits massive screening of several strains of microorganisms. If the goal is to select microorganisms with high capabilities of natural metabolite production and to develop these natural products for commercial application, then prescreening for antibiosis may be appropriate. For example, a collaborative study between Agriculture and Agri-Food Canada and the University of Saskatchewan (Saskatoon, Saskatchewan, Canada) involved the evaluation of several bacterial strains from canola and wheat plants for growth inhibition of phytopathogenic fungi (J.R. de Freitas, S.M. Boyetchko, J.J. Germida, and G.G. Khachatourians, unpublished data). The objective of this project is to discover novel, natural active metabolites against canola pathogens by screening bacteria for antibiosis *in vitro* (Figure 1). Out of 581 bacterial strains tested, 165 exhibited biocontrol activity against *R. solani* (AG2-1 and AG4 are treated as two different pathogens), *L. maculans, S. sclerotiorum*, and *Pythium* sp. The most effective bacteria were 12 strains that were inhibitory to all 5 fungal pathogens, followed by 28 that were inhibitory to 4 of the fungal pathogens (Figure 2). The possible use of one or more strains with activity against a wide range of pathogens is attractive from a commercial standpoint. Also, many of these bacterial strains were isolated from canola grown under various crop production systems, which may result in obtaining bacterial strains that are adapted to agroecosystem practices. The ability of these bacteria to produce novel metabolites and characterization of these metabolites are currently being determined.

Many potential antagonists, however, may be inadvertently disregarded because they demonstrate no inhibition in agar bioassays and this may exclude the discovery of antagonists that control plant pathogens through other mechanisms (Elad and Chet, 1995; Kloepper, 1991). Kloepper (1991) described two *in vivo* screening bioassays: a radicle bioassay as a screening method for pre-emergence damping-off and a hypocotyl bioassay for post-emergence damping-off caused by *R. solani*. The advantages of these bioassays include the ease of use and inclusion of the host in the evaluation. Inclusion of the host plant provides selection of biocontrol agents with modes of action such as induced resistance, competition, parasitism, as well as antibiosis. Nonetheless, selection of the most appropriate screening method should be part of the initial evaluation in a biocontrol program.

5.2. Formulation and Delivery

One of the major considerations in developing a viable biocontrol product is the production, formulation, and delivery of propagules that are viable and effective (Elad and Chet, 1995; Jacobsen and Backman, 1993; Lumsden *et al.*, 1995). Both solid state and liquid fermentation technology have been developed, but liquid fermentation is more readily available to industry for scale up production (Lumsden *et al.*, 1995). Bacteria are easily mass-produced in liquid fermentation, but careful selection of appropriate nutrients for enhanced efficacy is required. Maximum production of propagules (bacterial cells) do not often confer optimum biological activity (Boyetchko, unpublished data). Reddy *et al.* (1993a) also demonstrated that growth factors such as pH, temperature, sources of carbon, nitrogen, and amino acids, have a significant impact on microbial antagonism. Some biocontrol products, especially fungi, may either lack the ability to sporulate in liquid culture or may not survive the liquid fermentation process and, therefore, part of the fermentation process must be carried out in a solid-state phase. In North America, solid-state fermentation is not extensively used except for mushroom production (Lumsden *et al.*, 1995). However, solid-state fermentation should be considered, particularly for those biocontrol agents that cannot be produced in liquid culture.

Many researchers have developed a variety of methods for production of fungal biocontrol agents such as *Trichoderma, Gliocladium*, and *Talaromyces* (Backman and Rodrigues-Kabana, 1975; Hadar *et al.*, 1979; Papavizas, 1985; Papavizas *et al.*, 1984; Sivan *et al.*, 1984). One of the most successful media for sporulation and growth of *T. harzianum* was shown to be wheat bran which acted as both a food base and carrier (Hadar *et al.*, 1979). Conidial viability and shelf life were significantly improved. By adding peat to the wheat bran at a 1:1 (v/v) ratio, the pH of the medium was more effectively controlled and was found

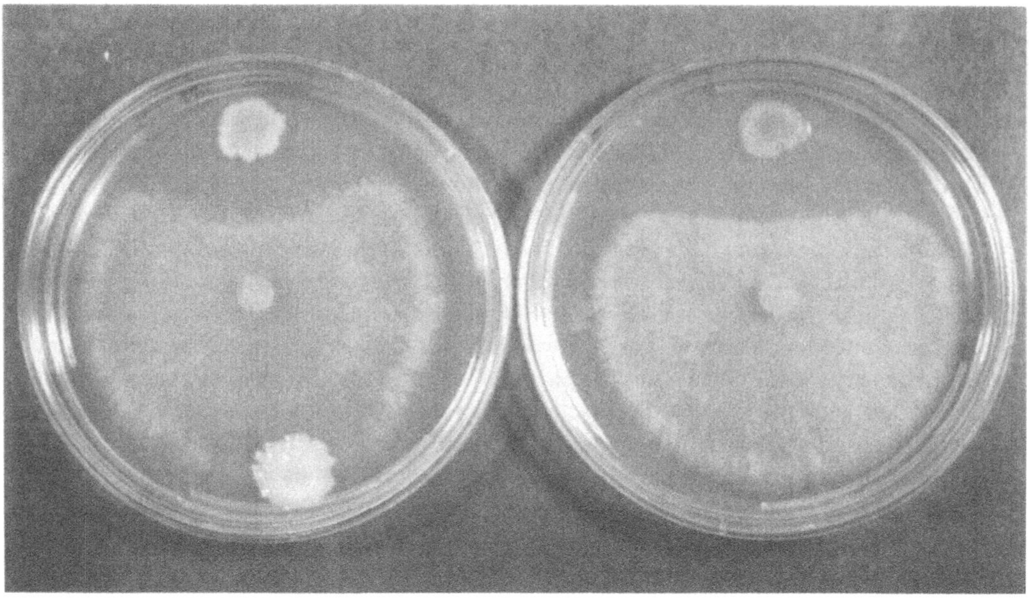

Figure 1. Bioassay used to screen bacterial strains for biological control activity *in vitro* against phytopathogenic fungi. Strain WPR3-16 (right) and strains WPR2-23 and WPR3-47 (left) are inhibitory to *Rhizoctonia solani* AG2-1.

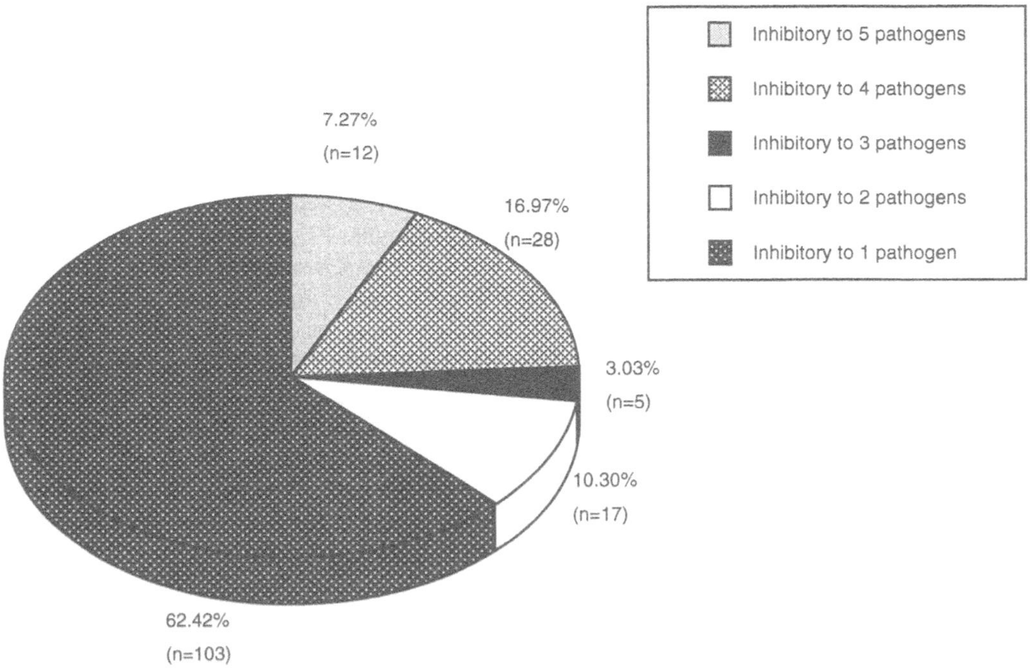

Figure 2. Efficiency of biocontrol agents to exhibit inhibitory activity *in vitro* against the phytopathogenic fungi *Rhizoctonia solani* AG2-1, *R. solani* AG4, *Sclerotinia sclerotiorum*, *Leptosphaeria maculans*, and *Pythium* sp.

to be a better carrier than wheat bran alone (Sivan *et al.*, 1984). It was also reported that various grain seeds and meals, straw, wheat bran, sawdust, and peat could be used as substrates for producing *Trichoderma, Gliocladium, Coniothyrium*, and *Talaromyces* (Papavizas, 1985; Papavizas *et al.*, 1984). Some of these fungal biocontrol agents could also be produced in 20-L liquid fermentors using relatively inexpensive ingredients such as molasses and either dried brewer's yeast, cotton seed flour, or corn-steeped liquor (Lewis and Papavizas, 1985).

Lumsden *et al.* (1995) indicated a number of important characteristics required in a formulated biocontrol product. It must be easy to prepare and apply, inexpensive, stable during transportation and storage, and provide adequate shelf-life of abundant and effective propagules. Processes that may affect the product are speed of drying and milling. Formulations that have been developed include granules, pellets, dusts, or wettable powders that are either applied directly or in suspension. Powder or dust formulations containing *Trichoderma* with Pyrax have been successfully developed. These types of formulations may be processed using spray drying, freeze-drying, or air drying either with or without the aid of a fluidized bed dryer. Inert carriers, such as fine clay, peat, vermiculite, alginate, or polyacrylamide, are often added to the dry formulations to facilitate delivery. Immobilization of bacteria with polymers, such as polyacrylamide and sodium alginate, was successfully used for bacteria such as *Pseudomonas* spp. (Bryant, 1994; Glass, 1993; Jha *et al.*, 1993; Lumsden *et al.*, 1995). Products such as *Pseudomonas fluorescens* (Dagger G") have been formulated as a peat granule and applied in-furrow for control of soilborne diseases. Storage conditions, such as temperature, will also have a significant impact on the stability and efficacy of the biocontrol products (Colvert *et al.*, 1997). For example, some bacterial strains were found to be very sensitive to temperature, while other bacterial strains were not affected by temperature. These factors should therefore be evaluated on a case-by-case basis.

5.3. Genetic Manipulation of Biocontrol Agents

Techniques in molecular biology offer additional methods for modifying or improving certain genetic characteristics of fungal and bacterial biocontrol agents (Elad and Chet, 1995; Kloepper, 1993). Understanding the molecular basis of the mechanism of action will provide researchers with tools for enhancing or transferring a specific trait into ecologically superior microbial agents or combining important traits from more than one biocontrol agent by techniques such as protoplast fusion. A practical application of molecular technology is to elucidate DNA sequences as strain specific probes for environmental risk assessment. It may also be possible to clone the chitinase gene from one biocontrol agent into another superior strain to enhance the antifungal activity (Elad and Chet, 1995). Mutations induced by chemical mutagens or irradiation may also be used to improve biocontrol agents. These methods can result in the development of fungicide-resistant mutants, and mutants that may grow, sporulate, and colonize the rhizosphere or phyllosphere more effectively.

6. FUTURE PROSPECTS

Over the years, several fungal and bacterial agents have been evaluated as potential biological control agents of plant diseases. The different approaches in biocontrol have included reducing initial inoculum, protecting the infection court, and reducing the virulence and/or spread of the pathogen. Despite these efforts, commercial development of biocontrol products has not been extensively realized and many of the biocontrol agents have not progressed past the initial discovery and evaluation phases. A systematic and pragmatic approach for developing biological control strategies must be considered for their implementation in a disease management program.

Researchers must understand which control strategy would be most the appropriate for a particular pathosystem. Extensive knowledge of the biology and epidemiology of the target pathogen must be clearly demonstrated. For example, if infection and spread by a plant pathogen is more prevalent by foliar inoculum than by soil inoculum, selection of a biocontrol strategy would favor a foliar-applied biocontrol agent. Study of the microbial ecology of the rhizosphere and phyllosphere is required to understand the complex interactions of biotic, physical, and chemical processes that occur in the ecosystem and what impact these factors have on the biocontrol agent. Multidisciplinary activities in plant pathology and applied microbiology, host-pathogen interactions, epidemiology, molecular technology, and fermentation and formulation technology will advance the development of biocontrol agents to the commercial stage.

Utkhede (1996) stressed the concern for the lack of appreciation by administrators for the effort and long-term funding required to develop biological control agents. Efforts are required to priorize target pests where biocontrol strategies will be effective. Resources for infrastructure, personnel, and technology development by government and university institutions, and industry are necessary and collaborations with groups with expertise in a number of disciplines will advance the development of biological control research. New technologies in fermentation and formulation are emerging and use of molecular tools may lead to strain improvement and enhanced efficacy. There is a need for public education in the benefits of biological control products, including the acceptability of genetically modified organisms.

Despite the numerous challenges that face researchers, biocontrol has tremendous potential. The success will remain in the hands of committed researchers who investigate various aspects in biocontrol, and adopt a holistic approach for implementation into a plant disease management program.

7. CONCLUSIONS

Biological control of plant diseases has been an area of intense study in plant pathology for several decades, but few biocontrol products have reached the marketplace. Incentives that have prompted more concerted efforts in this area include development of pesticide resistance, spray drift issues, de-registration of fungicides, and lower costs associated with the discovery of microbial biopesticide products. Emerging technologies in fermentation and formulation make biocontrol products attractive alternatives to chemical fungicides. Research in biocontrol of canola pathogens is replete with examples of fungal and bacterial agents that have been evaluated for their potential to control diseases such as sclerotinia stem rot, seedling blight and root rot, alternaria blackspot, and blackleg. Some fungi that have potential include *Trichoderma* spp., *Gliocladium virens*, *Coniothyrium minitans*, and *Talaromyces flavus*. Athough the number of studies focussing on bacteria are not as extensive, some prospective agents include *Bacillus* spp., *Pseudomonas fluorescens*, and *Burkholderia cepacia*.

REFERENCES

Abawi, G.S. and Grogan, R.G. 1975, Source of primary inoculum and effects of temperature and moisture on infection of beans by *Whetzelinia sclerotiorum*, Phytopathology 65:300-309.

Abawi, G.S. and Grogan, R.G. 1979, Epidemiology of diseases caused by *Sclerotinia* species, Phytopathology 69:899-904.

Adams, J.H. and Ayers, W.A. 1981, *Sporidesmium sclerotivorum*: distribution and function in natural biological control of sclerotial fungi, Phytopathology 71:90-93.

Adams, J.H. and Ayers, W.A. 1983, Histological and physiological aspects of infection of sclerotia of two

Sclerotinia species by two mycoparasites, Phytopathology 73:1072-1076.
Adams, P.B. 1990, The potential of mycoparasites for biological control of plant diseases, Annu. Rev. Phytopathol, 28:59-72.
Akhtar, M. 1969, Biological control of *Rhizoctonia solani* causing seed decay and damping-off of mustard, West Pakistan J. Agric. Res. 7:29-52.
Andrews, J.H. 1992, Biological control in the phyllosphere, Annu. Rev. Phytopathol, 30:603-635.
Backman, P.A. and Rodrigues-Kabana, R. 1975, A system for the growth and delivery of biological control agents to the soil, Phytopathology 65:819-821.
Bailey, K.L. 1997, IPM practices for reducing fungicide use in field crops, in : Techniques for Reducing Pesticide Use, D. Pimentel, ed., John Wiley & Sons, Chichester, pp.293-316.
Baker, K.F. and Cook, R.J. 1974, Biological Control of Plant Pathogens, W.H. Freeman and Company, San Francisco.
Baker, R. and Dickman, M.B. 1993, Biological control with fungi, in : Soil Microbial Ecology, F. Blaine Metting, Jr., ed., Marcel Dekker, Inc., New York, pp. 275-305
Boland, G.J. and Hall, R. 1994, Index of plant hosts of *Sclerotinia sclerotiorum*, Can. J. Plant Pathol. 16:93-108.
Boyko, L.A. 1995, Interactions of *Gliocladium roseum* and *Myrothecium verrucaria* with *Alternaria brassicae*, M.Sc. Thesis, University of Alberta, Edmonton, 266 p.
Bryant, J.E. 1994, Commercial production and formulation of *Bacillus thuringensis*, Agric. Ecosys. and Environ. 49:31-35.
Calman, A.I. 1990, Canola seedling blight in Alberta: pathogens, involvement of *Pythium* spp. and biological control of *Rhizoctonia solani*, M.Sc. Thesis, University of Alberta, Edmonton. 108 pp.
Campbell, R 1986, The search for biological control agents against plant pathogens : a pragmatic approach, Biol. Agric. Hort. 3:317-327.
Cardosa, J.E. and Echandi, E. 1987, Biological control of *Rhizoctonia* root rot of snap bean with binucleate *Rhizoctonia*-like fungi, Plant Dis. 71:167-170.
Chahal, A.S. and Kang, M.S. 1979, Some aspects of seed-borne infection of *Alternaria brassicae* in rape and mustard cultivars in the Punjab, Indian J. Mycol. Plant Pathol. 9:51-55.
Chet, I., Ordentlich, A., Shapira, R. and Oppenheim, A. 1990, Mechanisms of biocontrol of soil-borne plant pathogens by Rhizobacteria, Plant and Soil 129:85-92.
Colvert, D.C., Pedersen, E.A., Funk, L.M. , Seresinhe, N. , Reddy, M.S. and Brown, G.L., 1997, Influence of storage conditions on the stability and shelf-life of bacterial inoculants with commercial potential and their activity towards fungal pathogens (Abstr), Can. J. Plant Pathol. 19:107-108.
Conn, K.L. 1991, Phytoalexin production in crucifers, Ph.D. Thesis, University of Alberta, Edmonton, 151pp.
Conn, K.L. and Tewari, J.P. 1992, Phytoalexins as control agents for pathogens, in : Recent Developments in Biocontrol of Plant Diseases, K.G. Mukerji, J.P. Tewari, D.K. Arora and G. Saxena, eds., Aditya Books Private Limited, New Delhi, pp.146-153.
Cook, R.J. 1993, Making greater use of introduced microorganisms for biological control of plant pathogens, Annu. Rev. Phytopathol. 31:53-80.
Cook, R.J. 1996, Assuring the safe use of microbial biocontrol agents: a need for policy based on real rather than perceived risks, Can. J. Plant Pathol. 18:439-445.
Cook, R.J. and Baker, K.F. 1983, The Nature and Practice of Biological Control of Plant Pathogens, 2nd ed., American Phytopathological Society, St. Paul.
Dahiya, J.S. and Woods, D.L. 1987, Control of *Rhizoctonia solani*, causal agent of brown girdling root rot of canola/rapeseed by *Pseudomonas fluorescens* (Abstr), Can. J. Plant Pathol. 9:275-276.
Dahiya, J.S., Woods, D.L. and Tewari, J.P. 1988, Control of *Rhizoctonia solani*, causal agent of brown girdling root rot of rapeseed, by *Pseudomonas fluorescens*, Bot. Bull. Academia Sinica 29:135-142.
Dennis, C. and Webster, J. 1971, Antagonistic properties of species-groups of *Trichoderma*, I, Production of nonvolatile antibiotics, Trans. Br. mycol. Soc. 57:25-39.
Di Pietro, A., Lorito, M. , Hayes, C.K. , Broadway, R.M. and Harman,G.E. 1993, Endochitinase from *Gliocladium virens*: Isolation, characterization and synergistic antifungal activity in combination with gliotoxin, Phytopathology 83:308-313.
Dube, H.C. and Podile, A.R. 1988. Biological control of fungal diseases of plants, in : Perspectives in Mycology and Plant Pathology, V.P. Agnihotri, A.K. Sarbhoy, and D. Kumar, eds., Malhotra Publishing House, New Delhi, pp. 496-509.
Dutton, M.V., Evans, C.S. Atkey, P.T. and Wood, D.A. 1993, Oxalate production by basidiomycetes, including the white-rot species *Coriolus versicolor* and *Phanerochute chrysosporium*, Appl. Microbiol, Biotechnol. 39:5-10.
Elad, Y. and Chet, I. 1995, Practical approaches for biocontrol implementation, in : Novel Approaches to Integrated Pest Management, R. Reuveni, ed., Lewis Publishers, London, pp. 323-338.

Fiddaman, P.J. and Rossall, S. 1993, The production of antifungal volatiles by *Bacillus subtilis*, J. Appl. Bacteriol. 74:119-126.

Fiddaman, P.J. and Rossall, S. 1995, Selection of bacterial antagonists for the biological control of *Rhizoctonia solani* in oilseed rape (*Brassica napus*), Plant Pathol. 44:695-703.

Glass, D.J. 1993, Commercialization of soil microbial technologies, in: Applications in Agricultural and Environmental Management, Soil Microbial Ecology, F., Jr., Blaine Metting, ed., Marcel Dekker Inc., New York, pp. 595-618.

Gugel, R.K., and Petrie, G.A. 1992, History, occurrence, impact, and control of blackleg of rapeseed, Can. J. Plant Pathol. 14:36-45.

Gugel, R.K., Verma, P.R. and Kaminski, D.A. 1986, Protection of canola seedlings from *Rhizoctonia solani* infection by a nonpathogenic, binucleate *Rhizoctonia*-like isolate, (Abstr) Can. J. Plant Pathol. 8:349.

Gugel, R.K., Yitbarek, S.M., Verma, P.R., Morrall, R.A.A. and Sadasivaiah, R.S. 1987, Etiology of the *Rhizoctonia* root rot complex in the Peace River region of Alberta, Can. J. Plant Pathol. 9:119-128.

Hadar, Y, Chet, I. and Henis, Y. 1979, Biological control of *Rhizoctonia solani* damping-off with wheat bran culture of *Trichoderma harzianum*, Phytopathology 69:64-68.

Hall, R. 1992, Epidemiology of blackleg of oilseed rape, Can. J. Plant Pathol. 14:46-55.

Handelsman, J. and Stabb, E.V. 1996, Biocontrol of soilborne plant pathogens, The Plant Cell 8:1855-1869.

Howell, C.R., and Stipanovic, R.D. 1983, Gliovirin, a new antibiotic from *Gliocladium virens*, and its role in the biological control of *Pythium ultimum*, Can. J. Microbiol. 29:321.

Huang, H.C. 1977, Importance of *Coniothyrium minitans* in survival of sclerotia of *Sclerotinia sclerotiorum* in wilted sunflower, Can. J. Bot. 55:289-295.

Huang, H.C. 1978, *Gliocladium catenulatum* : hyperparasite of *Sclerotinia sclerotiorum* and *Fusarium* species, Can. J. Bot. 56:2243-2246.

Huang, H.C. and Hoes, J.A. 1976, Penetration and infection of *Sclerotinia sclerotiorum* by *Coniothyrium minitans*, Can. J. Bot. 54:406-410.

Huang, H.C. and Kokko, E.G. 1987, Ultrastructure of hyperparasitism of *Coniothyrium minitans* on sclerotia of *Sclerotinia sclerotiorum*, Can. J. Bot. 65:2483-2489.

Huang, H.C. and Kokko, E.G. 1988, Penetration of hyphae of *Sclerotinia sclerotiorum* by *Coniothyrium minitans* without the formation of appressoria, J. Phytopathol. 123:133-139.

Humpherson-Jones, F.M. 1986, The occurrence of virulent pathotypes of *Leptosphaeria maculans* in brassica seed crops in England, Plant Pathol. 35:224-231.

Jacobsen, B.J. and Backman, P.A. 1993, Biological and cultural plant disease controls: alternatives and supplements to chemicals in IPM systems, Plant Dis. 77:311-315.

Jha, P.K., Nair, S. and Babu, S. 1993, Encapsulation of seeds of *Sesbania sesban* with polyacrylamide and alginate gel entrapped rhizobia leads to effective symbiotic nitrogen fixation, Indian J. Expt. Biol. 31:161-167.

Jones, D. and Watson, D. 1969, Parasitism and lysis by soil fungi of *Sclerotinia sclerotiorum* (Lib.) de Bary a phytopathogenic fungi, Nature 224:287-288.

Jones, D., Gordon, A.H. and Bacon, J.S.D. 1974, Co-operative action by endo- and exo-ß-(1-3)-glucanases from parasitic fungi in the degradation of cell wall glucans of *Sclerotinia sclerotiorum* (Lib.) de Bary, Biochem. Jour. 140:47-55.

Kaminski, D.A. and Verma, P.R. 1985, Cultural characteristics, virulence, and *in vitro* temperature effect on mycelial growth of *Rhizoctonia* isolates from rapeseed, Can. J. Plant Pathol. 7:256-261.

Kharbanda, P.D. and Tewari, J.P. 1996, Integrated management of canola diseases using cultural methods, Can. J. Plant Pathol. 18:168-175.

Kharbanda, P.D., Yang, J., Beatty, P., Jensen, S. and Tewari, J.P. 1997, Potential of a *Bacillus* sp., to control blackleg and other diseases of canola (Abstr), Phytopathology 87(Suppl):S51.

Kloepper, J.W. 1991, Development of *in vivo* assays for prescreening antagonists of *Rhizoctonia solani* on cotton, Phytopathology 81:1006-1013.

Kloepper, J.W. 1993, Plant growth promoting rhizobacteria as biological control agents, in: Microbial Ecology, F.B. Metting Jr., ed., Marcel Dekker Inc., New York, pp.255-273.

Kloepper, J.W., Leong, J., Teinze, M. and Schroth, M.N. 1980, Enhanced plant growth by siderophores produced by plant growth-promoting rhizobacteria, Nature 286:885-886.

Kloepper, J.W., Tuzun, S. and Kuc, J. 1992, Proposed definitions related to induced disease resistance, Biocontrol Sci. Technol, 2:349-351.

Kolte, S.J. 1985, Diseases of Annual Edible Oilseed Crops, Vol. II, Rapeseed-mustard and Sesame Diseases, CRC Press Inc., Boca Raton. 135 pp.

Krüger, W. 1979, Verbreitung der Wurzelhals- und Stengelf Sule (verursacht durch *Phoma lingam*) bei Raps in der Bundersrepublik Deutschland, Nachrichtenbl, Deut. Pflanzenschutzd. 31:145-148.

Lacoste, L., Louvet, J., Anselme, C., Alabouvette, C., Brunin, B. and Pierre, J.G. 1969, Rôle de *Phoma lingam* (Tode) Desm. et de sa forme parfaite, *Leptosphaeria maculans* (Desm.) Ces. et de Not. dans les épidémies de nécrose du collet de colza (*Brassica napus* L. var. *oleifera* Meltzer), C.R. Hebd. Séances Acad. Agric. Fr. 55:981-989.

Lawton, M.A. and Lamb, C.J. 1987, Transcriptional activation of plant defense genes by fungal elicitor, wounding and infection, Mol. Cell. Biol. 7:335-341.

Lewis, J.A. and Papavizas, G.C. 1985, Characteristics of alginate pellets formulated with *Trichoderma* and *Gliocladium* and their effect on the proliferation of the fungi in soil. Plant Pathol. 34:571-577.

Lewis, J.A. and Papavizas, G.C. 1992, Potential of *Lactisaria arvalis* for the biocontrol of *Rhizoctonia solani*, Soil Biol. Biochem. 24 : 1075-1079.

Lorita, M., Harman, G.E., Hayes, C.K., Broadway, R.M., Tronsmo, A., Woo, S.L. and Di Pietro, A. 1993, Chitinolytic enzymes produced by *Trichoderma harzianum*: Antifungal activity of purified endochitinase and chitobiosidase, Phytopathology 83:302-307.

Lorita, M., Peterbauer, C., Hayes, C.K. and Harman, G.E. 1994, Synergistic interaction between fungal cell wall degrading enzymes and different antifungal compounds enhances inhibition of spore germination. Microbiology 140:623-629.

Lumsden, R.D., Lewis, J.A., and Fravel, D.R. 1995, Formulation and delivery of biocontrol agents for use against soilborne plant pathogens, in : Biorational Pest Control Agents, Formulation and Delivery, F.R. Hall and J.W. Barry, eds., ACS Symposium Series 595, Washington, pp. 166-182

LYth, P., Schuylz, R.R. and Pfeffer, H. 1993, The influence of bacterial antagonists on the infestation of a soil as well as on the yield of winter oilseed rape affected by *Sclerotinia sclerotiorum*, Zentralbl. Mikrobiol. 148:32-38.

McCredie, T.A., and Sivasithamparam, K. 1985, Fungi mycoparasitic on sclerotia of *Sclerotinia sclerotiorum* in some Western Australian soils, Trans. Br. mycol. Soc. 84:736-739.

McGee, D.C. and Emmett, R.W. 1977, Blackleg (*Leptosphaeria maculans* (Desm.) Ces. et de. Not.) of rapeseed in Victoria, Crop losses and factors which affect disease severity, Aust. J. Agric. Res. 28:47-51.

McLaren, D.L., Huang, H.C. Kozub, G.C. and Rimmer, S.R. 1994, Biological control of sclerotinia wilt of sunflower with *Talaromyces flavus* and *Coniothyrium minitans*, Plant Dis. 78:231-235.

McLaren, D.L., Huang, H.C. Rimmer, S.R. and Kokko, E.G. 1989, Ultrastructural studies on infection of sclerotia of *Sclerotinia sclerotiorum* by *Talaromyces flavus*, Can.J. Bot. 67: 2199-2205.

McQuilken, M.P., Mitchell, S.J., Budge, S.P., Whipps, J.M., Fenlon, J.S., and Archer, S.A. 1995, Effect of *Coniothyrium minitans* on sclerotial survival and apothecial production of *Sclerotinia sclerotiorum* in field-grown oilseed rape, Plant Pathol. 44:883-896.

Milner, J.L., Silo-Suh, L.A., Lee, J.C., He, H. , Clardy, J. and Handelsman, J. 1996, Production of kanosamine by *Bacillus cereus* UW85, Appl. Environ. Microbiol. 62:3061-3065.

Moreau, C. 1979, Moulds, Toxins, and Food, John Wiley and Sons, New York, 477 pp.

Morrall, R.A.A. and Dueck, J. 1982, Epidemiology of Sclerotinia stem rot of rapeseed in Saskatchewan, Can. J. Plant Pathol. 4:161-168.

Pachenari, A. and Dix, N.J. 1980, Production of toxins and wall degrading enzymes by *Gliocladium roseum*, Trans. Br. mycol. Soc. 74:561-566.

Papavizas, G.C. 1985, *Trichoderma* and *Gliocladium* : biology, ecology, and potential for biocontrol, Annu. Rev. Phytopathol. 23:23-54.

Papavizas, G.C., Dunn, M.T., Lewis, J.A. and Bengle-Ristaino, J. 1984, Liquid fermentation technology for experimental production of biocontrol fungi, Phytopathology 74:1171-1175.

Pastula, C.L., Neiser, C.D., Funk, L.M., Pedersen, E.A. and Reddy, M.S. 1997, Preliminary investigations on the modes of action of biological control agents : *Burkholderia cepacia* strains Ral-3 and Ral-3R and *Pseudomonas aureofaciens* strain 63-28 (Abstr), Can. J. Plant Pathol. 19:114.

Pedras, M.S., and Séguin-Swartz, G. 1992, The blackleg fungus: phytotoxins and phytoalexins, Can. J. Plant Pathol. 14:67-75.

Petrie, G.A. 1986, Consequences of survival of *Leptosphaeria maculans* (blackleg) in canola stubble residue through an entire crop rotation sequence, Can. J. Plant Pathol (Abstr), 8:353.

Phillips, A.J.L. 1986, Factors affecting the parasitic activity of *Gliocladium virens* on the sclerotia of *Sclerotinia sclerotiorum* and a note on its host range, J. Phytopathol. 116:212-220.

Punja, Z.K., and Jenkins, S.F. 1984, Light and scanning electron microscopic observations of calcium oxalate crystals produced during growth of *Sclerotium rolfsii* in culture and infected tissue, Can. J. Bot. 62:2028-2032.

Purdy, L.H. 1979, *Sclerotinia sclerotiorum* : History, diseases and symptomatology, host range, geographic distribution, and impact, Phytopathology 69:875-880.

Rao, D.V. and Tewari, J.P. 1987, Production of oxalic acid by *Mycena citricolor*, causal agent of the American leaf spot of coffee, Phytopathology 77:780-785.

Reddy, M.S., Colvert, D.C., Hynes, R.K. and Rennie, R.J. 1993a, Effect of nutritional factors on antagonism of soil-borne plant pathogens by select pseudomonads (Abstr), in: Proc. 6th International Congress of Plant Pathology, Montreal, July 28-August 6, 1993, p.270

Reddy, M.S., Colvert, D.C., Craig, K.A., Young, S.E., Rennie, R.J., Verma, P.R. and Mortensen, K. 1993b, Rhizobacterial influence on healthy stand establishment of canola grown in *Rhizoctonia solani* infested fields (Abstr), in : Proc. 6th International Congress of Plant Pathology, Montreal, July 28-August 6, 1993, p. 279

Reddy, M.S., Colvert, D.C., Craig, K.A., Rennie, R.J., McKenzie, D., Verma, P.R. and Mortensen, K. 1994, Suppression of preemergence damping-off of canola, caused by *Rhizoctonia solani*, by rhizobacteria, in : Improving Plant Productivity with Rhizosphere Bacteria, Proc. Third International Workshop on Plant Growth-Promoting Rhizobacteria, M.H. Ryder, P.M. Stephens, and G.D. Bowen, eds., Adelaide, March 7-11, 1994, pp. 80-82.

Seidel, D., Daebeler, F., Amelung, D., Engel, K.H. and Lücke, W. 1984, Auftreten, Schadwirkkung und BekSmpfung von *Phoma lingam* an Winterraps, Nachrichtenbl. Pflanzenschutz DDR 38:120-123.

Shinners, T.C., and Tewari, J.P. 1997, Diversity in crystal production by some bird's nest fungi (Nidulariaceae) in culture, Can. J. Chem. 75:850-856.

Silo-Suh, L.A., Lethbridge, B.J., Raffel, S.J., He, H., Clardy, J. and Handelsman, J. 1994, Biological activities of two fungistatic antibiotics produced by *Bacillus cereus* UW85, Appl. Environ. Microbiol. 60:2023-2030.

Sivan, A., Elad, Y. and Chet, I. 1984, Biological control effects of a new isolate of *Trichoderma harzianum* on *Pythium aphanidermatum*, Phytopathology 74:498-501.

Teo, B.K., Yitbarek, S.M., Verma, P.R. and Morrall, R.A.A. 1988, Influence of soil moisture, seeding date, and *Rhizoctonia solani* isolates (AG 2-1 and AG 4) on disease incidence and yield in canola, Can. J. Plant Pathol. 10:151-158.

Tewari, J.P. 1985, Diseases of canola caused by fungi in the Canadian prairies, Agri. For. Bull., Univ. Alberta 8:13-20.

Tewari, J.P. 1993, Biochemical basis of resistance to *Alternaria brassicae* in crucifers, in : Adv. Pl. Biotech. Biochem., M.L. Lodha, S.L. Mehta, S.Ramagopal, and G.P. Srivastava, eds., Indian Soc. Agil. Biochemists, Kanpur, pp. 33-38.

Tewari, J.P., Calman, A.I. and Furuya, H. 1987, Pathogens of seedling blight of canola in Alberta, in : Proc. 7th Intl. Rapeseed Cong., Poznan, Poland, May 11-14, pp. 1248-1253.

Tewari, J.P., and Briggs, K.G. 1995, Field infestation of canola stubble by a bird's nest fungus (Abstr), Can. J. Plant Pathol. 17:291.

Tewari, J.P., Shinners, T.C. and Briggs, K.G. 1997, Production of calcium oxalate crystals by two species of *Cyathus* in culture and infested plant debris, Z. Naturforsch. 52:421-425.

Tronsmo, A., and Hjeljord, L.G. 1998, Biological control with *Trichoderma* species, in : Plant-Microbe Interactions and Biological Control, G.J. Boland and L.D. Kuykendall eds., Marcel Dekker Inc., New York, pp. 111-126

Trutmann, P., Keane, P.J. and Merriman, P.R. 1980, Reduction of sclerotial inoculum of *Sclerotinia sclerotiorum* with *Coniothyrium minitans*, Soil Biol. Biochem. 12:461-465.

Trutmann, P., Keane, P.J. and Merriman, P.R. 1982, Biological control of *Sclerotinia sclerotiorum* on aerial parts of plants by the hyperparasite *Coniothyrium minitans*, Trans. Br. mycol. Soc. 78:521-529.

Tsuneda, A., and Skoropad, W.P. 1977, The *Alternaria brassicae-Nectria inventa* host-parasite interface, Can. J. Bot. 55:448-454.

Tsuneda, A., and Skoropad, W.P. 1978, The behavior of *Alternaria brassicae* and its mycoparasite *Nectria inventa* on intact and on excised leaves of rapeseed, Can. J. Bot. 56:1333-1340.

Tsuneda, A. and Skoropad, W.P. 1980, Interactions between *Nectria inventa*, a destructive mycoparasite, and fourteen fungi associated with rapeseed, Trans. Br. mycol. Soc. 74:501-507.

Tsuneda, A., Skoropad, W.P. and Tewari, J.P. 1976, Mode of parasitism of *Alternaria brassicae* by *Nectria inventa*, Phytopathology 66:1056-1064.

Tu, J.C. 1980, *Gliocladium virens*, a destructive mycoparasite of *Sclerotinia sclerotiorum*, Phytopathology 70:670-674.

Tu, J.C. 1984, Mycoparasitism by *Coniothyrium minitans* on *Sclerotinia sclerotiorum* and its effect on sclerotial germination, Phytopath. Z. 109:261-268.

Turhan, G. 1993, Mycoparasitism of *Alternaria alternata* by an additional eight fungi indicating the existence of further unknown candidates for biological control, J. Phytopathol. 138:283-292.

Ulhoa, C.J. and Peberdy, J.F. 1991, Regulation of chitinase synthesis in *Trichoderma harzianum*, J. Gen. Microbiol. 137:2163-2169.

Utkhede, R.S. 1996, Potential and problems of developing bacterial biocontrol agents, Can. J. Plant Pathol. 18:455-462.

Vannacci, G. and Harman, G.E. 1987, Biocontrol of seed-borne *Alternaria raphani* and *A. brassicicola*, Can. J. Microbiol. 33:850-856.

Verma, P.R. 1996, Oilseed rape and canola diseases incited by *Rhizoctonia* species, in : *Rhizoctonia* Species: Taxonomy, Molecular Biology, Ecology, Pathology and Disease Control, B. Sneh, S. Jabaji-Hare, S.

Neate, and G. Dijst, eds., Kluwer Academic Publishers, Netherlands, pp.249-258.

Weller, D.M. 1985, Application of fluorescent pseudomonads to control root diseases, in : Ecology and Management of Soilborne Plant Pathogens, C.S. Parker, A.D. Rovira, K.J. Moore, P.T.W. Wong and J.F. Kollmorgan, eds., American Phytopathological Society, St. Paul, pp.137-140.

Weller, D.M. 1988, Biological control of soilborne plant pathogens in the rhizosphere with bacteria, Annu. Rev. Phytopathol. 26:379-407.

Whipps, J.M. and Budge, S.P. 1990, Screening for sclerotial mycoparasites of *Sclerotinia sclerotiorum*, Mycol. Res. 94:607-612.

Whipps, J.M. and Gerlagh, M. 1992, Biology of *Coniothyrium minitans* and its potential for use in disease control, Mycol. Res. 96:897-907.

Whipps, J.M., Budge, S.P. and Mitchell, S.J. 1993, Observations on sclerotial mycoparasites of *Sclerotnia sclerotiorum*, Mycol. Res. 97:697-700.

Williams, J.R. and Stelfox, D. 1980, Influence of farming practices in Alberta on germination and apothecium production of sclerotia of *Sclerotinia sclerotiorum*, Can. J. Plant Pathol. 2:169-172.

Wu, W.S. and Lu, J.H. 1984, Seed treatment with antagonists and chemicals to control *Alternaria brassicicola*, Seed Sci. & Technol. 12:851-862.

Yang, J., Kharbanda, P.D. and Tewari, J.P. 1997a, Inhibitory effects of a biocontrol agent (*Bacillus* sp.) against *Leptosphaeria maculans in vitro* and *in vivo* (Abstr), Can. J. Plant Pathol. 19:119-120.

Yang, J., Kharbanda, P.D. and Tewari, J.P. 1997b, Inhibitory effects of a biocontrol agent (*Bacillus* sp.) against *Leptosphaeria maculans* and DNA fingerprinting of the biocontrol agent using PCR-RAPD, in : Advances in Biological Control of Plant Diseases, T. Wenhua, R.J. Cook and A. Rovira, eds., China Agricultural University Press, Beijing, p. 99.

Yang, J., Kharbanda, P.D. and Tewari, J.P. 1998, Detection of a biocontrol agent (*Bacillus* sp.) of *Leptosphaeria maculans* by Dig-labeled probes (Abstr), Can. J. Plant Pathol. (in press).

Yitbarek, S.M., Verma, P.R. and Morrall, R.A.A. 1987, Anastomosis groups, pathogenicity, and specificity of *Rhizoctonia solani* isolates from seedling and adult rapeseed/canola plants and soils in Saskatchewan, Can. J. Plant Pathol. 9:6-13.

Yitbarek, S.M., Verma, P.R., Gugel, R.K. and Morrall, R.A.A. 1988, Effect of soil temperature and inoculum density on pre-emergence damping-off of canola caused by *Rhizoctonia solani*, Can. J. Plant Pathol. 10:93-98.

INNOVATIVE APPLICATIONS OF MICROBIAL AGENTS FOR BIOLOGICAL WEED CONTROL

Susan M. Boyetchko

Agriculture and Agri-Food Canada, Saskatoon Research Centre, 107 Science Place, Saskatoon, Saskatchewan, Canada S7N 0X2

1. INTRODUCTION

The world chemical pesticide market was valued at $26.8 billion in 1991, with herbicides accounting for 44% of the market (Powell and Jutsum, 1993). Annual yield losses due to weeds were $984 million in 58 commodities grown in Canada and $1,060 million in vegetable, fruit and nut crops grown in the United States (Chandler *et al.*, 1984; Charudattan and DeLoach, 1988; Swanton *et al.*, 1993). Rising economic, environmental, and social costs associated with agricultural inputs, spray drift, pesticide residues, government legislation for reduced pesticide use, and development of herbicide-resistance in weeds make biopesticide products attractive alternatives to chemicals (Jacobsen and Backman, 1993; Pimental *et al.*, 1991; Swanton *et al.*, 1993). Countries such as Denmark, Sweden, and the Netherlands have implemented programs to reduce pesticide use by 50% (Pimental *et al.*, 1991). In Ontario, Canada, a government initiative was proposed to reduce pesticide applications by 50% by the year 2002 (Swanton *et al.*, 1993).

Biological control, by definition, is the deliberate use of living organisms to suppress, reduce, or eradicate a pest population. Inundative biological control, often referred to as the bioherbicide approach, is characterized by the artificial increase (mass-production) of high inoculum levels of a host-specific pathogen to a target weed (Boyetchko, 1997; Charudattan, 1991; TeBeest *et al.*, 1992). The term biopesticide refers to a biocontrol product used to control pests such as weeds, insects, and plant diseases and mycoherbicide refers to fungal pathogens used as weed biocontrol agents. Although exotic as well as indigenous pathogens have been used, the majority of biocontrol agents using this strategy have been indigenous pathogens. These pathogens are often applied as post-emergence bioherbicides, similar to chemical herbicides and they may be applied at least one or more times during the growing season. Weed control with bioherbicides is relatively short-term and they are not expected to persist in the environment for long periods. Despite the variety of microbial biocontrol agents

Biotechnological Approaches in Biocontrol of Plant Pathogens
Edited by Mukerji *et al.*, Kluwer Academic / Plenum Publishers, New York, 1999

investigated, there are still very few commercial successes. The perceived limitations or weaknesses of bioherbicides may, in part, be attributed to the comparisons made with chemical herbicides. Several constraints in the development of these biocontrol products include biological, environmental, technological, and economical challenges (Auld and Morin, 1995; Mortensen, 1998).

Researchers perceive biopesticide technologies will be implemented as a component of an integrated pest management program rather than act as replacements for chemical pesticides (Boyetchko, 1996b; Jacobsen and Backman, 1993). Although there is an emerging awareness of the benefits of biological control as another pest management strategy, less than 1% of the market has been penetrated by biopesticides; the majority being dominated by bio-insecticides. While synthetic herbicides have been in development for over 60 years, bioherbicide products have only been pursued for approximately 20 years (Jacobsen and Backman, 1993). Therefore, the few products that are commercially available should not be surprising since funding for pesticide development has been directed to a greater extent for chemicals and infrastructure required for biopesticides has not been readily available. Moreover, the scientific expertise and endeavors directed towards biological control are only now being realized by the private sector.

The objective of this chapter is to discuss the status of some bioherbicide projects using fungal and bacterial agents, their utility in weed management programs, and topics related to the implementation and enhancement of bioherbicidal activity. Some discussion on microbial/weed/crop relationships and how they relate to weed control strategies will be covered.

2. STATUS OF BIOHERBICIDES

More than 100 fungal and bacterial agents have been or are being evaluated as potential bioherbicides (Boyetchko, 1997; Charudattan, 1991; Greaves and MacQueen, 1992; Makowski and Mortensen, 1992; Mortensen, 1988, 1998; TeBeest, 1996; TeBeest and Templeton, 1985). The majority of projects investigating bioherbicides have primarily focused on the application of foliar fungal pathogens, with soilborne fungi and bacterial agents receiving less attention. Over the last decade or so, research efforts have included soil bacteria (Boyetchko, 1996b; Jones and Hancock, 1990; Kremer and Kennedy, 1996). The registration and/or commercialization of organisms such as *Colletotrichum, Phytophthora*, and *Puccinia* has generated more interest by researchers and industry in the utility of biological control as a weed management tactic.

2.1. Foliar Biological Control Agents

The majority of prospective mycoherbicides belong to the Hyphomycetes or Coelomycetes and their application has been primarily as foliar applied agents (Charudattan, 1991). A variety of fungi, including *Colletotrichum, Fusarium, Alternaria, Cercospora, Phoma* and *Phomopsis* have been explored as pathogens causing foliar and stem lesions, blights, wilts, and anthracnose symptoms (Table 1). Some examples of mycoherbicides that are being investigated will be discussed.

Table 1. Examples of foliar fungal pathogens as bioherbicides

Pathogen	Trade Name	Target Weed	References
Alternaria cassiae	CASST	sicklepod, coffee senna, showy crotalaria	Walker and Boyette 1985; Waker and Riley 1982
A. crassa		jimsonweed	Boyette, 1986
Cercospora rodmanii		waterhyacinth	Charudattan, 1991
Chondrostereum purpureum	ECOCLEAR™	red alder, black cherry, white birch, aspen	DeJong *et al.*, 1990; Wall 1990; Dumas *et al.*, 1997
Colletotrichum gloeosporioides f.sp. *aeschynomene*	Collego	northern jointvetch	TeBeest and Templeton, 1985; Smith, 1986
C. gloeosporioides f.sp. *malvae*	BioMal	round-leaved mallow	Makowski, 1987; Mortensen ,1988
C. gloeosporioides	Lubao No. 1	dodder	Yang, 1990
Colletotrichum coccodes		velvetleaf	Gotlieb *et al.*, 1987; Wymore *et al.*, 1987, 1988; Wymore and Watson, 1989
C. orbiculare		Bathurst burr	Auld *et al.*, 1990
C. truncatum		hemp sesbania	Boyette 1991a, 1991b
Dactylaria higginsii		purple nutsedge	Kadir *et al.*, 1997a,b
Nectria ditissima	PFC-ALDERKIL	red alder	Dorworth, 1995
Phomopsis amaranticolor		pigweed	Rosskopf, 1997
P. convovulus		field bindweed	Morin *et al.*, 1989, 1990
Puccinia canaliculata	DR. BIOSEDGE	yellow nutsedge	Phatak *et al.*, 1983; Beste *et al.*, 1992

2.1.1. Foliar Fungal Agents

Colletotrichum gloeosporioides (Penz.) Penz. and Sacc. f.sp. *aeschynomene* is a foliar fungal pathogen registered for use under the name Collego® for control of northern jointvetch (*Aeschynomene virginica* [L.] B.S.P.), a leguminous weed in rice and soybeans (TeBeest and Templeton, 1985). Collego was formulated as a wettable powder comprised of dried spores (Smith, 1986). The host-specific pathogen produces anthracnose symptoms and when applied onto the foliage provides up to 90 to 100% weed control. The product was on the market from 1982 to 1992, but Ecogen Inc. (Langhorne, PA) discontinued production because it was felt the small market size could not justify their anticipated costs involved in making necessary changes in fermentation technology to make the product. The University of Arkansas licensed the agent to Encore Technologies (Minnetonka, MN) and in 1997, Collego was sold in test-market amounts. It is anticipated that the product will be sold on a large-scale in 1998 and marketing and distribution will be conducted through a dealer in the rice growing region of the U.S. (D.R. Johnson, 1998, Encore Technologies, personal communication).

Puccinia canaliculata (Schw.) Lagerh. is an endemic rust that was registered as a bioherbicide, under the name DR. BIOSEDGE® for control of yellow nutsedge (*Cyperus esculentus* L.) (Greaves and MacQueen, 1992). When applied in early spring, *P. canaliculata* inhibited flowering and reduced yellow nutsedge stand and new tuber formation by 46% and 66%, respectively (Phatak *et al.*, 1983). Urediniospores of the rust applied in mid-June did not reduce the population of yellow nutsedge, but decreased its tuber biomass by 32% (Beste *et al.*, 1992). Mass production of the fungus continues to be a problem because the rust fungus is an obligate pathogen and inoculum production is achieved only on live plants. DR. BIOSEDGE is not commercially viable, but EPA registration for the rust is valid, indicating that the product may be commercially developed and sold under its present label (R. Charudattan, 1998, University of Florida, personal communication).

Colletotrichum gloeosporioides f.sp. *malvae* was registered in 1992 as BioMal®, the first bioherbicide in Canada for control of round-leaved mallow (*Malva pusilla* Sm.) in eight field crops (Makowski, 1987; Mortensen, 1988; Mortensen and Makowski, 1997). A dry formulation of spores at a concentration of 2 X 10^6 spores/ml provided excellent weed control, but a minimum dew period of 20 hours was required for infection and disease development to occur (Makowski, 1993; Mortensen and Makowski, 1990). Philom Bios (Saskatoon, Saskatchewan), the industrial partner involved in the development of this bioherbicide, has not commercially sold BioMal because of the relatively small market potential in field crops at this time and the cost involved to mass-produce the product given the current technologies.

In China, an isolate of *C. gloeosporioides* was developed as a mycoherbicide for control of dodder (*Cuscuta* spp.) under the trade name 'Lubao No. 1' (Mortensen, 1998; Wang, 1990). The mycoherbicide provides greater than 80% control of dodder, resulting in yield loss reductions of 30-80%. The fungus is formulated as a granular mixture but difficulties in fermentation resulted in a decrease in its commercial use. Recently, a new strain of the fungus was discovered and it is currently being produced in two county factories in China (Wang, 1990).

The basidiomycetous fungus, *Chondrostereum purpureum* (Pers. ex Fr.) Pouzar, is being developed as a mycoherbicide for control of stump sprouting and re-growth of hardwood tree species and forest vegetation management (Shamoun *et al.*, 1996; Wall, 1994). This fungus is a common saprophyte that will invade the cambium of hardwood trees upon injury or wounding. It has potential for biological control of red alder (*Alnus rubra* Bong.), black cherry (*Prunus serotina* Ehrh.), white birch (*Betula papyrifera* Marsh.), and aspen (*Populus* spp.) (DeJong *et al.*, 1990; Dumas *et al.*, 1997; Wall, 1990). Commercial application of this mycoherbicide is being developed for vegetation management in utility rights-of-way, under the trade name ECO-CLEAR™, in collaboration with the University of Victoria, MycoLogic Inc., and the Canadian Forest Service, Pacific Forestry Centre, Victoria, British Columbia. Two formulations of the fungus have been developed: ECO-CLEAR™ paste is a wet paste formulation and ECO-CLEAR™ EC is a spray emulsion formulation for application with a back-pack sprayer (W.E. Hintz, 1998, University of Victoria, personal communication).

Another promising mycoherbicide being developed is *Colletotrichium truncatum* (Schw.) Andrus and Moore for control of hemp sesbania (*Sesbania exaltata* [Rydb] ex. A.W. Hill) in soybean and cotton (Boyette, 1991a,b). This host-specific fungal pathogen is highly virulent but a major challenge has been to reduce the requirement for several hours of dew necessary for infection to occur. Research in formulation has resulted in significant improvements in the activity of *C. truncatum* by using unrefined corn oil as an adjuvant that reduced the dew period and spray volume requirements (Boyette, 1994; Egley and Boyette, 1995). Also, production of microsclerotia instead of conidia provided a viable method for producing the bioherbicide in submerged fermentation (Jackson and Schisler, 1995; Jackson *et al.*, 1996).

Biological control of sicklepod (*Cassia obtusifolia* L.), coffee senna (C. *occidentalis* L.), and showy crotalaria (*Crotalaria spectabilis* Roth) with *Alternaria cassiae* Jurair and Khan, resulted in leaf-blight symptoms (Walker and Boyette, 1985; Walker and Riley, 1982). The pathogen has a narrow host-range and can cause stunting and death of inoculated sicklepod seedlings. Weed control of 60 to 100% is possible when the fungus is applied as a post-emergence bioherbicide (Charudattan *et al.*, 1986; Walker, 1982). At least 8 hours of free water at 20 to 30°C are required for infection, however, older plants exhibit higher levels of resistance to this pathogen (Walker, 1982; Walker and Riley, 1982). In regional field trials, the fungus gave similar or better levels of control than toxaphene, a chemical used for control of the weed in some states of U.S. (Charudattan *et al.*, 1986). Commercial development of this fungus under the trade name "CASST" was pursued by Mycogen Corporation (San Diego, CA), but the company relinquished the licence to this product (Bannon, 1988; Powell and Jutsum, 1993). *Alternaria cassiae* is now being explored for development as a mycoherbicide in Brazil (R. Charudattan, 1998, University of Florida, personal communication). Progress in formulation has demonstrated that germination and infection of sicklepod by *A. cassiae* is achieved by modifying pH, nutrients, and surfactants which may improve bioherbicidal activity (Daigle and Cotty, 1991).

Biological control of velvetleaf (*Abutilon theophrasti* Medic.) has been relatively successful with *Colletotrichum coccodes* (Wallr.) Hughes (Gotlieb *et al.*, 1987; Wymore and Watson, 1989; Wymore *et al.*, 1987, 1988). The pathogen produces gray-brown lesions on velvetleaf and eventually causes desiccation and premature shedding of leaves. A 24 hour dew period at 24°C provides optimal disease development, but disease will occur at cooler temperatures and with shorter dew periods (Wymore *et al.*, 1988). However, velvetleaf plants will continue to grow after defoliation of inoculated leaves if environmental conditions for disease development are less than optimum. The potential of this mycoherbicide is deemed to be very good but, as of 1998, there is no commercial partner for further development. Since the initial discovery of the pathogen, new effective herbicides have become available that control velvetleaf, but it is anticipated that interest in *C. coccodes* could surface in the future (A.K. Watson, 1998, McGill University, personal communication).

Recently, a new fungal species, *Phomopsis amaranthicola* Rosskopf, Charudattan, and Shabana, was evaluated as a mycoherbicide for control of pigweeds and amaranthus (Rosskopf, 1997). The fungus causes leaf and stem blight on the plants and its host range is limited to *Amaranthus* spp. Application of spores at concentrations of 1.5×10^6 to 1.5×10^7 conidia/ml at the two to four-leaf stage provided high levels of mortality. The potential for development of this fungus as a mycoherbicide is considered extremely good and additional research on this fungus is continuing.

2.1.2. Foliar Bacterial Agents

Although fungi have been primarily investigated as potential foliar-applied bioherbicides, a pathogenic strain of *Xanthomonas campestris* pv. *poannua* was shown to control annual bluegrass (*Poa annua* L.) in bermudagrass (Table-2) (Johnson, 1994; Johnson *et al.*, 1996). The bacterial agent causes systemic wilting, yellowing and death of the grass and the host-range of this pathovar is restricted to *P. annua* ssp. *annua* under field conditions. Key considerations of timing and rate of application were factors important for providing a level of control that would be suitable in most golf courses. The application technique involves mowing the grass to provide wounds for entry of the bacteria which is very effective for enhancing the bioherbicidal activity of this bacterial agent. Additional research resulted in the isolation of several new *X. campestris* strains indigenous to Japan and an application for registration of the bacterial strain was submitted (Imaizumi *et al.*, 1997). Based on host specificity and bacteriological properties, the strain *X. campestris* pv. *poannua* identified in the U.S. was found to be identical to *X. campestris* pv. *poae* (Nishino *et al.*, 1995). Japan Tobacco has registered the bacterial agent under the name Camperico®. The U.S. strain, that was isolated in San Diego from symptomless bluegrass is highly effective in golf course greens overseeded with bermudagrass during the winter. The U.S. licensing rights belong to EcoSoil, who may market *X. campestris* pv. *poae* in the future (D.R. Johnson, 1998, Encore Technologies, personal communication).

Pseudomonas syringae pv. *tagetis* is being pursued as a possible bioherbicide for Canada thistle (*Cirsium arvense* L.) control in soybean (Table 2) (Johnson *et al.*, 1995). The bacterium was first reported on marigold, sunflower, common ragweed, and Jerusalem

bacterium was first reported on marigold, sunflower, common ragweed, and Jerusalem artichoke and causes apical chlorosis of Canada thistle (Johnson and Wyse, 1991; Johnson *et al.*, 1996). Application of the bacterium in the spring reduced the number of Canada thistle stems/m^2 by 78%, compared to the controls, and no disease occurred on soybean. Mycogen Corporation was interested in the project and after considerable progress in the development of the bacterial agent, the company decided to no longer pursue its commercial development. Encore Technologies has licensed *P. syringae* pv. *tagetis* from the University of Minnesota and commercial development is proceeding (D.R. Johnson, 1998, Encore Technologies personal communication).

Table 2. Examples of bacterial agents as bioherbicides

Pathogen	Target Weed	References
Foliar agents		
Xanthomonas campestris pv. *poae* (Camperico)	annual bluegrass	Johnson 1994; Johnson *et al.* 1996; Imaizumi *et al.*, 1997
Pseudomonas syringae pv. *tagetis*	Canada thistle	Johnson and Wyse 1991; Johnson *et al.*, 1995,
P. syringae pv. *phaseolicola*	kudzu	Zidack and Backman, 1996
Soilborne agents		
Pseudomonas fluorescens	downy brome jointed goatgrass	Kennedy *et al.*, 1991; Kennedy *et al.*, 1995
Pseudomonas *Xanthomonas* *Enterobacter*	downy brome Japanese brome jointed goatgrass	Harris and Stahlman, 1992, 1993
Pseudomonas	downy brome green foxtail wild oats	Boyetchko and Mortensen, 1993; Boyetchko, 1997
Pseudomonas *Enterobacter* *Erwinia herbicola*	velvetleaf giant and green foxtail; redroot pigweed, leafy spurge, jimsonweed	Kremer, 1987; Kremer *et al.*, 1990; Souissi and Kremer, 1994
Pseudomonas	marsh reedgrass	Dorworth ,1992

2.2. Soilborne Biological Control Agents

Hasan and Ayres (1990) suggested that foliar or stem applied pathogens constitute the more successful bioherbicides because they are host-specific and can be applied to shoots quite easily. Grassy weeds are considered to be more difficult to control because the meristem in grasses is covered by a leaf sheath which protects the plant from infection by foliar fungal pathogens (Greaves and MacQueen, 1992). It was also suggested that soilborne pathogens are poor candidates for biological control because they are not very host-specific. However, there are several prospective fungal and bacterial agents that show tremendous potential as pre-emergence bioherbicides (Table-3) and they are discussed below.

Table 3. Examples of soilborne fungal pathogens as bioherbicides

Pathogen	Target Weed	References
Phytophthora palmivora (DeVine)	stranglervine	Burnett *et al.*, 1974; TeBeest and Templeton, 1985
Fusarium lateritium	spurred anoda, velvetleaf, prickly sida	Boyette *et al.*, 1984; Boyette and Walker ,1986
F. solani f.sp. *cucurbitae*	Texas gourd	Weidemann and Templeton ,1988
Sclerotinia sclerotiorum	Canada thistle, dandelion, diffuse and spotted knapweeds	Brosten and Sands, 1986; Riddle *et al.*, 1991; Mortensen and Hogue, 1985
Gliocladium virens	pigweed	Howell and Stipanovic, 1984; Jones and Hancock, 1990

2.2.1. Soilborne Fungi

Although *Phytopthora palmivora* Butler, a soilborne fungus, was the first commercially available mycoherbicide, soilborne fungi constitute a small proportion of the targeted microorganisms under investigation as bioherbicides (Jones and Hancock, 1990). This group of fungi is capable of reducing the competitive ability and reproductive capacity of weeds by causing root decay, reducing weed emergence through pre- or post-emergence damping-off, producing lesions that girdle the stem at the soil line and causing internal necrosis or wilting. One of the major benefits of these fungi is their ability to germinate and cause infection without relying on environmental conditions, such as high humidity. Moreover, the use of granular formulations allows the application of soilborne fungi, similar to granular fertilizers or herbicides, and the various formulation components can provide the biocontrol agent with a substrate necessary for growth and proliferation.

Phytophthora palmivora is an endemic fungal pathogen that was registered as a mycoherbicide in 1981 under the tradename DeVine® for control of stranglervine (*Morrenia odorata* [H and A] Lindl.) in citrus orchards (Burnett *et al.*, 1974; Charudattan, 1991; TeBeest, 1996; TeBeest and Templeton, 1985). The fungus was originally isolated from stranglervine plants and the product consists of a liquid preparation of chlamydospores that can provide up to 96% weed control within 10 weeks when applied as a postemergence spray. Abbott Laboratories (Abbott Park, Ill.) markets DeVine on a small scale, predominantly in the citrus-growing area of Florida (Charudattan, 1991).

Many fusaria species have been explored for biological control as either pre-emergence or post-emergence pathogens (Jones and Hancock, 1990; Mortensen, 1998). Some *Fusarium* spp. are host-specific because of their specialization on various hosts as forme speciales. *Fusarium lateritium* (Nees.) ex Fr. is restricted to members of the Malvaceae [spurred anoda (*Anoda cristata* [L.] Schlecht), velvetleaf (*Abutilon theophrasti* Medic.), and prickley sida (*Sida spinosa* L.) (Boyette *et al.*, 1984; Boyette and Walker, 1986). It has been suggested that survival of these specialized fungi is limited because they do not infect plants in the absence of their host (Jones and Hancock, 1990).

Sclerotinia sclerotiorum (Lib.) De Bary has been explored as a possible biocontrol agent of Canada thistle (*Cirsium arvense* L.), dandelion (*Taraxacum officinale* Weber), and diffuse and spotted knapweeds (*Centaurea diffusa* Lam. and *C. maculosa* Lam.) (Brosten and Sands, 1986; Mortensen, 1998; Mortensen and Hogue, 1995; Riddle *et al.*, 1991). One of the major limitations is that this pathogen has a wide host-range, but molecular strategies may be used to create bioengineered isolates that are non-sporulating or have restricted growth requirements to limit their spectrum of activity (Sands *et al.*, 1990a).

More recently, soilborne fungal pathogens have been investigated for control of wild oats (*Avena fatua* L.) (Boyetchko *et al.*, 1997, 1998). Screening methods included use of growth pouches where mycelial suspensions were applied directly onto the seeds of wild oats. Dose-response studies indicated that emergence of wild oats could be reduced over a three-week period. Further investigations will focus on continued screening efforts and development of effective delivery systems and inoculum levels that reflect practical rates of application in the field.

Control of pigweed with *Gliocladium virens* Miller, Giddings, and Foster, a broad-spectrum mycoherbicide, was reported (Howell and Stipanovic, 1984; Jones and Hancock, 1990). The fungus has also demonstrated pre-emergence activity on more than 20 plant species where significant reductions in seedling emergence and dry weights were observed (Jones and Hancock, 1990). The wide host-range of this mycoherbicide makes it an attractive biocontrol agent for enhancing crop stand establishment while reducing weed competition early in the growing season. The ability of this fungus to produce phytotoxins (*e.g.* viridiol) results in necrosis and death of emerging seedlings. Also, *G. virens* is easy to mass-produce, grows rapidly and persists for relatively long periods of time. Moreover, this fungus has both bioherbicidal and biofungicidal activities by inhibiting plant pathogens such as *Sclerotinia sclerotiorum* (Lib.) de Bary, *Rhizoctonia solani* Kühn, and *Pythium ultimum* Trow (Howell, 1982; Tu, 1980).

2.2.2. Deleterious Rhizobacteria

Several researchers in the U.S. and Canada have considered deleterious rhizobacteria (DRB) as potential biological weed control agents (Table-2) (Boyetchko, 1996b; Boyetchko and Mortensen, 1993; Dorworth, 1992; Harris and Stahlman, 1992, 1993; Kennedy *et al.*, 1991, 1995; Kremer, 1987; Kremer *et al.*, 1990; Souissi and Kremer, 1994). These bacteria are easily overlooked because symptoms such as root discoloration or plant wilting, under field conditions, are not obvious. The majority of DRB associated with weeds are fluorescent and other pseudomonads, *Flavobacterium* spp., *Erwinia herbicola*, and *Alcaligenes* spp. (Kremer *et al.*, 1990). As of 1996, 8 out of 11 projects with DRB focused on the control of annual weeds in cereal and row crops, with several of the weeds targeted belonging to the Poaceae (Kremer and Kennedy, 1996). Many of these DRB are applied either to the soil as a drench, granular formulation or seed treatent, or directly onto crop residues (Boyetchko, 1996b; Kremer and Kennedy, 1996; Mooney *et al.*, 1996). This approach reduces weed competition by decreasing weed emergence and delaying weed growth and development to provide a competitive advantage to the crop. It may also be possible to target weed species where herbicide-resistance has been demonstrated.

Pseudomonas fluorescens, strain D7, exhibiting inhibitory activity against downy brome (*Bromus tectorum* L.), was capable of reducing weed populations in winter wheat fields resulting in yield increases of 18 to 35% (Johnson *et al.*, 1993; Kennedy *et al.*, 1991). These results support the concept that reduced weed competition through suppression can lead to higher yielding crops and that eradication of weeds may not be necessary. Root colonization by *P. fluorescens* occurred to a greater extent in moist, cool soils and it was suggested that lack of percolating water in soil limited the distribution of the bacteria along the roots (Johnson *et al.*, 1993). Colonization of the downy brome roots and plant growth suppression were also greatest at cool temperatures such as 10 to 15°C.

In western Canada, DRB are being investigated as potential biological control agents (Boyetchko, 1997; Boyetchko and Mortensen, 1993). Over 2000 rhizobacteria isolated from prairie soils have been screened for their inhibitory effects against annual grassy weeds. Many of these DRB and their secondary metabolites are host-specific and the amount of root reduction was found to be concentration-dependent (Boyetchko and Holmström-Ruddick,

1995). Other researchers have also indicated that the effect of rhizobacteria can be host-specific (Cherrington and Elliott, 1987; Elliott and Lynch, 1985; Kennedy *et al.*, 1991). Host-range tests on non-target hosts (i.e., spring and winter cereal crops) showed little or no detrimental effects to root and shoot growth, with some degree of plant growth promotion occurring in some cases, depending on the crop cultivar and rhizobacterial isolate (Boyetchko and Holmström-Ruddick, 1995). It may be possible to use DRB to reduce the competitive ability of weeds while using the same strain as a plant-growth promoting rhizobacterium in crops to improve crop stand establishment and provide a competitive advantage.

3. PRACTICAL APPROACHES FOR ENHANCING EFFICACY

In view of the number of prospective bioherbicides and some of the challenges that researchers have encountered in their evaluation and development, research activities in plant pathology, formulation and application technology, and agronomy may provide methods for implementation and enhancing efficacy. One area that requires more attention is knowledge of the infection process and whether phytotoxic metabolites are produced. Mode of pathogenicity will elucidate methods for fermentation, formulation, application, and delivery. Some physiological and biochemical mechanisms by which bioherbicides affect plant growth have been studied. This discussion will focus on a few mechanisms by which fungal and bacterial bioherbicides affect growth and development of weeds. Chemical and molecular strategies for enhancing their efficacy, as well as formulation and delivery as components of an integrated weed management system will be discussed.

3.1. Pathogenicity and Induced Resistance

Limited research has been conducted on the infection process of foliar fungal bioherbicides. More complete understanding of the mode of infection may provide clues on pathogenicity, virulence, and host-specificity. For example, application of high inoculum levels of conidia of *C. gloeosporioides* f.sp. *malvae* onto round-leaved mallow resulted in poor germination (Morin *et al.*, 1996). Self-inhibition of conidia was attributed to crowding of conidia on the leaves and to self-inhibitors present in the extracellular matrix from the acervulus. However, it may be possible to release self-inhibition by adding different surfactants to the mycoherbicide (Zhang *et al.*, 1998). For example, Tween 40 and Tween 80 released self-inhibition of *C. gloeosporioides* f.sp. *malvae* conidia. Similar effects were demonstrated with gelatin on *Phoma* spp.

Following application, conidia of *C. gloeosporioides* f.sp. *malvae* germinated on round-leaved mallow leaves within 3 to 4 hours and appressoria developed after 20 hours (Wei *et al.*, 1997). After an infection peg formed, a short biotrophic phase was demonstrated by the formation of intracellular infection vesicles. It was suggested that a biotrophic phase in some *Colletotrichum* spp. may determine host-specificity of the pathogen and that it may be a useful tool for selecting potential mycoherbicides with a limited host-range. The biotrophic phase was followed by a necrotrophic phase shown by the development of thin secondary hyphae that produced necrotic lesions. Morin *et al.* (1996) also studied the infection process of *C. gloeosporioides* f.sp.*malvae* on several malvaceae weeds, including moderately resistant and highly suceptible hosts. In susceptible hosts, infection was similar to that previously described (Wei *et al.*, 1997). However, in resistant hosts, infection was quickly followed by a hypersensitive reaction. It was suggested that the biochemical determinants of incompatible and compatible reactions occurred shortly after the short biotrophic phase.

Induced systemic resistance of sicklepod to *A. cassiae* was reported (Weete, 1992). Very high inoculum doses of the bioherbicide and age of the plant were two factors that limited the efficacy of this bioherbicide. Also, in jimsonweed infected with *Alternaria crassa* (Sacc.) Sands, phenylalanine ammonia-lyase (PAL) activity was detected in more resistant red- and green-stemmed jimsonweed and it was suggested that various degrees of resistance to the mycoherbicide may exist (Hoagland and Boyette, 1994). Induced plant defense mechanisms through the elicitation and accumulation of PAL and phytoalexins (low molecular weight antimicrobial compounds) may account for some inconsistent performance of specific bioherbicides (Hoagland, 1996; Hoagland and Boyette, 1994; Weete, 1992). If plant defense mechanisms are impaired, increased susceptibility to plant diseases (such as bioherbicides) may result in enhanced efficacy of the biological weed control agent.

Three chemical categories of phytoalexins (phenolics, isoprenes, and polyacetylenes) have been described (Hoagland, 1996). It was suggested that enhanced efficacy of bioherbicides with phytoalexins may be used in two ways: 1) to interfere with the chemical pathway leading to phytoalexin accumulation resulting in enhanced pathogenicity and efficacy of the bioherbicide; and 2) to enhance the amount of phytoalexins produced to levels that are phytotoxic to the plant but are not detrimental to the bioherbicide. For example, glyphosate was shown to interfere with the conversion of shikimate-3-phosphate to 5-enolpyruvylshikimate-3-phosphate by inhibiting 5-enolpyruvylshikimate-3-phosphate synthase (Amrhein, 1986; Hoagland, 1996). This may lower the plant's defense mechanisms leading to greater susceptibility to plant pathogens. This was demonstrated with the application of a sublethal dose of glyphosate that led to an increase in susceptibility of sicklepod to the mycoherbicide *A. cassiae* (Sharon *et al.*, 1992).

Other examples of synergy between mycoherbicides and chemicals include the application of *Puccinia canaliculata* in combination with the herbicide paraquat (1,1'dimethyl-4,4'-bipyridiniumion) which gave 99% control of nutsedge when compared to only 60% control with the rust alone (Phatak, 1984). Tank mixing *C. gloeosporioides* f.sp. *aeschynomene* with propanil [N-(3,4-dichlorophenyl propanamide]-2-benzimidazolecarbamate] inhibited infection by the fungal pathogen, but if the pathogen was applied before the chemical herbicide, infection and disease development were not affected (Smith, 1986). Experiments with split applications of several herbicides with *C. gloeosporioides* f.sp. *malvae* resulted in enhance efficacy when compared to the application of the mycoherbicide alone (Grant *et al.*, 1990b). More than 50% of the herbicides applied at 0.1 to 2 times the recommended rate did not interfere with the growth of the mycoherbicide (Grant *et al.*, 1990a). However, a study on pesticide and adjuvant compatibility with *Chondrostereum purpureum* showed that the majority of chemical pesticides (fungicides, insecticides, and herbicides) and adjuvants were fungitoxic to the mycoherbicide at 0.01 and 0.1% (v/v) concentrations (Prasad, 1994). The majority of results suggest that applications of sublethal doses of chemicals and synergism between herbicides and biocontrol agents may play significant roles in the enhancement of bioherbicide activity.

3.2. Microbially-Derived Phytotoxic Metabolites

Several review articles have described the merits of microbial natural products as herbicides (Duke, 1986; Hoagland, 1990; Kenfield *et al.*, 1988; Strobel *et al.*, 1991). Some phytotoxins that have been extensively studied include curvulins, eremophilanes, maculosins, ophiobolins, and bialaphos (Hoagland, 1990). Microbially-derived compounds may be pursued either as templates for new synthetic chemical herbicides or as pathogens applied directly to the target weed. With the growing awareness of herbicide-resistance, novel compounds from microorganisms may provide new chemistries for weeds that may otherwise be difficult to control. Moreover, understanding the precise mode of action can provide clues into how the bioherbicides are formulated and delivered to the target weed. Discussion will focus on a few examples of microbial metabolites and/or phytotoxins that have been investigated.

Colletotrichum gloeosporioides f.sp. *malvae* produces several metabolic fractions that have phytotoxic activity on round-leaved mallow, common mallow, small flowered mallow, and some extracts show necrosis and chlorosis on velvetleaf. Most fractions were characterized as complex components but one was identified as a single component by NMR and GC/MS techniques. The difficulty in working with complex fractions is to develop sensitive bioassays that are rapid, consistent, and easy to use. For *C. gloeosporioides* f.sp. *malvae*, protoplast techniques were developed which increased the sensitivity of the assay as compared to the droplets containing the active fraction applied to leaves on intact plants (K.L. Bailey and J. Derby, 1998, Agriculture and Agri-Food Canada, personal communication).

Alternaria cassiae produced two phytotoxins, ferricrocin and coprogen (Ohra *et al.*, 1995). Both compounds were extracted from the aqueous fraction and produced necrotic lesions with brown spots using a leaf-wounding assay. These phytotoxic compounds contained iron and it was concluded that they are siderophores (ferric iron chelating compounds). Confirmation of siderophore activity was provided by enhancement of ferricrocin phytotoxicity when iron was removed and it was suggested that the mechanism of activity of this compound is related to chelating activity.

Pseudomonas syringae pv. *tagetis*, produces tagetitoxin that causes apical chlorosis in Canada thistle in the field (Gronwald *et al.*, 1998; Johnson *et al.*, 1996). Injection of tagetitoxin into cotyledons of susceptible hosts (sunflower) reduced chlorophyll content in developing leaves by 90% compared to less than 50% reduction in chlorophyll content in soybean (nonhost). It was suggested that little injury to soybean through the inundative application of *P. syringae* pv. *tagetis* will occur.

A phytotoxin produced by *P. fluorescens* strain D7 was partially purified and caused root discoloration and suppressed root elongation (Tranel *et al.*, 1993b). Root elongation was restored if the plants were removed from the cell-free culture filtrates or crude preparation of the toxin resulting in the conclusion that a toxin was involved. Gurusiddaiah *et al.* (1992) partially purified the toxin and found it to be complex, consisting of at least two polypeptides, fatty acid esters, and lypopolysaccharides. However, further purification of the toxic complex resulted in almost complete loss of the phytotoxicity. It is unclear or unknown if the complex was a single compound or if the toxic fraction consisted of two or more compounds that were necessary for activity.

Tranel *et al.* (1993a) investigated the mode of action of the partially purified toxin of the bacterial agent *P. fluorescens* strain D7. The toxin had no effect on cell division, respiration, or protein synthesis, but lipid synthesis and membrane integrity were disrupted by the D7 toxic fraction. The authors were not certain if the concentrations of the toxin used in the bioassays could be produced at physiological levels in the rhizosphere where colonization of *P. fluorescens* D7 takes place.

Some bacterial strains produce excessive amounts of auxin (indole-3-acetic acid [IAA]), particularly when L-tryptophan is added as a precursor (Kremer and Sarwar, 1995; Sarwar and Kremer, 1995a,b). The high auxin-producing bacterium, *Enterobacter taylorae* inhibited the root length of field bindweed (*Convolvulus arvensis* L.) when L-tryptophan was added to a minimal salts medium. It was proposed that high auxin producing DRB can be applied with L-tryptophan to control weed seedling growth *in situ*.

Microbial cyanide produced by DRB may affect root metabolism and development by inhibiting enzymes such as cytochrome oxidase, an enzyme involved in respiration (Schippers *et al.*, 1987, 1990). Another enzyme sensitive to hydrogen cyanide is peroxidase, which has been associated with lignification and hypersensitve reactions of plant tissues after infection by pathogens (Schippers *et al.*, 1990). Amino acids such as glycine and proline have been found in root exudates, with glycine being a precursor of microbial cyanide production.

Development of phytotoxic metabolites produced by fungal and bacterial pathogens provide weed control options that can complement other weed management tactics. It may be possible to utilize these natural products derived and produced through fermentation, use them as templates for new synthetic herbicides, or develop these products as novel compounds with new modes of action. For example, a strain of DRB was capable of suppressing or inhibiting weed emergence of both susceptible and resistant wild oats, suggesting the potential for discovering bacterial strains that produce metabolites with modes of action differing from existing chemical herbicides (Figure-1). The addition of precursors to the phytotoxins in a formulated product may increase the toxin production in the fermentation process and may enhance the bioactivity of the microorganism.

Figure 1. Comparison of the effects of chemical herbicide, sethoxydim, (middle), bacterial metabolites (right), and untreated control (left) on germination and growth of herbicide resistant and susceptible wild oats. Each agar plate contains 2 rows of resistant (R) wild oats on the left and 2 rows of susceptible (S) wild oats on the right. Note the suppression of growth in S wild oats populations with the chemical herbicide and bacterial metabolites while R wild oats is only suppressed by bacterial metabolites.

3.3. Molecular Technologies

The development and application of recombinant DNA technology is an exciting area of research that has received very little attention until recently, yet it offers opportunities for management of agricultural pests. Genetic manipulation of fungi and bacteria used in biological weed control has potential for monitoring introduced microbes, enhancing efficacy and virulence, constructing microbes with desirable or unique traits, modifying specificity, and developing methods of microorganism containment to reduce risk (Kennedy, 1996; TeBeest *et al.*, 1992; Turgeon and Yoder, 1985).

Molecular identification techniques have been used to characterize strains of mycoherbicides such as *C. purpureum* (ECOCLEAR) and *C. gloeosporioides* isolated from northern jointvetch (Collego) (Ramsfield *et al.*, 1996; TeBeest *et al.*, 1992). Research on *C.*

purpureum as a mycoherbicide in forestry has indicated that there is gene flow of several isolates across North America (Ramsfield *et al.*, 1996). Two nuclear types were detected, with type I being predominant in eastern North America and type II occuring in the west. Mitochondrondial recombination in the fungus is rare and it was suggested that the chance of introducing a single isolate of *C. purpureum* is relatively low because of the freedom of gene flow in *C. purpureum* populations. Further, PCR-based technologies provided highly reliable and predictable methods for differentiating between the strain of *C. purpureum* applied as a mycoherbicide and other isolates of the fungus (W.E. Hintz, 1998, University of Victoria, personal communication). Strains of *C. gloeosporioides* from winged water primose do not mate with the strain from pecan but sexual compatibility between various strains has not been ruled out (TeBeest *et al.*, 1992). Genetic exchange between introduced bioherbicides and indigenous pathogens must therefore be clearly defined to assess the risk of applying these agents into the environment.

Development of reliable tracking systems for environmental fate assessment can be readily achieved through molecular technology. In past studies, development of antibiotic resistant mutants have been used but the reliability and sensitivity of this technique is not very high. Molecular markers may be used to monitor specific strains of microbes using PCR amplication to detect specific DNA probes with high precision (Kennedy, 1996). For example, an effective extraction method that allows the isolation of DNA from microorganisms can be used to fingerprint mixed species DNA by restriction endonuclease analysis of PCR-amplified fragments (Porteous *et al.*, 1994). Biological diversity using this technique facilitated the characterization of microbial populations. *Lux* genes as bioluminescence markers have also been used to monitor pseudomonad bacteria (Kennedy, 1996). The *lux* gene codes for enzymes involved in bioluminescence and bacteria expressing this gene can be easily enumerated. Another method for detecting microbes is use of the green fluorescent protein which allows visual inspection for fluorescence activity (Poppenborg *et al.*, 1997). Green fluorescent proteins are easy to use, stable, and can form functional products with other proteins. Applications of this technique include the identification of genetically transformed organisms, tracking, and promoter studies.

The identification of genes responsible for virulence, enzyme or phytotoxin production, and competitive ability can be utilized to enhance the bioherbicidal activity of the organism. Insertion of cutinase genes into pathogens have been used to alter and therefore enhance fungal pathogenicity (Dickman *et al.*, 1989). Cutin is commonly found in the cuticle of aerial plant parts (Dickman, 1993). Providing pathogens with the mechanism to breach the cuticle layer and penetrate into the host by introducing the cutinase gene can enhance the bioactivity of bioherbicides. With growing evidence of the role of phytotoxins in bioherbicides, it may be possible to clone toxin-production genes into bacterial and fungal pathogens (Turgeon and Yoder, 1985). However, the genetics of toxin production in fungal pathogens is not well understood, but knowledge of the genetics of pathogenicity offers researchers with options for developing superior strains for biological control.

Potential mycoherbicides such as *S. sclerotiorum*, a fungus with a very wide host range, may be genetically altered to reduce survival and host-specificity (Sands *et al.*, 1990a,b). Nonsclerotial forming mutants and auxotrophs were developed to reduce virulence and host range. Genetic modification through induced mutagenesis using ultraviolet light or chemical mutagens led to *S. sclerotiorum* mutants that lacked the ability to produce sclerotia for overwintering (Sands *et al.*, 1990a). Auxotrophic mutants with a dependence for exogenous sources of cytosine and pyrimidine were used to reduce the spread and survival of the fungal pathogen. Genetic containment may also be achieved with suicide genes to limit the spread of the agent in the environment by constructing microbes with strict substrate requirements. The suicide vector is turned off in the presence of a particular substrate, but once the substrate becomes limiting, the microbe carrying the gene self-destructs (Sands *et al.*, 1990b).

Molecular strategies for detection and enhancement of bioherbicidal properties offer unique opportunities in biocontrol research. Improvements in risk assessment and monitoring introduced strains and determining the mechanisms of action at a genetic level can increase our understanding of the physiological basis for pathogenicity.

3.4. Formulation and Delivery

Formulation of bioherbicides has received a great deal of attention and critical review over the last decade or more (Boyette *et al.*, 1996; Green *et al.*, 1998). It involves the preparation of the biologically active ingredient (i.e. microbial agent), a carrier, and other adjuvants to provide a viable and effective delivery system. DeVine was formulated as a liquid preparation of chlamydospores of *P. palmivora* and only had six weeks of shelf life (Burnett *et al.*, 1974; TeBeest and Templeton, 1985). Collego and BioMal were both formulated as wettable powders that consisted of the fungal spores that are dried from the liquid ferment and mixed with a carrier (Mortensen, 1988; TeBeest and Templeton, 1985).

Formulation may be used to enhance the bioherbicidal activity of the active ingredient. Environmental factors that may affect efficacy of foliar applied mycoherbicides are temperature and moisture. Moisture required for disease development is often dependent on the amount of dew period. Formulations may be used to extend the dew period to allow the fungal pathogen to infect the weed, reduce the dew period requirement, extend shelf life of the biocontrol product, and enhance stability and bioactivity (Green *et al.*, 1998). In liquid formulations, adjuvants are used to assist or modify the activity of the active ingredient, while surfactants are used to modify the surface tension of the spray droplets in order to improve the spread and wetting properties of the formulated microbial agent.

Invert emulsions (water-in-oil) are known to enhance fungal germination and infection because of moisture-retaining properties that overcome long dew period requirements (Auld and Morin, 1995; Boyette *et al.*, 1996; Connick *et al.*, 1990). The efficacy of *C. truncatum* was significantly improved when applied with an invert emulsion (Boyette *et al.*, 1993). However, these emulsions are very viscous, phytotoxic in some target plants, and are expensive. Connick *et al.* (1991b) developed an invert emulsion with lower viscosity and improved water-retention properties. In addition, *A. cassiae* and *Alternaria crassa* (Sacc.) Rands formulated an invert emulsion that led to a reduction in the level of spore inoculum per droplet from 10s and 100s of spores to 1 spore per droplet while maintaining effective control of sicklepod (Amsellen *et al.*, 1990). Formulation of *Colletotrichum orbiculare* for control of spiny cocklebur using vegetable oils enhanced the efficacy and improved the spread of the invert emulsion, with no phytotoxicity to the host being exhibited (Auld, 1993). A polar oil-based invert emulsion containing soybean oil, paraffinic oil, paraffin wax and emulsifying agents (Arlacel 780 and Arlacel 989) was found to be a very good formulation with low viscosity and moisture-retaining properties (Womack *et al.*, 1996). In addition, germ-tube elongation was enhanced and it was indicated that this may reduce the time required for initial infection of the weed to occur.

Several adjuvants and other amendments can be used to increase spore germination, improve pathogen stability and modify the environmental requirements or expand the host-range of various mycoherbicides (Boyette *et al.*, 1996). Although the lack of free moisture has limited the bioherbicidal potential of *C. truncatum* (Walker and Boyette, 1985), unrefined corn oil as an adjuvant improved its bioactivity and reduced its dew period requirement from 12 hours to 2 hours (Boyette, 1994). Spray volume requirements were also reduced from 500 L/ha to 5 L/ha. The water-corn oil emulsion increased spore germination from 30% in water alone to 92% (Egley and Boyette, 1995). Moreover, following an initial 4-hour dew period, delays of 24 to 72 hours of dew resulted in 89% control of hemp sesbania or higher, while similar delays in dew period with water-conidia provided 11% control or less. Although the water-unrefined corn oil was inferior to the invert emulsion, stimulation of conidial germination was deemed a desirable characteristic for enhancing the bioherbicidal activity.

Eight hydrophilic polymers were evaluated for their ability to retain moisture for long periods of time, enhance germination and viability of fungal spores, and improve pathogenicity (Shabana *et al.*, 1997). All gels were able to hold moisture for 6 days and there was an 18 to 22% water loss per day. Germination in all gels was very high (over 95%) and viability was over 50%, with Kelzan® xanthan gum being the most effective. Enhanced pathogenicity was demonstrated, particularly with Gallan gum and Kelgin®-HV, followed by Evergreen® 500 polyacrylamide and Kelgin®-LV. The ability of these hydrophilic polymers to retain moisture and enhance bioactivity indicates their potential as components in formulation for microbial herbicides.

Surfactants have been utilized as ingredients in formulations because they help wet the plants by reducing surface tension and improving the dispersal of the fungal spores in the spray droplet mix. Several surfactants that have been used are Tween 20 with *Fusarium lateritum* , nonoxynol with *Alternaria macrospora* Zimm. *C. truncatum*, and *A. cassiae*, and sorbitol with *Colletotrichum coccodes* (Boyette *et al.*, 1996). Selection of appropriate surfactants must consider their inhibitory or stimulatory effects on spore germination, infection, etc.

Phytopathogenic bacteria have a requirement of free water for dispersal and the presence of wounds or natural openings for bacteria to enter into the plant (Johnson *et al.*, 1996; Zidack *et al.*, 1992). Cutting or mowing of turfgrass for control of annual bluegrass with *X. campestris* pv. *poae* will permit the bacteria to enter into the plant (Imaizumi *et al.*, 1997). In addition, bacteria applied at a rate of 10^9 cfu/ml at high water volumes (400 ml/m2) showed over 90% disease severity in the annual bluegrass.

Silwet L-77 (OSi Specialties, Tarrytown, NY) is an organosilicone surfactant that has been used to faciliate the penetration and entry of bacteria into plant stomata and hydathodes (Zidack *et al.*, 1992). A low surface tension of 30 dynes cm^{-1} or less is required for delivery of the liquid into the stomata of a leaf. This surfactant reduces the water surface tension to 20 dynes cm^{-1}. Application of *Pseudomonas syringae* pv. *tagetis* with Silwet L-77 enhanced the penetration and entry of the bacteria into stomata and hydathodes, resulting in significant increases in disease severity and incidence in Canada thistle when compared to plants sprayed with the bacteria minus the surfactant (Johnson *et al.*, 1996). It has also been suggested that delivery of the bacteria into these natural openings protects them from UV irradiation and dessication. Similar results were obtained with *Pseudomonas syringae* pv. *phaseolicola* and Silwet L-77 for biocontrol of kudzu (*Pueraria lobata* [Willd.] Ohwi) (Zidack and Backman, 1996). Formulation of *P. syringae* pv. *phaseolicola* and Silwet L-77 can be applied under hot, dry, and sunny conditions when the stomata are open but when applied at dusk or dawn when the stomata are closed and the light levels are significantly reduced, there is no entry of the bacteria into the plant (Zidack *et al.*, 1992).

Application of spray droplets and their distribution onto the leaf surface must also be considered when measuring the efficacy of foliar-applied mycoherbicides. Effective desposition of the inoculum on the target has often not been given much attention (Lawrie *et al.*, 1997). For example, droplets that were 290 µm in diameter contained more conidia of two mycoherbicides (*Stagonospora* and *Mycocentrospora* spp.) than droplets of 240 or 150 µm in diameter. If the droplets were less than 150 µm, at least 20% of the droplets contained no conidia, while 60% of the droplets contained at least 20 propagules. Small droplets also dehydrated quickly, which may affect viability of the mycoherbicide. Conversely, larger droplets will persist longer under high humidity. Evidently, greater attention to optimization of spray droplet size should be given when evaluating the formulation and delivery of bioherbicides. It was further suggested that research on spray application of appropriate inoculum under field and laboratory conditions requires the cooperation of a multidisciplinary team of researchers in plant pathology, formulation chemists, and spray application technologists.

Liquid formulations have been mostly used for foliar applied, post-emergence mycoherbicides, while solid-based formulations have been developed for pre-emergence mycoherbicides that infect the weeds at or below the soil surface (Boyette *et al.*, 1996). These formulations can provide a substrate for the pathogen and act as a buffer in environmental extremes. Pesta is a wheat-gluten matrix (liquid inoculum, semolina wheat flour and kaolin) that has been used to formulate mycoherbicides such as *C. truncatum, A. crassa* and *Fusarium lateritium* (Connick *et al.*, 1991a). It can be soil-applied as a pre-emergence treatment and shelf-life of the product can be improved by manipulating the water activity (moisture content of the granule) and sucrose content (Connick *et al.*, 1996). Bran, wheat kernels, cornmeal/sand, and vermiculite are other solid substrates used to formulate mycoherbicides (Auld and Morin, 1995; Boyette *et al.*, 1996). For example, mycelium, micro- and macro-conidia and chlamydospores of *Fusarium solani* f.sp. *cucurbitae* (Snyd.) Snyd. and Hans. formulated in cornmeal-sand for control of Texas gourd provided 96% control (Boyette *et al.*, 1985).

Polymers such as polyacrylamide and sodium alginate have been used to immobilize bacteria in granular formulations (Digat, 1989; Jha *et al.*, 1993). However, slow hydration and release of the active ingredient are major impediments to this technology. A variety of carriers such as Pyrax or powdered wheat bran, that provide a food base, have been added with the bacterial biomass and alginate. A new encapsulation method for bacterial inoculants that were suspended in their nutrient broth resulted in a high concentration of bacteria (107 cfu) in a 6 mm granule (Digat, 1991). This method caused less nutritional stress and it was suggested that the system enables the formulation of several microbial agents or strains (i.e. a cocktail mix). Research at Agriculture and Agri-Food Canada in Saskatoon is currently underway to evaluate the shelf life and efficacy of bacteria formulated into several solid-based products, including alginate prills, pesta, and pregelatinized starches. The efficacy of bacteria is dependent on each combination of strain and formulation and some formulations offer potential for application of DRB as pre-emergence, slow release biocontrol agents. Future research will focus on placement of granules (e.g. within-furrow application, broadcasting, side-row banding), and granular size, water activity, and stability. The application method should be compatible with existing agricultural practices and should not require new equipment that would preclude the farmer from using the biopesticide product. Bacteria may be delivered to the soil but the type of delivery system will depend on the type of crop and farming practice in use.

3.5. Weed Ecology and Integrated Weed Management

Introduction of microbial agents as part of a weed management program should consider the compatibility of the agents with agronomic practices. After all, biological control is not intended to replace chemicals, but is used to control weeds where chemical herbicides are not available, and to complement cultural and chemical control measures that can lead to reductions in chemical pesticide use. Chemical pesticides may either enhance, reduce, or have a neutral effect on the efficacy of fungal and bacterial agents. Therefore, compatibility with various chemical herbicides, fungicides, insecticides, and fertilizers will be a desirable characteristic in a bioherbicide agent.

Two factors in weed ecology that are considered in the evaluation of crop losses due to weeds are time of emergence and critical period (Aldrich, 1987; Boyetchko, 1997; O'Donovan *et al.*, 1985; Swanton and Weise, 1991; Weaver, 1984). The competitive ability of crops is affected by the relative time of emergence of the weed. The earlier weed emerges in the growing season, the more competitive that weed is in the crop and the greater the potential for increasing crop yield loss (O'Donovan *et al.*, 1985). Critical period defines the minimum amount of time that suppression of weeds must occur in order to prevent crop yield loss (Swanton and Weise, 1991; Weaver, 1984). Two important concepts to consider in critical period are : 1) the length of time weed control is required to prevent crop yield loss and 2) the length of time a crop can tolerate the presence of weeds before the weeds interfere with crop growth and yield (Weaver, 1984).

Time of emergence and critical period will vary with the weed species and density and the crop and crop cultivar. They provide useful information on the suitability of a bioherbicide candidate for weed control in any particular crop (Boyetchko, 1997). For example, a bioherbicide that takes too long to kill or suppress a weed may be considered unsuitable in a crop that cannot tolerate the presence of weeds for very long without experiencing significant yield losses. However, the same bioherbicide may be appropriate in a crop that is more competitive and can withstand the presence of the weed before yield losses occur. New and improved formulations may enhance the efficacy and speed of weed control, thus making the bioherbicide a more efficacious candidate. In addition, a weed that emerges early in the growing season, compared to the crop, may require a biocontrol agent that either controls the weed very quickly or is applied as a pre-emergence rather than as a post-emergence bioherbicide.

Prospective biocontrol candidates that were initially considered not suitable or effective may be highly efficacious when weeds are grown in a competitive environment. *Colletotrichum coccodes* had only limited bioherbicidal activity on velvetleaf in pure stands, but when grown in competition with soybean in the field, its bioherbicidal effects were substantially increased (DiTommaso and Watson, 1995; DiTommaso *et al.*, 1996). Application of *C. coccodes* onto velvetleaf grown in soybean fields resulted in significantly higher reproductive losses in velvetleaf compared to velvetleaf grown in pure stands. Conversely, yield losses in soybean were lower in plots containing velvetleaf treated with the mycoherbicide, compared to control plots. At high plant densities, inoculated velvetleaf plants were not able to recover from initial reductions in plant growth and development, thus decreasing its competitive ability and possible reproductive capacity that would lead to fewer viable seeds returning to the seed bank (DiTommaso and Watson, 1995). These results indicate the importance of interspecific plant competition while evaluating potential biocontrol agents, since these conditions may alter the candidate's suitability and effectiveness.

DRB that have been used to control downy brome in winter wheat, reduced weed competition and provided an increase in crop yield (Kennedy *et al.*, 1991). Although complete control was not achieved, suppression of the weed reduced its competitive ability. Similar results with jointed goatgrass in winter wheat were demonstrated (Kennedy *et al.*, 1995). DRB screened and evaluated for control of leafy spurge decreased weed emergence, seedling vigor, and seed production, compared to the un-inoculated control (Kremer and Kennedy, 1996; Souissi and Kremer, 1994). These bacteria applied as a pre-emergence treatment reduced weed competition in velvetleaf and reductions of weed seeds in the seed bank may provide long-term reductions in weed populations (Kremer, 1993). Host-specific microbes and their metabolites may be utilized to accelerate the depletion of weed seeds that are dormant or persistent in the soil. Some DRB in weeds may also have plant-growth promoting activity in crops (Boyetchko and Holmström-Ruddick, 1995). Therefore, it may be possible to utilize these biocontrol agents to improve the competitive ability of crops while reducing weed competition, suggesting that eradication of weeds by biocontrol agents may not be required.

Crop production systems are subject to a variety of farm management practices, including application of fertilizers and pesticides and tillage operations. To date, the impact of agroecosystem management on biocontrol agents has rarely been evaluated. For the most part, many bioherbicide candidates are still in the developmental phases, including evaluation of formulations, delivery, and chemical compatibility. The next phase will have to consider how to integrate these biocontrol agents under different crop management systems. For example, evolving research in conservation tillage may provide opportunities in the application of bioherbicides (Derksen *et al.*, 1996). Collego overwinters on infected seed and was recovered from unburied infected plant debris (TeBeest, 1982). It may be possible to obtain residual control of northern jointvetch in conservation tillage systems via secondary spread of inoculum. Similarly, DRB populations were greater in crops sown in zero-tillage plots than those seeded into tilled plots (Stroo *et al.*, 1988). It may be conjectured that with reduced tillage practices, crop residues may provide less invasive habitats for biocontrol agents, and provide a substrate (i.e. the crop residue) for overwintering which extends the period of control (Derksen *et al.*, 1996).

4. FUTURE PROSPECTS

Research into microbial pathogens for biological weed control has been receiving a great deal of activity and attention. Several constraints in the development of bioherbicides have been classed as biological, environmental, technological and commercial (Auld and Morin, 1995; Mortensen, 1998). But in reality, these are challenges and not constraints in the evolving discipline of biological control. Researchers, industry, and farmers must evaluate their perceptions in what is considered adequate weed control from an ecologically balanced perspective (Makowski, 1997). Complete kill or eradication of weeds that is often used to evaluate chemical herbicides may not be the appropriate measure for the suitability of bioherbicides. Reduced competition and suppression and/or inhibition of weeds may also lead to improvements in crop yield which is the primary objective.

The debate on market size and niche continues to present controversy in the development of bioherbicides. Large companies appear willing to develop agents with world market potential, whereas small companies have been investing in the development and commercialization of several prospective bioherbicides for niche markets. Regardless of market size, researchers should no longer rely on the chance discovery of a plant pathogen on a weed, but take a pragmatic approach in the selection of target weeds and screening of potential microbial candidates.

A number of potential bioherbicides are being developed, including a proprietary organism for dandelion control in turf delivered as a pellet and/or granular to plant/soil interface (G.J. Boland, 1998, University of Guelph, personal communication), a *Fusarium* that is added to seed at planting for striga control (A.K. Watson, 1998, McGill University, personal communication), and *C. purpureum* for vegetation management in utility rights-of-ways (W.E. Hintz, 1998, University of Victoria, personal communication). In addition, patent protection has been obtained for *Phomopsis amaranthicola* for control of several pigweed species and *Dactylaria higginsii* (Luttrell) M.B. Ellis for control of purple nutsedge (R. Charudattan, 1998, University of Florida, personal communication). New innovations in application of multiple pathogens for control of grass weeds with three fungi native to Florida resulted in highly efficacious activity (Chandramohan and Charudattan, 1997; Chandramohan *et al.*, 1997). A patent application for this process has been submitted to the U.S. Patent Office (R. Charudattan, 1998, personal communication).

Innovative technologies in fermentation and formulation are providing new options for application and enhancement of bioherbicide activity. However, it should be borne in mind that discovery of a pathogen is only one step in the development of a bioherbicide. The final biocontrol product is one where all processes in production and formulation, delivery, and information on the pathogenicity and mode of action have been evaluated. These activities will involve a committed team of researchers in a number of disciplines including plant pathology, weed ecology, agronomy, biochemistry, and spray application technology. Understanding the biochemical and physiological processes of infection and resistance, the role of phytotoxic metabolites, and molecular strategies for enhancing efficacy will be effective tools for bioherbicide development. Formulation research has often been conducted by trial and error and often based on each bioherbicide candidate. Perhaps a systematic approach in the selection of formulations based on fungal and bacterial characteristics will lead to a better understanding and selection process of the interactions between the microbial agent and formulation components (Zhang *et al.*, 1998).

These examples underscore the potential utility of bioherbicides for weed management. Moreover, the emergence of companies becoming involved in microbial product development and fermentation and formulation expertise indicates that microbial pest control products are an emerging technology. Their success will come from multidisciplinary and collaborative partnerships amongst researchers at public and private institutions.

5. CONCLUSIONS

Biological weed control with microbial agents, often referred to as bioherbicides, provides an additional weed management strategy for eradication, suppression, or delay of plant growth and development. A variety of fungal and bacterial agents applied to the foliage and/or soil have been explored as potential bioherbicides, but there are few commercial products available. The majority of fungal pathogens belong to the Hyphomycetes and Coelomycetes, while bacterial agents, most of which belong to *Pseudomonas* species, have only recently been evaluated for biocontrol. A number of challenges that researchers have encountered in the development of microbial agents are formulation and application technology, while physiological and biochemical processes regulating pathogenicity have received little attention. Understanding the physical and biochemical barriers for infection, such as induced resistance, may provide clues for enhancing the efficacy of prospective bioherbicides. Microbially-derived phytotoxic compounds offer opportunities for developing natural products as templates for synthetic herbicides or for optimization of microbial activity. Research in molecular biology and formulation technology are providing tools for enhancing bioherbicidal activity and knowledge of weed ecology contributes to the understanding of weed/crop competition. This discusses innovative strategies on application and enhancement of bioherbicides.

REFERENCES

Aldrich, R.J. 1987, Predicting crop yield reductions from weeds, Weed Technol. 1:199-206.

Amrhein, N. 1986, Specific inhibitors as probes into the biosynthesis and metabolism of aromatic amino acids, Recent Adv. Phytochem. 20:83-117.

Amsellem, Z., Sharon, A., Fressel, J. and Quimby, P.C., Jr. 1990, Complete abolition of high inoculum threshold of two mycoherbicides (*Alternaria cassiae* and *A. crassa*) when applied in invert emulsion, Phytopathology 80:925-929.

Auld, B.A. 1993, Vegetable oil suspension emulsions reduce dew dependence of a mycoherbicide, Crop Prot. 12:477-479.

Auld, B.A., and Morin, L. 1995, Constraints in the development of bioherbicides, Weed Technol. 9:638-652.

Auld, B.A., Say, M.M., Ridings, H.I. and Andrews, J. 1990, Field applications of *Colletotrichum orbiculare* to control *Xanthium spinosum*, Agric. Ecosystems Environ. 23:315-323.

Beste, C.E., Frank, J.R., Bruckart, W.L., Johnson, D.R. and Potts, W.E. 1992, Yellow nutsedge (*Cyperus esculentus*) control in tomato with *Puccinia canaliculata* and pebulate, Weed Technol. 6:980-984.

Boyetchko, S.M. 1996a, Formulating bacteria for use as biological control agents, in : Proc. National Meeting of Expert Committee on Weeds, Victoria, B.C., December 9-12, 1996, pp. 85-88.

Boyetchko, S.M. 1996b, Impact of soil microorganisms on weed biology and ecology, Phytoprotection 77:41-56.

Boyetchko, S.M. 1997, Efficacy of rhizobacteria as biological control agents of grassy weeds, in : Proc. Soils and Crops Workshop '97, Saskatoon, Saskatchewan, February 20-21, 1997, pp. 460-465.

Boyetchko, S.M., and Holmström-Ruddick, B. 1995, Host-range of rhizobacteria effective as biocontrol agents of downy brome, Can. J. Plant Pathol. 17, (Abstr).

Boyetchko, S.M., Bailey, K.L., Mortensen, K., Wolf, T.M. and Zhang, W.M. 1997, Survey and evaluation of fungal pathogens for biological control of grass weeds (Abstr), Phytopathology 87 (Suppl):S11.

Boyetchko, S.M., and Mortensen, K. 1993, Use of rhizobacteria as biological control agents of downy brome, in : Proc. Soils and Crops Workshop '93, Saskatoon, Saskatchewan, February 25-26, 1993, pp.443-448.

Boyetchko, S.M., Wolf, T.M., Bailey, K.L., Mortensen, K. and Zhang, W.M. 1998, Survey and evaluation of fungal pathogens for biological control of grass weeds, in : Proc. Soils and Crops Workshop, Saskatoon, Saskatchewan, February 19-20, 1998 (in press).

Boyette, C.D. 1986, Evaluation of *Alternaria crassa* for biological control of Jimsonweed: host range and virulence, Plant Sci. 45:223-228.

Boyette, C.D. 1991a, Host range and virulence of *Colletotrichum truncatum*, a potential mycoherbicide for hemp sesbania (*Sesbania exaltata*), Plant Dis. 75:62-64.

Boyette, C.D. 1991b, Control of hemp sesbania with a fungal pathogen, *Colletotrichum truncatum*, U.S. Patent No. 5,034, 328.

Boyette, C.D. 1994, Unrefined corn oil improves the mycoherbicidal activity of *Colletotrichum truncatum* for hemp sesbania (*Sesbania exaltata*) control, Weed Technol. 8:526-529.

Boyette, C.D., Templeton, G.E. and Oliver, L.R. 1984, Texas gourd (*Cucurbita texana*) control with *Fusarium solani* f.sp. *cucurbitae*, Weed Sci. 32:649-655.

Boyette, C.D., Quimby, P.C. Jr., Caesar, A.J., Birdsall, J.L., Connick, W.J., Jr., Daigle, D.J., Jackson, M.A., Egley, G.H. and Abbas, H.K. 1996, Adjuvants, formulations, and spraying systems for improvement of mycoherbicides, Weed Technol. 10:637-644.

Boyette, C.D., and Walker, H.L. 1986, Evaluation of *Fusarium lateritium* as a biological herbicide for controlling velvetleaf (*Abutilon theophrasti*), and prickly sida (*Sida spinosa*), Weed Sci. 34:106-109.

Boyette, C.D., Quimby, P.C., Jr., Bryson, C.T. , Egley, G.H. and Fulgham, F.E. 1993, Biological control of hemp sesbania (*Sesbania exaltata*) under field conditions with *Colletotrichum truncatum* formulated in an invert emulsion, Weed Sci. 41:497-500.

Brosten, B.S. and Sands, D.C. 1986, Field trials of *Sclerotinia sclerotiorum* to control Canada thistle (*Cirscium arvense*), Weed Sci. 34:377-380.

Burnett, H.C., Tucker, D.P.H. and Ridings, W.H. 1974, *Phytophthora* root and stem rot of milkweed vine, Plant Dis. Rep. 58:355-357.

Chandler, J.M., Hamill, A.S. and Thomas, A.G. 1984, Crop losses due to weeds in Canada and the United States, Spec. Rep. Weed Sci Soc. Am., Champaign, Ill.

Chandramohan, S. and Charudattan, R. 1997, Bioherbicidal control of grassy weeds with a pathogen mixture, WSSA Abstracts 37:56.

Chandramohan, S., Charudattan, R. and Singh, M. 1997, Field testing of a -multiple-pathogen strategy' for bioherbicidal control of grassy weeds (Abstr), Phytopathology 87(Suppl.):S17.

Charudattan, R. 1991, The mycoherbicide approach with plant pathogens, in : Microbial Control of Weeds, D.O. TeBeest, ed., Chapman and Hall, Inc., New York, p. 24-57.

Charudattan, R., and DeLoach, C.J. Jr. 1988, Management of pathogens and insects for weed control in agroecosystem, in : Weed Management in Agroecosystems: Ecological Approaches, M.A. Altieri and M. Liebman, eds., CRC Press, Boca Raton, Florida. p.245-264

Charudattan, R., Walker, H.L., Boyette, C.D., Ridings, W.H., TeBeest, D.O., Van Dyke, C.G. and Worsham, A.D. 1986, Evaluation of *Alternaria cassiae* as a mycoherbicide for sicklepod (*Cassia obtusifolia*) in regional field tests, Southern Coop. Ser. Bull. 317. Alabama Agric. Exp. Sta., Auburn University, Auburn, AL.

Cherrington, C.A. and Elliott, L.F. 1987, Incidence of inhibitory pseudomonads in the Pacific Northwest, Plant Soil. 101:159-165.

Connick, Jr. W.J., Boyette, C.D. and McAlpine, J.R. 1991a, Formulation of mycoherbicides using a pasta-like process, Biol. Control 1:281-287.

Connick, Jr. W.J., Daigle, D.J. and Quimby, P.C., Jr. 1991b, An improved invert emulsion with high water retention for mycoherbicide delivery, Weed Technol. 5:442-444.

Connick, Jr. W.J., Daigle, D.J., Boyette, C.D., Williams, K.S. and Vinyard, B. 1996, Water activity and other factors that affect the viability of *Colletotrichum truncatum* conidia in wheat flour-kaolin granules ('pesta'), Biocontrol Sci. Technol. 6:277-284.

Connick, Jr. W.J., Lewis, J.A. and Quimby, P.C., Jr. 1990, Formulation of biocontrol agents for use in plant pathology, UCLA Symp. Mol. Cell Biol. pp. 345-372.

Daigle, D.J. and Cotty, P.J. 1991, Factors that influence germination and mycoherbicidal activity of *Alternaria cassiae*, Weed Technol. 5:82-86.

Daigle, D.J. and Cotty, P.J. 1992, Production of conidia of *Alternaria cassiae* with alginate pellets, Biol.

Control 2:278-281.
De Jong, M.D., Scheepens, P.C. and Zadoks, J.C. 1990, Risk analysis for biological control: a Dutch case study in biocontrol of *Prunus serotina* by the fungus *Chondrostereum purpureum*, Plant Dis. 74:189-194.
Derksen, D.A., Blackshaw, R.E. and Boyetchko, S.M. 1996, Sustainability, conservation tillage and weeds in Canada, Can. J. Plant Sci. 76:651-659.
Dickman, M.B. 1993, Molecular biology of plant-parasite relations, in : Soil Microbial Ecology: Applications in Agricultural and Environmental Management, F. Blaine Metting, Jr., ed., Marcel Dekker, New York, pp. 177-202.
Dickman, M.B., Podila, G.K. and Kolattukudy, P.E. 1989, Insertion of cutinase gene into a wound pathogen enables it to infect intact host, Nature 342:446-448.
Digat, B. 1989, Strategies for seed bacterization, Acta Hortic. 253:121-130.
Digat, B. 1991, A new encapsulation technology for bacterial inoculants and seed bacterization, in : Plant Growth-Promoting Rhizobacteria Progress and Prospects, Int'l Workshop on Plant Growth-Promoting Rhizobacteria, (2nd: 1990: Interlaken, Switzerland), Bulletin SROP, No. 14, pp. 383-391.
DiTommaso, A., and Watson, A.K. 1995, Impact of a fungal pathogen, *Colletotrichum coccodes* on growth and competitive ability of *Abutilon theophrasti*, New Phytol. 131:51-60.
DiTommaso, A., Watson, A.K. and Hallett, S.G. 1996, Infection by the fungal pathogen *Colletotrichum coccodes* affects velvetleaf (*Abutilon theophrasti*)-soybean competition in the field, Weed Sci. 44:924-933.
Dorworth, C. 1992, The case for biological control - Deleterious rhizobacteria as biocontrol agents for *Calamagrostis canadensis*, Northern Forest Vegetation Manager, pp. 9-11.
Duke, S.O. 1986, Naturally occurring chemical compounds as herbicides, Rev. Weed Sci. 2:15-44.
Dumas, M.T., Wood, J.E., Mitchell, E.G. and Boyonoski, N.W. 1997, Control of stump sprouting of *Populus tremuloides* and *P. grandidentata* by inoculation with *Chondrostereum purpureum*, Biol. Control 10:37-41.
Egley, G.H. and Boyette, C.D. 1994, Water-corn oil emulsion enhances conidia germination and mycoherbicidal activity of *Colletotrichum truncatum*, Weed Sci. 43:312-317.
Elliott, L.F., and Lynch, J.M. 1985, Plant growth-inhibiting pseudomonads colonizing winter wheat (*Triticum aestivum* L.) roots, Plant and Soil 84:57-65.
Gotlieb, A.R., Watson, A.K. and Poirier, C. 1987, First report of *Colletotrichum coccodes* on velvetleaf, Plant Dis. 71:281.
Grant, N.T., Prusinkiewicz, E., Makowski, R.M.D., Holmström-Ruddick, B. and Mortensen, K. 1990a, Effect of selected pesticides on survival of *Colletotrichum gloeosporioides* f.sp.*malvae*, a bioherbicide for round-leaved mallow (*Malva pusilla*), Weed Technol. 4:701-715.
Grant, N.T., Prusinkiewicz, E., Mortensen, K. and Makowski, R.M.D. 1990b, Herbicide interactions with *Colletotrichum gloeosporioides* f.sp. *malvae* a bioherbicide for round-leaved mallow (*Malva pusilla*) control, Weed Technol. 4:716-723.
Greaves, M.P., and MacQueen, M.D. 1992, Bioherbicides: their role in tomorrow's agriculture. in : Resistance: Achievements and Developments in Combating Pesticide Resistance, Sci. Symposium, Harpenden, England, UK, pp. 295-306.
Green, S., Stewart-Wade, S.M., Boland, G.J., Teshler, M.P. and Liu, S.H. 1998, Formulating microorganisms for biological control of weeds, in : Plant-Microbe Interations and Biological Control, G.J. Boland and L.D. Kuykendall, eds., Marcel Dekker Inc., New York, pp. 249-281
Gronwald, J.W., Johnson, D.R., Plaisance, K.L. and Wyse, D.L. 1998, Factors influencing the efficacy of *Pseudomonas syringae* pv. *tagetis* on host and nonhost species (Abstr), WSSA Abstracts, 38:46.
Gurusiddaiah, S., Gealy, D.R., Kennedy, A.C. and Ogg, A.G. Jr. 1992, Production, isolation, and characterization of phytotoxic and fungistatic compounds for biocontrol of downy brome (*Bromus tectorum* L.) and plant pathogenic fungi (Abstr), Weed Sci. Soc. Am. 32:84.
Harris, P.A. and Stahlman, P.W. 1992, Biological weed control using deleterious rhizobacteria, Abstr. Weed Sci. Soc. Am. 32:50.
Harris, P.A. and Stahlman, P.W. 1993, Soil bacteria selectively inhibit winter wheat annual grass weeds in winter wheat, in : Agronomy Abstracts (Soil Science Soc. Amer.), Cincinnati, Ohio, November 7-12, 1993, p. 250.
Hasan, S. and Ayres, P.G. 1990, The control of weeds through fungi:Principles and prospects, New Phytol. 115:201-222.
Hoagland, R.E. 1990. Microbes and microbial products as herbicides : An overview, in : Microbes and Microbial Products as Herbicides, R.E. Hoagland, ed., ACS Symposium Series 439, American Chemical Society, Washington, pp.2-52.
Hoagland, R.E. 1996, Chemical interactions with bioherbicides to improve efficacy, Weed Technol. 10:651-674.
Hoagland, R.E. and Boyette, C.D. 1994, Pathogenic interaction of *Alternaria crassa* and phenolic metabolism

in jimsonweed (*Datura stramonium* L.) varieties, Weed Sci. 42:44-49.
Howell, C.R. 1982, Effect of *Gliocladium virens* on *Pythium ultimum, Rhizoctonia solani*, and damping-off of cotton seedlings, Phytopathology 72:496-498.
Howell, C.R. and Stipanovic, R.D. 1984, Phytotoxicity to crop plants and herbicidal effects on weeds of viridiol produced by *Gliocladium virens*, Phytopathology 74:1346-1349.
Imaizumi, S., Nishino, T., Miyabe, K., Fujimori, T. and Yamada, M. 1997, Biological control of annual bluegrass (*Poa annua* L.) with a Japanese isolate of *Xanthomonas campestris* pv. *poae* (JT-P482), Biol. Control 8:7-14.
Jackson, M.A. and Schisler, D.A. 1995, Liquid culture production of microsclerotia of *Colletotrichum truncatum* for use as bioherbicidal propagules, Mycol. Res. 99:879- 884.
Jackson, M.A., Schisler, D.A., Slininger, P.J., Boyette, C.D., Silman, R.W. and Bothast, R.J. 1996, Fermentation strategies for improving the fitness of a bioherbicide, Weed Technol. 10:645-650.
Jacobsen, B.J. and Backman, P.A. 1993, Biological and cultural plant disease controls: Alternatives and supplements to chemicals in IPM systems, Plant Dis. 77:311-315.
Jha, P.K., Nair, S. and Babu, S. 1993, Encapsulation of seeds of *Sesbania sesban* with polyacrylamide and alginate gel entrappted rhizobia leads to effective symbiotic nitrogen fixation, Ind. J. Expt. Biol. 31:161-167.
Johnson, B.J. 1994, Biological control of annual bluegrass with *Xanthomonas campestris* pv. *poannua* in bermudagrass, Hort Science, 29:659-662.
Johnson, B.N., Kennedy, A.C. and Ogg, A.G. 1993, Suppression of downy brome growth by a rhizobacterium, in controlled environments, Soil Sci. Soc. Am. J. 57:73-77.
Johnson, D.R. and Wyse, D.L. 1991, Use of *Pseudomonas syringae* pv. *tagetis* for control of Canada thistle, Proc. North Cent. Weed Sci. Soc. 46:14-15.
Johnson, D.R., Wyse, D.L. and Jones, K.J. 1995, Efficacy of spring and fall applications of *Pseudomonas syringae* pv. *tagetis* for Canada thistle (*Cirsium arvense* L.) control in soybean, WSSA Abstracts, 35:61.
Johnson, D.R., Wyse, D.L. and Jones, K.J. 1996, Controlling weeds with phytopathogenic bacteria, Weed Technol. 10:621-624.
Jones, R.W., and Hancock, J.G. 1990, Soilborne fungi for biological control of weeds, in : Microbes and Microbial Products as Herbicides, R.E. Hoagland, ed., American Chemical Society Symposium Series, Washington, D.C.
Kadir, J, Charudattan, R., Berger, R.D., Stall, W.M. and Brecke, B.J. 1997a, Field efficacy of *Dactylaria higginsii* for control of purple nutsedge (Abstr), Phytopathology 87(Suppl):S49, pp.277-286.
Kadir, J., Charudattan, R., Stall, W.M. and Bewick, T.A. 1997b, Effect of *Dactylaria higginsii* on the interference of purple nutsedge with tomato and pepper (Abstr), Phytopathology 87(Suppl):S50.
Kennedy, A.C. 1996, Molecular biology of bacteria and fungi for biological control of weeds, in : Molecular Biology of the Biological Control of Pests and Diseases of Plants, M. Gunasekaran and D.J. Weber, eds., CRC Press Inc., Boca Raton, Florida, pp.155-172.
Kennedy, A.C., Elliott, L.F.,Young, F.L. and Douglas, C.L. 1991, Rhizobacteria suppressive to the weed downy brome, Soil Sci. Soc. Am. J. 55:722-727.
Kennedy, A.C., Stubbs, T.L. and Fuerst, E.P. 1995, Biological control of jointed goatgrass (*Aegilops cylindrica*), WSSA Abstracts. 35:61.
Kenfield, D., Bunkers, G., Strobel, G.A. and Sugawara, F. 1988, Potential new herbicides - phytotoxins from plant pathogens, Weed Technol. 2:519-524.
Kremer, R.J. 1987, Identity and properties of bacteria inhabiting seeds of selected broadleaf weed species, Microb. Ecol. 14:29-37.
Kremer, R.J. 1993, Management of weed seed banks with microorganisms, Ecol. Appl. 3:42-52.
Kremer, R.J. and Kennedy, A.C. 1996, Rhizobacteria as biocontrol agents of weeds, Weed Technol. 10:601-609.
Kremer, R.J., Begonia, M.F.T., Stanley, L. and Lanham, E.T. 1990, Characterization of rhizobacteria associated with weed seedlings, Appl. Environ. Microbiol. 56:1649-1655.
Kremer, R.J., and Sarwar, M. 1995a, Microbial metabolites with potential applications in weed management, Proc. Plant Growth Reg. Soc. Am. 21:48-51.
Lawrie, J., Greaves, M.P., Down, V.M. and Chassot, A. 1997, Some effects of spray droplet size on distribution, germination of and infection by mycoherbicide spores, Aspects Appl. Biol. 48:175-182.
Makowski, R.M.D. 1987, The evaluation of *Malva pusilla* Sm. as a weed and its pathogen *Colletotrichum gloeosporioides* (Penz.) Sacc. f.sp. *malvae* as a bioherbicide, Ph.D. Dissertation, University of Saskatchewan, Saskatoon, 225 pp.
Makowski, R.M.D. 1993, Effect of inoculum concentration, temperature, dew period, and plant growth stage on disease of round-leaved mallow and velvetleaf by *Colletotrichum gloeosporioides* f.sp. *malvae*, Phytopathology 83:1229-1234.

Makowski, R.M.D. 1997, Foliar pathogens in weed biocontrol: ecological and regulatory constraints, in : Ecological Interactions and Biological Control, D.A. Andow, D.W. Ragsdale, and R.F. Nyvall, eds., Westview Press, Colorado, pp.87-103.

Makowski, R.M.D., and Mortensen, K. 1992, The first mycoherbicide in Canada: *Colletotrichum gloeosporioides* f.sp. *malvae* for round-leaved mallow control, in : Proceedings of the First International Weed Control Congress, Monash University, Melbourne, Australia, February 17-21, 1992, pp. 298-300

Mooney, H.D., Boyetchko, S.M. and Punja, Z.K. 1996, Development of application techniques for biological weed control using rhizobacteria, in : Proc. IX International Symposium on Biological Control of Weeds, V.C. Moran and J.H. Joffmann eds., 19-26 January, 1996, Stellenbosch, South Africa, University of Cape Town, South Africa, pp.297-299.

Morin, L., Derby, J.L. and Kokko, E.G. 1996, Infection process of *Colletotrichum gloeosporioides* f.sp. *malvae* on Malvaceae weeds, Mycol. Res. 100:165-172.

Morin, L., Watson, A.K. and Reeleder, R.D. 1989. Efficacy of *Phomopsis convolvulus* for control of field bindweed (*Convolvulus arvensis*), Weed Sci. 37:830-835.

Morin, L., Watson, A.K. and Reeleder, R.D. 1990, Effect of dew, inoculum density, and spray additives on infection of field bindweed by *Phomopsis convolvulus*, Can. J. Plant Pathol. 12:48-56.

Mortensen, K. 1988, The potential of an endemic fungus *Colletotrichum gloeosporioides*, for biological control of round-leaved mallow (*Malva pusilla*) and velvetleaf (*Abutilon theophrasti*), Weed Sci. 36:473-478.

Mortensen, K. 1998, Biological control of weeds using microorganisms, in : Plant-Microbe Interactions and Biological Control, G.J. Boland and L.D. Kuykendall, eds., Marcel Dekker Inc., New York, pp. 223-248.

Mortensen, K. and Hogue, E.J. 1995, *Sclerotinia sclerotiorum* as a potential biological control agent for diffuse knapweed on dry rangeland in interior British Columbia, in : Proceedings of VIII Int. Symp. Biol. Contr. Weeds, E.S.Delfosse and R.R. Scott, eds., 1992,Lincoln University, Cantebury, New Zealand, DSIR/CSIRO, Melbourne, Australia, pp. 397-406.

Mortensen, K. and Makowski, R.M.D. 1990, Field efficacy at different rates of *Colletotrichum gloeosporioides* f.sp. *malvae* as a bioherbicide for round-leaved mallow (*Malva pusilla* Sm.) in : Proceedings of the VIIth International Symposium on the Biological Control of Weeds, E.S. Delfosse, ed., Rome, 1988, pp. 523-530

Mortensen, K., and Makowski, R.M.D. 1997, Effects of *Colletotrichum gloeosporioides* f.sp. *malvae* on plant development and biomass of non-target field crops under controlled and field conditions, Weed Res. 37:351-360.

Nishino, T., Imaizumi, S., Yamada, M. and Goto, M. 1995, *Xanthomonas campestris* pv. *poae* as the causal agent of wilt symptoms on annual bluegrass in Japan, Ann. Phytopathol. Soc. Jpn. 61:555-561.

O'Donovan, J.T., de St. Remy, E.A., O'Sullivan, P.A., Dew, D.A. and Sharma, A.K. 1985, Influence of the relative time of emergence of wild oat (*Avena fatua*) on yield loss of barley (*Hordeum vulgare*) and wheat (*Triticum aestivum*), Weed Sci. 33:498-503.

Ohra, J., Morita, K., Tsujino, Y., Fujimori, T., Goering, M., Evans, S. and Zorner, P. 1995, Production of two phytotoxic metabolites by the fungus *Alternaria cassiae*, Biosci. Biotech. Biochem. 59:1782-1783.

Phatak, S.C. 1984, Knock out nutsedge, Amer. Veg. Grower 32:44-46.

Phatak, S.C., Sumner, D.R., Wells, H.D., Bell, D.K. and Glaze, N.C. 1983, Biological control of yellow nutsedge with the indigenous rust fungus *Puccinia canaliculata*, Science 219:1446-1447.

Pimental, D., McLaughlin, L., Zepp, A., Lakitan, B., Kraus, T., Kleinman, P., Vancini, F., Roach, W.J., Graap, E., Keeton, W.S. and Selig, G. 1986, Environmental and economic effects of reducing pesticide use, BioScience 41:402-409.

Poppenborg, L., Friehs, K. and Flaschel, E. 1997, The green fluorescent protein is a versatile reporter for bioprocess monitoring, J. Biotechnol. 58:79-88.

Porteous, L.A., Armstrong, J.L., Seidler, R.J. and Watrud, L.S. 1994, An effective method to extract DNA from environmental samples for polymerase chain reaction amplication and DNA fingerprint analysis, Curr. Microbiol. 29:301-307.

Powell, K.A. and Jutsum, A.R. 1993, Technical and commercial aspects of biocontrol products, Pesticide Sci. 37:315-321.

Prasad, R. 1994, Influence of several pesticides and adjuvants on *Chondrostereum purpureum* - a bioherbicide agent for control of forest weeds, Weed Technol. 8:445-449.

Ramsfield, T.D., Becker, E.M., Rathief, S.M., Tang, Y., Vrain, T.C., Shamoun, S.F. and Hintz, W.E. 1996, Georgraphic variation of *Chondrostereum purpureum* detected by polymorphisms in the ribosomal DNA, Can. J. Bot. 74:1919-1929.

Riddle, G.E., Burpee, L.L. and Boland, G.J. 1991, Virulence of *Sclerotinia sclerotiorum* and *S. minor* on dandelion, Weed Sci. 39:109-118.

Rosskopf, E.N. 1997, Evaluation of *Phomopsis amaranthicola* sp. nov. as a biological control agent for *Amaranthus* spp., Ph.D. Thesis, University of Florida, Gainesville, Florida.

Sands, D.C., Ford, E.J. and Miller, R.V. 1990a, Genetic manipulation of broad host-range fungi for biological control of weeds, Weed Technol. 4:471-474.

Sands, D.C., Miller, R.V. and Ford, E.J. 1990b, Biotechnological approaches to control of weeds with pathogens, in : Microbes and Microbial Products as Herbicides, ACS Symposium Series, R.E. Hoagland, ed., American Chemical Society, Washington, pp.184-190.

Sarwar, M. and Kremer, R.J. 1995a, Enhanced suppression of plant growth through production of L-tryptophan-derived compounds by deleterious rhizobacteria, Plant and Soil 172:261-269.

Sarwar, M. and Kremer, R.J. 1995b, Determination of bacterially derived auxins using a microplate method, Letters Appl. Microbiol. 20:282-285.

Schippers, B., Bakker, A.W. and Bakker, P.A.H.M. 1987, Interactions of deleterious and beneficial rhizosphere microorganisms and the effect of cropping practices, Annu. Rev. Phytopathol. 25:339-358.

Schippers, B., Bakker, A.W., Bakker, P.A.H.M. and Van Peer, R. 1990, Beneficial and deleterious effects of HCN-producing pseudomonads on rhizosphere interactions, Plant and Soil 129:75-83.

Shabana, Y.M., Charudattan, R., Devalerio, J.T. and Elwakil, M.A. 1997, An evaluation of hydrophilic polymers for formulating the bioherbicide agents *Alternaria cassiae* and *A. eichhorniae*, Weed Technol. 11:212-220.

Shamoun, S.F., Ramsfield, T.D., Shrimpton, G. and Hintz, W.E. 1996, Development of *Chondrostereum purpureum* as a mycoherbicide for red alder (*Alnus rubra*) in utility rights-of-way (Abstr), in: Proc. National Meeting of the Expert Committee on Weeds, P. Comeau and G. Harper, eds., December 9-12, 1996, p. 199.

Sharon, A., Amsellem, Z. and Gressel, J. 1992, Glyphosate suppression of an elicited defence response, Plant Physiol. 98:654-659.

Smith, R.J. 1986, Biological control of northern jointvetch (*Aeschynomene virginica*) in rice (*Oryza sativa*) and soybeans (*Glycine max*) - A researchers view, Weed Sci. 34(Suppl. 1):17-23.

Souissi, T. and Kremer, R.J. 1994, Leafy spurge (*Euphorbia esula*) cell cultures for screening deleterious rhizobacteria, Weed Sci. 42:310-315.

Strobel, G., Kenfield, D., Bunkers, G., Sugawara, F. and Clardy, J. 1991, Phytotoxins as potential herbicides, Experientia 47:819-826.

Stroo, H.F., Elliott, L.F. and Papendick, R.I. 1988, Growth, survival and toxin production of root-inhibitory pseudomonads on crop residues, Soil Biol. Biochem. 20:201-207.

Swanton, C.J., Harker, K.N. and Anderson, R.L. 1993, Crop losses due to weeds in Canada, Weed Technol. 7:537-542.

Swanton, C.J. and Weise, S.F. 1991, Integrated weed management: the rationale and approach, Weed Technol. 5:657-663.

TeBeest, D.O. 1982, Survival of *Colletotrichum gloeosporioides* f.sp. *aeschynomene* in rice irrigation water and soil, Plant Dis. 66:469-472.

TeBeest, D.O. 1996, Biological control of weeds with plant pathogens and microbial pesticides, in: Advances in Agronomy, D.L. Sparks ed., Volulme 56, Academic Press, Toronto, pp.115-137.

TeBeest, D.O. and Templeton, G.E. 1985, Mycoherbicide : progress in the biological control of weeds, Plant Dis. 69:6-10.

TeBeest, D.O., Yang, X.B. and Cisar, C.R. 1992, The status of biological control of weeds with fungal pathogens, Annu. Rev. Phytopathol. 30:637-657.

Tranel, P.J., Gealy, D.R. and Irzyk, G.P. 1993a, Physiological responses of downy brome (*Bromus tectorum*) roots to *Pseudomonas fluorescens* strain D7 phytotoxin, Weed Sci. 41:483-489.

Tranel, P.J., Gealy, D.R. and Kennedy, A.C. 1993b, Inhibition of downy brome (*Bromus tectorum* L.) root growth by a phytotoxin from *Pseudomonas fluorescens* strain D7 phytotoxin, Weed Technol. 7:134-139.

Tu, J.C. 1980, *Gliocladium virens*, a destructive mycoparasite of *Sclerotinia sclerotiorum*, Phytopathology 70:670-674.

Turgeon, G. and Yoder, O.C. 1985, Genetically engineered fungi for weed control, in : Biotechnology Applications and Research, P.N. Cheremisinoff and R.P. Ouellette, eds., Technomic Publishing Co. Inc., Lancaster pp. 221-230,

Walker, H.L. 1982, A seedling blight of sicklepod caused by *Alternaria cassiae*, Plant Dis. 66:426-428.

Walker, H.L. and Boyette, C.D. 1985, Biocontrol of sicklepod (*Cassia obtusifolia*) in soybeans (*Glycine max*) with *Alternaria cassiae*, Weed Sci. 33:212-215.

Walker, H.L. and Riley, J.A. 1982, Evaluation of *Alternaria cassiae* for the biocontrol of sicklepod (*Cassia obtusifolia*), Weed Sci. 30:651-654.

Wall, R.E. 1990, The fungus *Chondrostereum purpureum* as a silvicide to control stump sprouting in hardwoods, North J. Appl. For. 7:17-19.

Wall, R.E. 1994, Biological control of red alder using stem treatments with the fungus *Chondrostereum purpureum*, Can. J. For. Res. 24:1527-1530.
Wang, R. 1990, Biological control of weeds in China : a status report. in : Proc. VII Int. Symp. Biol. Contr. Weeds, E.S. Delfosse, ed., Rome, Italy, 1988, 1st Sper, Patol. Veg. (MAF), Rome, Italy, pp.689-693.
Weete, J.D. 1992, Induced systemic resistance to *Alternaria cassiae* in sicklepod, Physiol. Molec. Plant Pathol. 40:437-445.
Weaver, S.E. 1984, Critical period of weed competition in three vegetable crops in relation to management practices, Weed Res. 24:317-325.
Wei, Y.D., Byer, K.N. and Goodwin, P.H. 1997, Hemibiotrophic infection of round-leaved mallow by *Colletotrichum gloeosporioides* f.sp. *malvae* in relation to leaf senescence and reducing reagents, Mycol. Res. 101:357-364.
Weidemann, G.J. and Templeton, G.E. 1988, Control of Texas gourd, *(Cucurbita texana)* with *Fusarium solani* f.sp. *cucurbitae*, Weed Technol. 2:271-274.
Womack, J.G., Eccleston, G.M. and Burge, M.N. 1996, A vegetble oil-based invert emulsion for mycoherbicide delivery, Biol. Cont. 6:23-28.
Wymore, L.A. and Watson, A.K. 1989, Interaction between the velvetleaf isolate of *Colletotrichulm coccodes* and thidiazuron for velvetleaf (*Abutilon theophrasti*) control in the field, Weed Sci. 37:478-483.
Wymore, L.A., Watson, A.K. and Gotlieb, A.R. 1987, Interaction between *Colletotrichum coccodes* and thidiazuron for control of velvetleaf (*Abutilon theophrasti*), Weed Sci. 35:377-383.
Wymore, L.A., Poirier, C., Watson, A.K. and Gotlieb, A.R. 1988, *Colletotrichum coccodes*, a potential bioherbicide for control of velvetleaf (*Abutilon theophrasti*), Plant Dis. 72:534-538.
Zhang, W.M., Wolf, T.M., Bailey, K.L., Mortensen, K. and Boyetchko, S.M. 1998, Principles for selection of adjuvants used in bioherbicide formulations (Abstr), WSSA Abstracts 38:44.
Zidack, N.K. and Backman, P.A. 1996, Biological control of kudzu (*Pueraria lobata*) with the plant pathogen *Pseudomonas syringae* pv. *phaseolicola*, Weed Sci. 44:645-649.
Zidack, N.K., Backman, P.A. and Shaw, J.J. 1992, Promotion of bacterial infection of leaves by an organosilicone surfactant: implications for biological weed control, Biol. Control 2:111-117.

BACTERIA AS BIOCONTROL AGENTS OF INSECTS

Surinder Kaur and K. G. Mukerji

Applied Mycology Laboratory
Department of Botany
University of Delhi
Delhi 110 007, INDIA

1. INTRODUCTION

Managing pest insects that affect economically important plants and that are vectors of human diseases is a major concern worldwide in food production and human health. It has been estimated that over 28% of the world food production is damaged by pest insects either in the field or during storage. Current pest control strategies rely heavily on chemical insecticides, however, has led to many harmful effects, such as acccumulation of potentially carcinogenic chemicals in the ecosystem, contamination of ground water, development of resistance insect population and indiscriminate destruction of beneficial insects. These undesired side effects warrant a search for an alternative, environmentally favourable control agent. Although biological control methods have been known for several decades, they did not gain much attention until the development of insect resistance to chemical pesticides was recognized.

Biological pest control agents are naturally occurring organisms, including viruses, bacteria, fungi, protozoa and nematodes. However, out of these, bacteria are the most potential and successful group of organisms identified as far for the effective control of insect pests and vector of diseases that have rapidly become resistant to chemical pesticides.

Many bacteria cause disease and death in insects. Four groups have been recognized based on their pathogenicity to insects: 1) obligate pathogens *e.g. Bacillus popillae*, *B. larvae*; 2) crystalliferous spore forming bacilli *e.g. B. thuringiensis*; 3) facultative pathogens *e.g. B. sphaericus*, *Pseudomonas aeruginosa*, *Achromobacter* spp.; 4) potential pathogens *e.g. Serratia marcescens*.

The best example of bacterial insecticides and one that has been investigated for almost 90 years is that of the members of the genus *Bacillus*. Four species are considered as primary insecticides: *B. thuringiensis*, *B. popillae*, *B. moritai* and *B.*

Biotechnological Approaches in Biocontrol of Plant Pathogens
Edited by Mukerji *et al.*, Kluwer Academic / Plenum Publishers, New York, 1999

sphaericus (Table 1). However, *Bacillus thuringiensis* is a commercially proved control agent for agronomically important insect pests. The advantages of using *B. thuringiensis* are its narrow insect specificity, environmentally friendly nature, and low cost of production, another advantage is the low level of resistance development that has been detected in insect populations. In recent years, the emphasis has been on understanding the mechanism of action of *B. thuringiensis*: endotoxins to pave the way for the construction of more potent and broader range second generation toxins.

Table 1. Bacteria used for insect pest management

Bacterium	Insect Host	Host Plant or Ecosystem	Country	Reference
Bacillus popilliae	*Popillia japonica*	turf grass	USA	Burges & Hussey, 1971
B. lentimorbus				Hutton & Burbitts, 1974
B. moritai	Diptera	unknown	Japan	Khachatourians, 1986
B. sphaericus	mosquito larvae	standing water	USA	DesRachers & Gareia, 1984 Hertlein *et al.*, 1979 Lacey & Undeen, 1986
B. thuringiensis var. *aizawai*	*Galleria mellonella*	beehives	USA	Payne, 1988
B. thuringiensis var. *israelensis*	mosquito larvae	water	USA France	Khachatourians, 1986 Lacey & Undeen, 1986
B. thuringiensis var. *kurstaki*	numerous Lepidoptera	forests, vegetables, home gardens, row crops	USA, France, USSR, Canada, China *etc.*	Hussey & Tinsley, 1981 Khachatourians, 1986 Kristiansen & Lewis, 1986
B. thuringiensis var. *san diego*, var. *tenebrionis*	beetles	crops, treees	USA, Europe	Roberts *et al.*, 1991

2. *BACILLUS THURINGIENSIS*

B. thuringiensis comprises a group of aerobic gram-positive spore forming bacteria closely related to *Bacillus cereus.* As a group, *B. thuringiensis* contains 58 recognized flagellar serotypes and several aflagellar biotypes. The crystalline inclusion (δ-endotoxin, or Cry toxin) of *B. thuringiensis* which is usually produced during sporulation, is toxic against economically important insect pests and vectors of human and animal diseases. These inclusion bodies acts as a gut poison to three orders of insects: Lepidoptera, Diptera and Coleoptera (Hofte and Whitely, 1989) as well as Nematodes (Feitelson *et al.*, 1992). Hofte and Whiteley (1984) classified 42 crystal proteins genes by their amino-acid sequence and host range, and subsequently other classes of toxic genes were added and analyzed for their evolutive relationships (Feitelson *et al.*, 1992). Table 2 lists some of the commercial products based on *B. thuringiensis.*

Berliner (1911) for the first time realised the use of *Bt* for insect control. He found the bacterium to be toxic so mediterranean flour moth larvae but not to mealworm larvae (*Tenebrio malitor*). In 1928, a pioneering project was started in Europe with spore-crystal preparations of *Bt* produced in small scale laboratory fermentors for control of the European corn borer (*Ostrinia nubilalis*) a severe corn pest. The earliest commercial production of *Bt* begain in France in approximately 1938, under the trade name Sporeine.

2.1. Insecticidal Crystal Proteins

Berliner (1911) noticed the presence of small inclusions in the spores in *Bt.* But these inclusions of crystals were not correlated with insecticidal activity until the 1950s (Hannay, 1953). Later it was shown that soluble proteins were the toxic factors in sporulated cultures of *Bt*. The final proof of the nautre of the insecticidal factor was provided by Schnepf and Whiteley (1989). They cloned a crystal-protein gene from *Bt* and transferred it to *Escherichia coli*. The transferred *E. coli* produced a protein that was highly toxic to tobacco to tobacco horn worm larvae.

Most ICP genes are located on large plasmids (Lereclus *et al.*, 1989). These plasmids can be naturally transferred between different *Bt* strains. This process is also exploited for the combination of different ICPs in single strains. Several of the genes have been found to be flanked by transposable elements which allow the genes to be excised and redistrubuted in the bacterium genome (Menou *et al.*, 1990).

In *Bt*, when ICPs are produced within the cytoplasm concomitantly with sporulation, they are tightly packed by hydrophobic bonds and disulphide bridges (Lambert and Peferoen, 1992). The crysrtals are released into the environment after cell lysis and are only stable for a couple of days in natural environments.

When crystals are ingested by sensitive larvae, they are solubilized in the alkaline environment of the larval gut. The solubilized protoxins are then converted by protein-digesting enzymes present in the gut juices into smaller toxic fragments of approximately 650 amino acids. The toxic, protease-resistant fragment is localized in one half of the protein. The other half is completeyl digested during the proteolytic activation process. Some of the smaller ICPs lack the digestable part. These proteins donot crystalize in the typical bipyramidal shape but have other forms.

Table 2. Commercial products based on *Bacillus thuringiensis* (after Gelemter 1990)

Target Insects	Target Markets	*Bt* Subspecies	Products	Companies
Lepidoptera (cabbage looper, imported cabbage worm, diamondback moth, gypsy moth, Indian meal moth)	Vegetabel and fruit crpos, forestry, stored products, corn	*kurstaki*	Bactospeine	Duphar
			Biobit	Novo Labs
			Condor*	Ecogen
			Cutlass*	Ecogen
			DiPel	Abbott Labs
			Javelin	Sandoz
			Larvo *Bt*	Knoll Labs
		encapsulated *kurstaki* toxin	MVP†	Mycogen
Lepidoptera (greater was moth)	Beekeepers	aizawai	Certan	Sandoz
Diptera (mosquitoes and blackfies)	Vector management agencies, mosquito abatement districts	*israelensis*	Skeetal	Novo Labs
			Teknar	Sandoz
			Vectobac	Abbott Labs
Coleoptera	Potato and vegetable crops, tree care	*san diego*	M-one	Mycogen
		tenebrionis	Trident	Sandoz
		tenebrionis/kurstaki	Foil*	Ecogen
		encapsulated san diego toxin	M-one Plus†	Mycogen

* Geneticallly improved, nonrecombinant.

† Genetically engineered, killed recombinants. Experimental use permit granted or registered

Since 1981, many ICPs have been discovered and cloned. To date, the nucleotide sequences and the deduced amino acid sequence of 27 different ICP genes have been discribed (Hofte and Whiteley 1989). They encode proteins of 640 to 1200 amino acids. The ICPs constitute a family of related proteins (Cry).

Over 100 crystal genes (Cry) have been cloned and sequenced from various *B. thuringiensis* strains. The cry genes are usually borne on large plasmids (Gonzalez and Carlton, 1984; Gonzalez *et al.*,1981), although exceptions are known in which the gene is located on the chromosome (Gonzalez *et al.*, 1982; Klier *et al.*, 1983) that can rapidly mobolize these genes throughout the population.

2.2. Insecticidal Activity and Target Specificity

Hofte and Whiteley (1989) classified the Cry proteins into four major classes based on their host range and primary structure. The four classes of toxins are lepidopteran specific (Cry I), lepidopteran and dipteran specific (Cry II), coleopteran specific (Cry III), and dipteran specific (Cry IV). The Cry I-type genes encodes a 130-K Da peptides in which the N-terminal region (60-65 K Da) represents the active (toxic) fragment (Hofte & Whiteley, 1989). The Cry I-type genes are further classified into several subclasses (Cry I A, Cry I B, Cry I C, Cry I E, Cry I F *etc.*) based on amino acid homolog and their degree of a activity toward lepidopteran insects. The cry II-type genes produce a protein of 68-71 K Da and are classified as lepidopteran and dipteran specific, the Cry II B and Cry II C toxins are active only against lepidopterans. The Cry III genes encode a 73-k Da coleopteran-specific protein and are classsified as Cry III A, Cry II B, and Cry II C. The Cry IV genes comprise Cry IVA, Cry IVB, Cry IVC and Cry IVD types, which produce 135, 128, 74 and 72-k Da proteins respectively (Rajamohan *et al.*, 1998).

Amino acid sequence alingnment of several Cry toxins has identified five highly conserved blocks of amino acids with a central hypervariable region (Geiser *et al.*, 1986). It has been speculated that these variable regions might be involved in insect specificity, and this was confirmed when the *Bombyx mori* activity of cry I Aa toxin was completely transfered to nontoxic cry I Ac by exchanging a hypervariable region between aminoacid residues 332 and 450 (Ge *et al.*, 1989). This experiment suggests that the *B. mori* specificity of Cry I Aa toxin is located within residues 335-450 on Cry I Ac are important for the activity of *Trichoplusia ni* (Ge *et al.*, 1991). However, a larger fragment (335-615) of Cry I Ac toxin is essential for the full activity against *Heliothis virescens* these results suggest that toxicity-determing regions are not restricted to hypervariable regions alone and might differ according to the type of insect receptor interacting with the toxin.

2.3. Structure of δ-Endotoxins

Li *et al.* (1991) reported the crystal structure of a coleopteran active δ-endotoxin, Cry III A, from *B. thuringiensis* subsp. *tenebrionis*, at a resolution of 25 Å. It composed of three structurally distinct domains. Domain I consists of seven α-helix bundles encompassing the N-terminal residues 63-290. Helix 5 is located at the center of domain I and is surrounded by six outer helices of varying sizes. Domain II (residues 291-500) is composed of three antiparallel ß-sheets connected to each other with surface exposed loops. Domain III (residues 501-644) is a sandwich of antiparallel β-sheets in a "jelly-roll" configuration. The structure of another crystal protein cry I Aa was resolved at 2.25 Å. The major difference between these two toxins lies in domain II. The loops connecting the domain II β-sheets in the Cry I Aa molecule are comparatively much larger than those of Cry III A. The functions of each domain with respect to insect toxicity have been the subject of speculation. In general, domain I has been assumed to be involved in membrane partitioning and ion-channel regulation Domain III has been recognized to have structure similar to that of immunoglobins and is proposed to be involved in insect specificity and in recognition of receptor molecules located in target insect mid gut epithelial cells. The function of domain III, although not clear is believed to be involved in ion-channel formation, receptor binding, and insect specificity.

2.4. Molecular Mode of Action of the δ-Endotoxins

The mechanism of action of *B. thuringienesis* Cry proteins consists of three major steps:

1. Solubilization and activation of the crystals in the insect mid gut.
2. Binding of the activated toxin to mid gut receptors.
3. Insertion of the activated toxin into the mid gut apical membrane to create ion channel or pores.

Most of the information regarding mode of action was obtained from studies of lepidopteran insects be cause of their lage size and easy maintenance under laboratory conditions.

2.4.1. Solubilization and Activation

As mentioned earlier, the Cry proteins are usually deposited as crystals that are primary composed of inactive protoxins. They must be consumed by susceptible larvae to have any lethal effect. The primary target of the action is the gut membrane, hence they are gut poisons on ingestion, the crystals are solubilized under the alkaline mid gut (Huber and Luthy, 1981). Following solubilization, the protoxins are digested by insect mid gut proteases producing an active toxin form (Tojo and Aizawa, 1983). The major lepidopteran insect mid gut proteases are trypsin like (Milne and Kaplan, 1993). Indeed, the protoxins can be activated *in vitro* by trypsin without losing their potencey. The 130 to 135 K Da protoxins, examplified by Cry I A, are digested progressively from the C terminus toward the N terminus, forming a 55 to 65 K Da protease-resistant toxic core (Chestukhina *et al.*, 1982). Because the carboxyl terminus is not essential for insect toxicity and contains many Cry residues, this region may be involved in crystallization of the protoxin and may also protect the toxin from promature protease cleavage.

2.4.2. Receptor Binding

Histopathological studies indicate that the toxin-binding target molecules are located in the microvillar brush border membranes which line the midguts of susceptible insects (Denalf *et al.*, 1993). *In vitro* binding studies have been performed with brush bordes membrane residues (BBMV) prepared from larval midguts, Several workers have found correlation between toxicity and the number of binding sites. However, some studies have failed to demonstrate a direct correlation between larvicidal activity and either the affinity or the number of binding site concentrations. For example, Cry IA toxins do bind saturably to the BBMV prepared from *Spodoptera* species even though the toxin does not kill the larvae (Van Rie *et al.*, 1990a).

The extensive study of specifity, binding, and structure studies indicates that domain II of toxin is involved in binding to the receptors. Recent experiments demonstrated that the loops of domain II are contact points in receptor in binding (Lu *et al.*, 1996) as predicted by Li *et al.* (1991). However, according to Wu and Aronson (1992), the mutation A92D (domain I) in the Cry IAc gene radially disrupts receptor binding, as assessed by heterologous competition, to *M. sexta* BBMV. These are results concerning Cry I Ab and IAc suggesting that bonding may also take place *via* residues in domain III. The amino acid difference between these two proteins lies soley in domain III in the region between conserved block three and four. Several toxin binding receptors molecules have been (receptors) proteins are generally larger and sometimes glycoproteins. Recently, the Cry IAc toxin binding protein for *M. sexta* has been identified as amino peptodase-N (Sangadala *et al.*, 1994). The amino peptidase - N is a glycoprotein of 120 K Da size (Knowles *et al.*, 1991).

2.4.3. Ion Channel (pore forming) Function

The α-helical structure of domain I has been proposed as the primary candidate for promoting membrane penetration by analogy with the pore-forming structure of colicin A and diptheria toxin. However, knowledge of the mechanism of channel formation of Cry toxins is still in its infancy. The ion-channel activity of several Cry toxins was studied using using whole insect midgut by voltage clamping (Harvey and Wolfresheryes, 1979, Liebig *et al.*, 1995), BBMV swelling assays, insect tissue culture cells (Knowles and Ellar, 1987; Schwartz *et al.*, 1991), phospholipid vesicles (Engiish *et al.*, 1991; Haider and Ellar, 1989) and lipid bilayers (Schwartz *et al.*, 1993).

Knowles and Ellar (1987) used CF-I cells and proposed a "colloidal osmotic lysis" theory as a model for the cytolytic toxicity of Cry toxins. According to their models, the Cry toxin make small holes in the midgut membrane leading to the influx of water and ions, resulting in cell swelling and eventually lysis.

2.5. Specificity

Understanding the mode of action of ICPs has enabled scientists to explain the high specificity of the ICPs composed to synthetic pesticides. In *Bt*, toxic activity requires specificity binding of toxic fragments to receptors in the midgut epithelium of susceptible insects. There are no reports of toxic ICPs that donot bind to midgut receptors. In addtion, no specific binding between midgut epithelia of insect larvae and nontoxic ICPs has been observed (Hofmann *et al.*, 1988; Van Rie *et al.*, 1989, 1990a). Marginal different ICPs that are toxic to the same insect may interact with different receptors on the midgut epithelium. It appears that specific binding is essential for specificity, but it is not sufficient for toxicity. Specificity is not only determined by the presence of receptors but also by the proteolytic activity and pH in the guts of the larvae (Haider *et al.*, 1986).

According to the current gene nomenclature, in the first level of classification, the roman numeral denotes host range, that is, the specificity of the Cry I group of toxins is against lepidopterans, as that of the Cry III toxins is against coleopterans (Rajamohan and Dean, 1996). However, there is a significant difference in the degree of specificity among toxins of the same classification. For example, Cry IAa is about 400 times more active against *Bombyx mori* than Cry IAc and Cry IAc is more active (10 times) against *Heliothis virescens* or *Tricoplusia ni* than Cry IAa (Ge *et al.*, 1989, 1991). In order to locate the specificity determing regions of the Cry IA toxins, Ge *et al.* (1989) constructed a series of structurally stable proteins by exchanging smaller segments of amino acids between Cry IAa and IAc toxins. They transferred the *B. mori* activity from Cry IAa to IAc by exchanging amino acid residues 332-450 of Cry IAa indicating that the *B. mori* specificity of Cry IAa toxin lies within the residues 332-450. Later, Ge *et al.* (1991) revealed that the amino acids 335-450 on CryIAc are associated with the activity against *T. ni.*, whereas, residues 335-615 on the same toxin are required to exchange full *H. virescens* specificity. Lee *et al.* (1992) established a direct correlation between the specificity-determining region and the receptor binding in the case of *B.mori*. However, the "specificity" of a toxin and its receptor binding properties are not always directly correlated.

2.6. Mechanism of Insect Resistance

The problem of resistance to synthetic insecticides has been a serious concern in insect control. Until recently, it was suggested that biological insecticides such as *Bt* might not have this problem. This suggestion was mainly based on the assumption that multiple effects were involved in the *Bt* toxicity. However, recently data show that *Bt* based bioinsecticides are also prone to resistance development both, in the field and in laboratory conditions. Table 3 lists some of the insects that have developed resistance to cry toxins (Rajamohan *et al.*, 1998).

Table 3. Examples of resistance to *Bacullus thuringiensis* δ-Endotixin (after Rajamohan *et al.*, 1998)

Insect	Cry toxin or source	Resistance ratio
Lepidoptera		
Heliothis virescens	Cry 1Ac	>10,000
Plutella xylostella	*Bacillus thuringiensis* var. *kurstaki*	2800
Plodia interpuctella	*Bacillus thuringiensis* var. *kurstaki*	140
Spodoptera exigua	Cry 1C	100
Trichoplusia ni	Cry 1Ab	26
Coleoptera		
Leptinotarsa decemlineata	Cry 3A	400
Diptera		
Aedes aegypti	*Bacillus thuringiensis* var. *israelensis*	1.1
Culex quinquefasciatus	*Bacillus thuringiensis* var. *israelensis*	6.0
Culex quinquefasciatus	Cry 10A	70

Studies of *B. thuringiensis* susceptible and resistant insects suggest several mechanisms by which the insects develop resistance. It was found that the midgut proteolytic activity of *B. thuringiensis* (*B. t. endomocidus* HD-198)- resistant *Ploidia interpintella* larvae is significantly reduced compared to midgut juice from susceptible insects. The data sugggest that altered or inadequate processing of Cry toxins is a possible factor in the resistance mechanism. Receptor binding studies with ^{125}I labelled toxins and BBMVs prepared from suceptable and resistant insects (*P. interpunctella* and *Plutella xylostella*) indicate that the toxin binding sites are altered or modified in the resistant strains (Van Rie *et al.*, 1990; Ferre *et al.*, 1991). Lee *et al.*, (1995) compared the receptor binding properties of Cry I Aa, Cry I Ab, and Cry I Ac forms for Cry I Ac susceptible and resistant (10,000-folds resistant) to laboratory selected *H. virescens* strains. The binding constants of Cry I Ab binding and cry I Ac toxins are similar for susceptable and resistant strains, but the binding of cry I Aa toxins in the resistant strain was totally abolished. It was proposed that atteration in one of the binding proteins that is shared by all three toxins (cCy I Aa, Cry I Ab, and Cry I Ac) is the major factor for resistant. However, in the case of another Cry toxin-resistant *H. virescens* strain, neither receptor binding affinity nor binding site concentration is affected. These data suggest that the probably a post binding event, such as integration into the membrane or ion channel activity, is altered in the above example.

A valuable strategy to delay resistance may be the combination of ICPs that are toxic to the same target insect but bind to different receptors. Multiple mutations in target sites would have to arise to avoid all binding and toxicity. In some cases however, combinations of toxins may select more rapid adaptation than single toxins (Gould, 1986).

Thus, multiple ICPs do not successfully avoid the prablem of resistance. Resistance management will have to combine different strategies, such as integrated pest management systems, use of refugia, and use of other insect control agents with independent modes of action, including multiple genes with unique receptor-binding activities (Lambert and Peferoen, 1992).

2.7. Genetic Manipulation of ICP Genes in Bacteria

Historically *E. coli* has been the host organism of choice for the expression of cloned ICP genes and subsequent characterization of their encoded crystal proteins (Adang *et al.*, 1985; Klier *et al.*, 1982; Schnepf and Whitely, 1981) although *Bacillus megaterium* (Donovan *et al.*, 1988, Shivkumar *et al.*, 1986) and *Bacillus sphaericus* have now been successfully used as host organisms for the heterologous expression of cloned ICP genes. Efficient expression of ICP genes *E. coli* has typically required the use of heterologous promoters such as Ptac (Oeda *et al.*, 1987), Plac (Haider *et al.*, 1986) as Lambda P_L and P_R (Hofte *et al.*, 1986) whereas significant ICP expression in the bacilli has typically not, presumbly because the native ICP promoters are utilized to some extent in other *Bacillus* species.

In the native host, the expression of crystal protein is both sporulation specific and most eficient crytal protein can account for upto 20 to 30% of the dry cell weight (Lilley *et al.*, 1980). However, there have been reports of vegetative rather than sporulation linked expression in *Bacillus* species, such as the vegetative expression of Cry IA (b) and Cry IAc in *B. subtilis* (Shivkumar *et al.*, 1986). This report conflicts with that of Klier *et al.*, (1982) in which the cry IAb gene from *Bt* strain *berliner* 1715 was reported to be developmentally regulated in *B. subtilis*. It was observed that the CryIAa and CryIAc genes of *Bt* strain HDI are differentially expressed in *B. subtilis* and *B. megaterium* with expression occurring during the vegetative growth phase of both organisms. For both genes, higher levels of expression were observed in *B. megaterium*. Cry III A expression was observed *B. megaterium* under growth conditions that prevented sporulation (Donovan *et al.*, 1988), suggesting that this gene is also expressed in the vegetative cells. In contract high-level expression of the cry II A gene appears to be restricted to the sporulation phase of *B. megaterium* (Donovan *et al.*, 1988).

2.8. Exploitation of *Bt* ICP Gene in Microbial Bioinsecticide Development

Microbial genetic approaches such as plasmid curing and conjugal transfer can be used for the construction of noval ICP gene combinations. These approaches can lead to *Bt* strains exhibiting significantly improved insecticidal activities for specific target pests. This strategy has been successfully employed by Ecogen Inc in their development of several bioinsecticide products including Candor® and Foil® bioinsecticides, marketed for soybean and potato crep aplication respectivily. Candor® bioinsecticide was among the first registered *Bt* bioinsecticides employing a genetically modified *Bt* strain.

The active ingradient of Candor® bioinsecticide, *Bt* var. *kurstaki* strain EG2348, was constructed by first curing a wild-type *kurstaki* isolate of certain plasmids (including those encoding ICP) and then introducing by conjugating an ICP gene present on a transferable plasmid from a different *Bt* strain. The ICP genes combined in EG2348 encode crystal proteins with insecticidal activity against specific soybean crop pests, such as *Pseudoplusia includens*, the soybean looper.

Conjugal plasmid transfer was also used to develop *Bt* strain EG2424, the active ingradient in Ecogens Foil® bioinsecticide product. In this case, however, conjugal transfer was employed to develop a *Bt* strain with a broader spectrum of insecticidal activity. More specifically plasmids encoding ICPs with activity against a lepidopteran insect, the European corn borer (*Ostrinia nubilalis*), were combined with a plasmid encoding an ICP with activity against the colorado potato beetle.

2.9. Construction of Recombinant *Bt* Strains

While consdiderable emphasis has been placed on the introduction of cloned ICP genes into alternative hosts for the production of insecticidal crystal proteins, the recent development of an efficient transformation system for *Bt*, together with the development of *Bt* cloning vectors, has made possible the construction of improved *Bt* strains for use as microbial insecticides with the use of recombinant DNA techniques. The recombinant DNA approache offers the maximum flexibility in deriving novel combinations of both native and for modified ICP genes in *Bt*.

The use of *Bt* as the host organism offers a number of advantages for the development of microbial-based bioinsecticides via recombinant DNA technology. (1) The addition to being efficient procedures of ICPs, native *Bt* strains can stably maintain and express several homologous ICP genes (Carlton and Gonzalez, 1985; Kronstad *et al.*, 1983). Finally, multiple ICPs, with differening modes of action or reception binding properties, may reduce the chance of developing insect resistance (McGaughey and Johnson, 1987; Van Rie *et al.*, 1990). (2) The spores of an ICP-producing *Bt* strain can have an effect on the insecticidal activity of that strain against certain insect pests (Burges *et al.*, 1976; Li *et al.*, 1987; McGaughey, 1978; Nishitsutsuji and Endo, 1980; Somervillae and Pockett, 1975). Thus depending upon the target pest, the indusion of *Bt* spores in the formulated product may neccessary for maximum potency. (3) Finally, environmental concern argue for the development of recombinant *Bt* bioinsecticides based on the native host organism for the ICP genes.

2.10. Cloning Vectors for *Bt*

A variety of plasmids, functional in other gram-positive bacteria, have been tested for their ability to transform *Bt* via electroporation.

A number of convenient shuttle vectors, functional in *E. coli* and *Bacillus* species have been constructed by inserting the large EcoRI fragment of the *B. cereus* plasmid pBC 16 into the *E. coli* cloning vectors pUC8 or pUC18/19. These shuttle vectors, pX 161, pEG 147 and pBU 4 also contain multiple cloning sites and have been successfully used to introduce and express cloned ICP genes in *Bt*. Chemeric derivatives of pBC 16 containing an *E. coli* replicon (*e.g.*, pUC, pBR 322). Shuttle vectors have also beeen consulted that contain replication origins derived from native plasmids of *Bt*.

The small cryptic plasmids of *Bt*, are a convenient of replication origins for vector development. Several of these plasmids have been cloned directly in *E. coli*, resulting in relatively large bifunctional plasmids exhibiting varying degree of segregational stability (Mahillon *et al.*, 1988).

A set of compatible shuttle vectors has been constructed based on the replication origins of large *Bt* plasmids. Vectors pEG 597, pEG 853 and pEG 854 contain (Table 4) replication origins derived from the 44-60 and 43-MD plasmids respectively of *Bt* strain HD 263 var. *kurstaki*.

Significant levels of ICP production can be achieved by the introduction of ICP-encoding recombinant plasmids into *Bt*.

2.11. Insecticidal Activity of Recombinant *Bt* Strains

The construction of recombinant *Bt* strains with noval ICP gene combinations may lead to the development of improved strains of *Bt* for use as commercial bioinsecticides. Numerous ICP genes have been cloned and their encoded crystal proteins purified and assessed for their insecticidal properties (Hofte and Whitely, 1989), *e.g.* insecticidal activity of *Bt* isolated (EG6345), was broadened against *Spodoptera exigua*.

A Cry IA c cntaining recombinant plasmid was introduced into the EG6345 background by electroporation. The resultant recombinant EG6345 rec, retained the insecticidal activity of the parental strain against *Spodptera exigua* and exhibited significant improvements in insecticidal activity against a variety of vegetable crop insect pests including *Plutella xylostella* and *Trichoplusia ni.*

Table 4. Examples of *Bt* Cloning Vectors (after Gawron-Burke and Baum, 1991)

Plasmid	Size (kb)	*Bt*	*E. coli*	*Bt*	*E. coli*	Comments (ref.)
p EG 147	5.8	pBC16	pUC18	Tet	Amp	Contains MCS[1]
pX161	5.6	pBC16	pUC8	Tet	Amp	Contains MCS (Schurter *et al.*, 1989)
pBU4	5.8	pBC16	pUC19	Tet	Amp	Contains MCS
pHT3101	6.6	pHT1030 *ori*	pUC18	Erm	Amp	Contains MCS, resident *Bt* plasmid replicon (Lereclus *et al.*, 1989)
pEG597	6.6	*ori 44*[2]	pTZ19u	Cm	Amp	Contains MCS, resident *Bt* plasmid replicon, phagemid vector
pEG853	6.6	*ori 60*[2]	pTZ19u	Cm	Amp	Contains MCS, resident *Bt* plasmid phagemid vector
pEG854	7.1	*ori 43*[2]	pTZ19u	CM	Amp	Contains MCS, resident *Bt* plasmid replicon, phagemid vector
pEG434	3.1	pBC16	-	Tet	-	Contains MCS, uses tet promoter for expression
pHV33	7.3	pC194	pBR322	Tet Cm	Amp	(Klier *et al.*, 1982)

[1] Multiple cloning site.
[2] *Ori 44*, *ori 60* and *ori 43* were derived from the 44, 60 and 43 MD plasmids of *Bt* strain HD263 var. *kurstaki.*

3. *BACILLUS SPHAERICUS*

Bacillus sphaericus is an aerobic, rod shaped, endospore forming saprophytic bacterium occurring in nature. Although commercial interest in *B. thuringiensis* subsp. *isralensis* is high on account of its broader spectrum of action (mosquito and black fly), its field stability has been low. The long-term persistence of *B. sphaerus* strains in the aquatic environment directed researchers to pay more attention to this species as a potential candidate for development of commercial mosquito larvicide.

Based on the mosquito larvicidal activity and serological cross-reactivity, de Barjac *et al.* (1985) have classified *B. sphaericus* (BS) strain. Among the pathogenic strains, some are "highly lethal" (1593, 2362, 2297) and others are not so lethal (SSII-I) to mosquito larvae. Krych *et al.* (1980) and Yousten (1984) have classified BS based on DNA homologies. Although the former, could not find any correlation between toxicity and DNA homology the latter, provided five homology groups with group II being divided into subgroups IIA and IIB Krych *et al.* (1989) observed marginal difference in the G + C content of the DNA among the BS strain investigation. Delta endotoxins of *B. thuringiensis* are generally associated with mega plasmids, which are reported to be present in low copy number (Gonzalez *et al.* 1982). On the other hand, the search for the presence of plasmids in BS has been incomplete and somewhat conflicting Kalfon *et al.* (1984) reported small plasmids in some BS strains, especially 2297, whereas Davidson (1982) mentioned the existence of a large plasmid in *B. sphaericus* 1593 and 1881 but not in *B. sphaericus* 1691 or 2362. Abe *et al.* (1983) found large plasmids in strains

1881 but not in 1593 or 1691. However, no direct evidence has been reported to link the mosquito larvicidal activity with plasmids in *B. sphaericus*. Most of the BS mosquito larvicidal genes have been cloned from the total cellular DNA.

3.1. Cloning of Mosquito Larvicidal Genes

Ganesh *et al.* (1983) for the first time reported the cloning of mosquito larvicidal gene from *B. sphaericus* 1593. The recombinant plasmid (pGspo3), containing a 3.7 kb insert, was toxic to *Culex pipiens* and *Anopheles stephensi* (LC50 of 1μg and 2 μg cellular protein per milliliter respectively). Subsequently, it was noted that the clone had lost its larvicidal potential, although the DNA insert appeared to remain of the same size. Moreover, as the restriction map of this gene does not match the sites of the other mosquito larvicidal genes of *B. sphaericus*, therefore, this gene is not considered to be a potential larvicidal gene. A 3.5 kb Hind III fragment was cloned from several hjghly toxic *B. sphaericus* strains (1593, 2362, 2297 and IAB5a) coding for the 42 and 51 kDa peptides by Baumann *et al.* (1987, 1988). Hindley and Berry (1987) cloned the gene coding only for the 42 kDa protein from *B. sphaericus* 1593 into *E. coli* and determined its sequence. The nucleotide sequence alignment of different clones showed that the nucleotides of the coding region are identical for the strains 1593, 2362 and 2317.3, whereas the strain 2297 and IAB5a differ from the previous strains by 25 and 7 nucleotides respectively.

The gene organisation of the 51 and 42 kDa toxins suggests that the genes coding for these two proteins are located in a single transcription unit in all the highly toxic strains and are separated by a spacer region of 174-176 nucleotides (Rajamohan and Dean, 1996). More recently, another toxin gene coding for a 100.6 kDa protein has been cloned and sequenced from a low toxic strain *B. sphaericus* SSII-I (Thanabalu *et al.*, 1991).

3.2. Location of BS Toxin

Untill recently, it was thought that *B. sphaericus* did not synthesize parasporal crystals like that of *B. thuringiensis*. However, Davidson and Myers (1985) and Kalfon *et al.* (1984) have identified some parasporal inclusions in some BS strains, notably the highly toxic strains. The polyhedral, mosquito larvicidal crystal is largely synthesized during sporulation except in strain SSII-I which synthesizes the toxin in the vegetative growth phase (Kalfon *et al.*, 1984). Unlike *B. thuringiensis* crystals the inclusion bodies of BS are held in close proximity to the spores by a membrane envelope because of a deficiency in autolytic enzymes (Holt *et al.*, 1975).

Two distinct toxins have been isolated from *B. sphericus*, the 51 and 42 kDa, binary highly active toxin and a 100 kDa weakly active toxin. The antigenically distinct 51 and 42 KDa toxins are produced by the highly active strains (1593, 2362, 2297) and both are necessary for mosquito larvicidal activity (Broadwell *et al.*, 1990a; Davidson et al., 1990). Broadwell *et al* (1990b) through quantitative bioassay suggested an equal ratio of 51 and 42 kDa toxins for maximum larvicidal activity. The 100 kDa toxin which is isolated from *B. sphaericus.* SS II-I, is weakly toxic to mosquito larvae and is produced during vegetative growth. Its potency is about 1000 times less than that of the highly toxic strains and is more unstable.

3.3. Mode of Action

To identify the precise region of 51 and 42 kDa proteins required for toxicity, several N- and C- terminal deletion mutants have been constructed. These results indicate that the amino acid residues between 32 to 395 and 17 to 353 of the 51 and 42 kDa proteins respectively, are essential to retain complete mosquito larvicidal activity (Broadwell *et al.*, 1990a; Clark and Baumann, 1990; Oei *et al.*, 1990, Sebo *et al.*, 1990)

The action of toxin to the susceptible larvae as given in Rajamohan and Dean, (1996) occurs in the following sequence

1) ingestion of the inclusion bodies by the filter-feeding susceptible larvae; 2) solubilization of the crystal in the alkaline pH of the midgut; 3) activation of the 51 and 42 kDa toxin to 43 and 39 kDa toxins, respectively, by larval gut enzymes; 4) binding of the activated toxins to specific receptor molecules located at the gastric cecum (GC) and posterior midgut (PMG); and 5) internalization of the toxin followed by mitochondrial swelling and cell lysis. Both the toxins are found to be essential for internalization. Neither 51 nor 42 kDa toxin can be internalized alone. Moreover, the specific binding and internalization process of the toxins is not found in nontoxic and toxic-resistant cell lines (Davidson, 1988, 1989; Oei *et al.*, 1992; Schroeder *et al.*, 1989 and Thanabalu *et al.*, 1993). This suggests an inevitable, conformational change induced by the 51 and 42 kDa toxin is essential for internalization of thje toxin and for toxicity.

The 100 kDa toxin of *B. sphaericus* SSII-I does not share homology with any other known insectidal toxin (Thanabalu *et al.*, 1993). However, the amino acid sequence of this toxin has an N terminal potential hydrophobic signal sequence and a transmembrane domain similar to that of SI subunit of pertussis toxin and the A subunit of cholera toxin. This strongly suggests that this 100 kDa toxin might exert its action by ADP-ribosylation like pertussis and diphtheria toxins. The mechanism of action of the 51, 42 kDa binary toxin and the 100 kDa toxin of *B. sphaericus* to susceptible mosquito larvae are different.

4. CONCLUSIONS

Insect pests can cause dramatic crop loss, either per se through direct destruction of plants or through wounds produced by the insect itself, that allow the infection and diffusion of other pathogens. Research on the use of natural insecticides received a great impulse over the past years because of the consciousness of the ecological damages caused by the chemical substances.

Biopesticides are rapidly substituting and replacing the conventional chemical pesticides used for control of agricultural pests and disese causing vectors. The attractiveness stems from their ecofriendliness and target specificity. Other advantages are the absence of insect resistance, absence of residue build-up in the environment and potential impact from biotechnological research and development. However, biological pest control agents have a number of limitations, but, it is hoped that biotechnology and a responsible environmental outlook will help us to change the limitations and make biological pest control agents products for the agriculture of today. For this, future research and development in the areas of identification of new etiological agents of insect disease, and mechanism and genetic basis of bioinsecticide-induced insect disease are needed.

It can be concluded that the biotechnology has a significant role in the production and manufacturing of biological pest control agents and ultimately lowering the input costs of agriculture. The challenge in this area has not been fully met. However, it is anticipated that with many firms producing or developing bioinsecticides, there will be greater activity in the area. Ultimately, the production of effective bioinsecticides in sufficient numbers will displace the synthetic chemical insecticides and the associated ecological disadvantages.

REFERENCES

Abe, K., Faust, R.M. and Bulla, L.A., Jr. 1983, Plasmid deoxyribonucleic acid in strains of *B. sphaericus* and in *Bacillus moritai*, J. Invertebr. Pathol. 41: 328.

Adang, M.J., Staver, M.J., Rocheleau, T.A., Leighton, J., Barker, R.F. and Thompson, D.V. 1985, Characterized full-length and truncated plasmid clones of the crystal protein of *Bacillus thuringiensis* subsp. *kurstaki* HD-73 and their toxicity to *Manduca sexta*, Gene 36: 289-300.

Baumann, L. and Baumann, P. 1991, Effects of the components of the *Bacillus sphaericus* toxin on mosquito larvae and mosquito-derived tissue culture grown cells, Curr. Microbiol. 23: 51.

Baumann, P., Baumann, L., Bowditch, R.D. and Broadwell, A.H. 1987, Cloning of the gene for the larvicidal toxin of *Bacillus sphaericus* 2362: evidence for a family of related sequences, J. Bacteriol. 170: 2045.

Baumann, L., Broadwell, A.H. and Baumann, P. 1988, Sequence analysis of the mosquitocidal toxin genes encoding 51.4 and 41.9 kilodalton proteins from *Bacillus sphaericus* 2362 and 2297, J. Bacteriol. 170: 2045.

Berliner, E. 1911, Uberdie Schlaffsucht der Mehlmottehraupe (*Ephestia kaechniella*), und inven Erreger *Bacillus thuringiensis* n. sp. Zeitschrift fur angewandttes, Entomol. 2: 29-56

Broadwell, A.H., Baumann, L., and Baumann, P. 1990, Larvicidal properties of the 42 and 51 kilodalton *Bacillus sphaericus* proteins expressed in different bacterial hosts: evidence for a binary toxin, Curr. Microbiol. 21: 361.

Broadwell, A.H., Baumann, P. 1990, The 42- and 51-kilodalton mosquitocidal proteins of 2362: construction of recombinants with enhanced expression and *in vivo* studies of processing and toxicity, J. Bacteriol. 172: 2217.

Burges, H.D. and Hussey, N.W. 1971, Past achievements and future prospopects, in: Biotechnology, Biological Pesticides and Noval Plant-Pest Resistance for Insect Pest Management, D.W. Roberts and R.R. Granados, eds., Boyce Thompson Institute, Ithaca, N.Y., pp 38-43.

Burges, H.D., Thomson, E.M. and Latchford, R.A. 1976, Importance of spores and δ-endotoxin protein crystals of *Bacillus thuringiensis* in *Galleria mellonella*, J. Invertebr. Pathol. 27: 87-94.

Carlton, B.C. and Gonzalez, J.M. 1985, Plasmids and delta-endotoxin production in different subspecies of *Bacillus thuringiensis*, in: Molecular Biology of Microbial Differentiation, J.A. Hoch and P. Setlow, eds., American Society for Microbiology, Washington D.C. p 246.

Chestakhina, G.G., Kostina, L.I., Mikhailova, A.L., Tyarin, S.A., Klepikova, F.S. and Stepanov, V.M. 1982, The main features of *Bacillus thuringiensis* delta-endotoxin molecular structure, Arch. Microbiol 132: 159-162.

Clark, M.A. and Baumann, P. 1990, Deletion analysis of the 51- kilodalton protein of the *Bacillus sphaericus* 2362 binary mosquitocidal toxin: construction of derivatives equivalent to the larva-processed toxin, J. Bacteriol. 172: 6759.

Davidson, E.W. 1988, Binding of the *Bacillus sphaericus* (Eubacteriales: Bacillaceae) toxin to midgut cells of mosquito (Diptera: culicidae) larvae: relationship to host range, J. Med. Entomol. 25: 151.

Davidson, E.W. 1989, Variation in binding of *Bacillus sphaericus* toxin and wheat germ agglutinin to larval midgut cells of six species of mosquitoes, J. Invertebr. Pathol. 53: 251.

Davidson, E.W. and Myers, P. 1985, Parasporal inclusions in *Bacillus sphaericus*, FEMS Microbiol. Lett. 10: 2.

Davidson, E.W., Oei, C., Meyer, M., Bieber, A.L., Hindley, J. and Berry, B. 1990, Interaction of the *Bacillus sphaericus* mosquito larvicidal proteins, Can. J. Microbiol. 36: 870.

Davidson, E.W., Spizizen, J. and Yousten, P. 1982, Recent advances in the genetics of *B. sphaericus*, Proc. Int. Colloq., Inverteb. Pathol. Brighton, England, 14.

de Barjac, H., Larget-thiery, I., Dumanior, V.C. and Ripouteau, H. 1985, Serological classification of *B. sphaericus* strains on the basis of toxicity to mosquito larvae, Appl. Microbiol. Biotechnol. 21: 85.

DesRochers, B. and Garcia, R. 1984, Evidence for persistence and recycling of *Bacillus sphaericus*, Mosquito News, 44: 160-165.

Donovan, W. P., Dankocsik, C., Gilbert, M. P., Gawron-burke, M. C., Groat, R. G. and Carlton, B. C. 1988, Amino acid sequence and entomocidal activity of the P2 crystal protein, An insect toxin from *Bacillus thuringiensis* var *kurstaki*, J. Biol. Chem. 263: 561.

English, L.H., Reddy, T.L. and Bastian, A.E. 1991, Delta-endotoxin-induced leakage of $^{86}Rb^+$ -K^+ and H_2O from phospholipid vesicles is catalyzed by reconstituted midgut membrane, Insect Biochem. 21: 177-184.

Feitelson, J. S., Payne, J. and Kim, L. 1992, *Bacillus thuringiensis*: insects and beyond, Biotech. 10: 271

Ferre, J., Real, M. D., Van Rie, J., Jansen, S. and Peferoen, M. 1991, Resistance to the *Bacillus thuringiensis* bioinsecticides in a field population of *Plutella xylostella* is due to a change in a midgut membrane receptor, Proc. Natl. Acad. Sci. 88: 5119-5123.

Ganesh, S., Kamdar, H., Jayaraman, K. and Szulmajster, J. 1983, Cloning and expression in *Escherichia coli* of a DNA fragment from *Bacillus sphaericus* coding for biocidal activity against mosquito larvae, Mol. Gen. Genet. 189: 181.

Gawron-Burke, C. and Baum, J.A. 1991, Genetic manipulation of *Bacillus thuringiensis* insecticidal crystal protein genes in bacteria, in: Genetic Engineering, Vol. 13, J.K. Setlow, ed., Plenum Press, New York, pp. 237-263.

Ge, A.Z., Rivers, D., Milne, R. and Dean, D.H. 1991, Functional domains of *Bacillus thuringiensis* insecticidal crystal proteins, J. Biol. Chem. 266: 17954.

Ge, A. Z., Shivarova, N.I. and Dean, D.H. 1989, Location of the *Bombyx mori* specificity domain on a *Bacillus thuringiensis* δ-endotoxin protein, Proc. Natl. Acad. Sci. 86: 4037.

Geiser, M., Schweitzer, S. and Grimm, C. 1986, The hypervariable region in the genes coding for entomopathogenic crystal proteins of *Bacillus thuringiensis*: nucleotide sequence of the Kurhd I gene of subsp. *kurstaki HDI*. Gene 48: 109.

Gelernter, W.D. 1990, *Bacillus thuringiensis*, bioengineering and the future of bioinsecticides, paper presented at the Brighton Crop Protection Conference on Pests and Diseases, Brighton, U.K.

Gordon, R. E., Hayens, W. E. and Pang, C. H. N. 1973, Agricultural Research Service USDA, Washington D.C.

Gonzalez, J. M., Dulmage, H. T. and Carlton, B. C. 1981, Correlation between specific plasmids and endotoxin production in *Bacillus thuringiensis*, Plasmid 5: 351.

Gonzalez, J. M., Brown, B. J. and Cartlon, B. C. 1982, Transfer of *B. thuringiensis* plasmids coding for delta-endotoxin among strains of *B. thuringiensis* and *B. cereus*, Proc. Natl. Acad. Sci. 79: 6951-6955.

Gonzalez, J. M. and Carlton, B. C. 1984, Correlation between specific plasmids and endotoxin production in *Bacillus thuringiensis*, Plasmid 11: 28

Gould, F. 1986, Simulation models for predicting durability of insect resistance germ plasm: a deterministic diploid, two-locus model. Environ. Entomol. 15: 1-10

Grochukski, P., Masson, L., Borisova, S., Pusztai-Carrey, M., Schwart, J. L., Brourseau, R. and Cyglies, M. 1995. J. Mol. Biol. 254: 447.

Haider, M. Z., Knowles, B. and Ellar, D. J. 1986, Specificity of *Bacillus thuringiensis* var. *calmeri* insecticidal delta-endotoxin by different processing of the rotoxin by larval gut proteases, Eur. J. Biochem. 156: 531-540

Haider, M. Z. and Ellar, D. J. 1989, Functional mapping of an entomocidal d-endotoxin, Single amino acid changes produced by site-directed mutagenesis influence toxicity and specificity of the protein, J. Mol. Biol. 208-183.

Harvey, W.R. and Wolferrberger, M.G. 1979, Mechanism of inhibition of active potassium transport in isolated midgut of *Manduca sexta* by *Bacillus thuringiensis* endotoxin, J. Exp. Biol. 83: 293-304.

Hertlein, B. C., Levy, R. and Miller, T. N. Jr. 1979, Recycling potencial and selective retrieval of *Bacillus sphaericus* from soil in a mosquito habitat, J. Inverteb. Pathol. 33: 217-221.

Hindley, J. and Berry, C. 1987, Identification, cloning and sequence analysis of the *Bacillus sphaericus* 1593 41.9 kDa larvicidal toxin gene, Mol. Microbiol. 1: 187.

Hofmann, C., Vanderbruggen, H., Hofte, H., Van Rie, J., Jansens, S. and Van Mellart, H. 1988, Specificity of *Bacillus thuringiensis* delta-endotoxins is correlated with the presence of high affinity binding sites in the brush border membrane of target insect midgut, Proc. Natl. Acad. Sci. 85: 7844-7848

Hofte, H. and Whiteley, H. R. 1989, Insecticidal crystal proteins of *Bacillus thuringiensis*, Microbiol. Rev. 53: 242-255.

Holt, S.C., Gauthier, J.J. and Tipper, D.J. 1975, Ultrastructural studies of sporulation in *B. sphaericus*, J. Bacteriol. 122: 1322.

Huber, H.E. and Luthy, P. 1981, *Bacillus thuringiensis* delta-endotoxin composition and activation, in : Pathogenesis of Invertebrate Microbial Diseases, E.W. Davidson, ed., Totowa, N.J., Allenheld, Osmun, pp. 209-234.

Hussey, N. W. and Tinsley, T. W. 1987, Impression of insect pathology in the people's Republic of China, in : Microbial Control of Pests and Plant Diseases, 1970-1980, H. D. Burger, ed., Academic Press, London, pp. 785-795.

Hutton, P.O. and Burbutis, P.B. 1974, Milky disease and Japanese beetle in Delaware, J. Econ. Entomol. 67: 247-248.

Kalfon, A., Charles, J.F., Bourgouin, C. and de Barjac, H. 1984, Sporulation of *B. sphaericus* 2297: an electron microscopic study of crystal-like inclusions biogenesis and toxicity to mosquito larvae, J.

Gen. Microbiol. 130: 893.
Khachatourians, G. G. 1986, Production and use of biological pest control agents, Trends in Biotech. 4: 120-124.
Klier, A., Fargette, F., Ribier, J. and Rapport, G. 1982, Cloning and expression of the crystal protein genes from *Bacillus thuringiensis* strain *berliner* 1715, EMBOT 1: 791-799.
Klier, A., Bourgowin, C. and Rapport, G. 1983, Mating between *Bacillus subtilis* and *Bacillus thuringiensis* and transfer of cloned crystal genes, Mol. Gen. Genet. 191: 257-262.
Knowles, B. H. and Elllar, D. J. 1987, Colloid osmotic lysis is a general feature of the mechanism of action of *Bacillus thuringiensis* δ-endotoxins with different insect specificity, Biochem. Biophys. Acta 924: 509.
Knowles, B.H., Knight, P.J.K. and Ellar, D.J. 1991, N-acetyl galactosamine is part of the receptor in insect gut epithelia that recognises an insecticidal protein from *Bacillus thuringiensis*, Proc. Roy. Soc. Lon. B295:31.
Kristianson, B. and Lewis, C. 1986, Bacterial insecticides: recent developments, Trends in Biotech. 4: 56-58.
Krych, V., Johnson, J.L. and Yousten, A.A. 1980, Deoxyribonucleic acid homologies among strains of *Bacillus sphaericus*, Int. J. Syst. Bacteriol. 30: 476.
Lacey, L. A. and Undeen, A. H. 1980, Microbial control of black flies and mosquitoes, Ann. Rev. Entomol. 31: 265-296.
Lambert, B. and Peferoen, M. 1992, Insecticidal promise of *Bacillus thuringiensis*: facts and mysteries abount a successful biopesticide, Bioscience 42: 112-122.
Lee, M.K., milne, R.E., Ge, A.Z. and Dean, D.H. 1992, Location of *Bombyx mori* receptor binding on *Bacillus thuringiensis* delta-endotoxin, J. Biol. Chem. 267: 3115.
Lereclus, D., Bourgouin, C., Leradet, M. M., Klier, A. and Rapport, G. 1984, Role, structure, and molecular organization of the genes coding for the parasporal delta-endotoxins of *Bacillus thuringiensis*, in : Regulation of Procaryote Development, I. Smith R. A., Siepecky and P. Setlow, eds., American Society for Microbiology, Washington D. C., pp. 255-276.
Li, J., Carrol, J. and Ellar, D. J. 1991, Crystal structure of insecticidal δ-endotoxin from *Bacillus thuringiensis* at 2.5 Å resolution, Nature 353: 815.
Liebig, B., Stetson, D.L., Dean, D.H. 1995, Quantification of the effect of *Bacillus thuringiensis* toxins on short circuit current in the midgut of *Bombyx mori*, J. Insect Physiol. 41(1): 17-22.
Lilley, M., Ruttel, R. N. and Somerville, H. J. 1980, Purification of the insecticidal toxin crystals of *Bacillus thuringiensis*, J. Gen. Microbiol. 118: 1-11.
Lu, Y.J., Adang, M.J. 1996, Conversion of *Bacillus thuringiensis* Cry I Ac-binding aminopeptides to a soluble form by endogenous phosphatidylinositol phospholipase C, Insect Biochem. Mol. Biol. 26(1): 33-40.
Mahillon, J., Hespel, F., Pierssens, A. M. and Delcour, J. 1988, Cloning and partial characterization of three small cryptic plasmids from *Bacillus thuringiensis*, Plasmid, 19: 169-173.
Maider, M. Z., Knowles, B. H. and Ellar, D. J. 1986, Specificity of *Bacillus thuringiensis* var. *colmeri* insecticidal δ-endotoxin is determined by differential proteolytic processing of the protoxin by larval gut proteases, Eur. J. Biochem. 156: 531-540.
McGaughey, W. H. 1978, response of *Ploidia interpunctella* and *Ephestia cautella* larvae to spores and parasporal crystals of *Bacillus thuringiensis*, J. Eco. Entomol. 71: 687-688.
McGaughey, W. H. and Johnson, P. E. 1987, Toxicity of different serotypes and toxins of *Bacillus thuringiensis* to resistant and susceptible Indian meal moths (Lepidoptera: Pyralidae), J. Econ. Environ., 80: 1122-1126.
Menou, G., Manillon, J., Lecader, M. M. and Leredus, D. 1990, Structural and genetic organization of IS 232, a new insertion sequence of *Bacillus thuringiensis*, J. Bacteriol. 172: 6689-6696.
Milne, R. and Kaplani, H. 1993, Purification and characterization of a trypsin-like digestive enzymes from spruce budworm (*Choristoneura fumiferana*) responsible for the activation of d endotoxin from *Bacillus thuringiensis*, Insect Biochem. Mol. Biol. 23(6): 663-673
Nicolas, L., Nielsen-Leroux, C., Charles, J.F. and Delecluse, A. 1993, Respective roles of the 42 and 51 kDa components of the *Bacillus sphaericus* toxin overexpressed in *Bacillus thuringiensis*, FEMS Microbiol. Lett. 106: 275.
Nishiitsutsuji-Uwo and Endo, Y. 1980, Mode of action of *Bacillus thuringiensis* δ-endotoxin: relative roles of spores and crystals in toxicity to *Pieris lymantria* and *Ephestia* larvae, Appl. Entomol. Zool. 15(4): 416-424.
Oei, C., Hindley, J. and Berry, C. 1990, An analysis of the genes encoding the 51.4 and 41.9 kDa toxins of *Bacillus sphaericus* 2297 by deletion mutagenesis: the construction of fusion proteins, FEMS Microbiol. Lett. 72: 265.
Oei, C., Hindley, J. and Berry, C. 1992, Binding of the purified *Bacillus sphaericus* binary toxin and its

deletion derivatives to *Culex quinquefasciatus* gut: elucidation of functional binding domains, J. Gen. Microbiol. 138: 1515.
Payne, C. C. 1988, Pathogens for the control of insects: where next? Phil. Trans. Royal Soc. 318: 225-298.
Perlak, F. J., Deaton, R.W., Armstrong, T.A., Fuchs, R.L., Simson, S.R., Greenplate, J.T. and Fishoff, D.A. 1990, Insect resistant cotton plants, Biotech. 8: 939-943.
Rajamohan, F. and Dean, D.H. 1996, Molecular biology of bacteria for the biological control of insects, in: Molecular Biology of the Biological Control of Pests and Diseases of Plants, M. Gunasekaran and D.J. Weber, eds., CRC Press, Boca Raton, Florida, pp 105-122.
Rajamohan, F., Lee, M. K. and Dean, D. H. 1998, *Bacillus thuringiensis* insecticidal proteins: molecular mode of action, Prog. Nucl. Acid Res. Mol. Bio. 60: 1-27
Roberts, D. W., Fuxa, J. R., Gaugles, R., Goettel, M., Jaques, R. and Maddox, J. 1991, Use of pathogens in insect control, in : Handbook of Pest Management in Agriculture, 2nd ed. Vol. II, D. Pimental, ed., CRC Press, Boca Raton, Florida pp. 243-278.
Sangadala, S., Walters, F.S., English, L.H. and Adang, M.J. 1994, A mixture of *Manduca sexta* aminopeptidase and phosphatase enhances *Bacillus thuringiensis* insecticidal cry IA (c) toxin binding and $^{86}Rb^{+}$ -K^{+} efflux *in vitro*, J. Biol. Chem. 269: 10088-10092.
Schnepf, H. E. and Whiteley, H. R. 1989, Cloning and expression of the *Bacillus thuringiensis* crystal protein genes in *Escherichia coli*, Proc. Natl. Acad. Sci. 78: 2893-2897
Schwartz, J. I., Garneau, I., Masson, Z. and Brousseau, R. 1991, Early response of cultured lepidopteran cells to exposure to δ-endotoxin from *Bacillus thuringiensis*: involvement of calcium and anionic channels, Biochem. Biophys. Acta, 1865: 250
Schwartz, J. L., Garneau, L., Savarla, D., Masson, L., Brousseau, R. and Rousseau, E. 1993, Lepidopteran-specific crystal toxin from *Bacillus thuringiensis* from cation and anion selective channels in planar lipid bilayer, J. Memb. Biol., 132: 53.
Schurter, W., Geiser, M. and Mathe, D. 1989, Efficient transformation of *Bacillus thuringiensis* and *B. cereus via* electroporation: Transformation of acrystalliferous strains with a cloned delta-endotoxin gene, Mol. Gen. Genet. 218: 177-181.
Schroeder, J.M., Chamberlain, C. and Davidson, E.W. 1989, Resistance to the *Bacillus sphaericus* toxin in cultured mosquito cells, *in vitro*, Cell. Dev. Biol. 25: 887.
Sebo, P., Bennardo, T., Torre, de la and Szulmajster, J. 1990, Deletion of the minimal portion of the *Bacillus sphaericus* 1593 toxin required for the expression of larvicidal activity, Eur. J. Biochem. 194: 161.
Shivkumar, A. G., Gundling, G. J., Benson, Y. A., Cusuto, D., Miller, M. F. and Spear, B. B. 1986, Vegetative expression of the δ-endotoxin genes of *Bacillus thuringiensis* subsp. *kurstaki* in *Bacillus subtilis*, J. Bacteriol. 166: 194-204.
Somervillae, H.J. and Pockett, H.V. 1975, An insect toxin from spores of *Bacillus thuringiensis* and *Bacillus cereus*, J. Gen. Microbiol. 87: 359-369.
Thanabalu, T., Berry, C. and Hindley, J. 1993, Cytotoxicity and ADP-ribosylating activity of the mosquitocidal toxin from *Bacillus sphaericus* SSII-I: possible roles of the 27 and 70 kilodalton peptides, J. Bacteriol. 175: 2314.
Thanabalu, T., Hindley, J., Jackson-Yap, J., and Berry, B. 1991, Cloning, sequencing, and expression of a gene encoding a 100 kilodalton mosquitocidal toxin from *Bacillus sphaericus* SSII-I, J. Bacteriol. 173: 2776.
Tojo, A. and Aizawa, K. 1983, Dissolution and degradation of *Bacillus thuringiensis* delta-endotoxin by gut juice protease of the silkworm *Bombyx mori*, Appl. Environ. Microbiol. 45: 576-580.
Vaeck, M., Reynaerts, A., Hofte, H., Jansens, S., Beuekeleer, M. D.; Dean, C., Zabeaw, M., Van Montagu, M. and Leemans, J. 1982, Transgenic plants protected from insect attack, Nature 327: 33-37
Van Rie, J., Jansens, S., Hofte, H., Degheele, D. and Van Mellaert, H. 1989, Specificity of *Bacillus thuringiensis* delta-endotoxins: importance of specific receptors on the brush border membranes of the midgut of target insects, Eur. J. Biochem. 186: 239-247.
Van Rie, J., Jansens, S., Hofte, H., Degheele, D. and Van Mellaert, H. 1990a, Receptors on the brush border membrane of the insect midgut as determinants of the specificity of *Bacillus thuringiensis* delta-endotoxins, Appl. Environ. Microbiol. 56: 1378-1385.
Van Rie, J., McGaughey, W. H., Johnsn, D. E., Barnett, B. D. and Van Mellaert, H. 1990b, Mechanism of insect resistance to the microbial insecticide *Bacillus thuringiensis*, Science 247: 72-74.
Yousten, A.A. 1984, *Bacillus sphaericus*: microbiological factors related to its potential as a mosquito larvicide, Adv. Biotechnol. Process 3: 315.

BIOCONTROL - PLANT GROWTH PROMOTING RHIZOBACTERIA: MECHANISM OF ACTION

K.V.B.R. Tilak[1], Geeta Singh[1] and K.G.Mukerji[2]

[1]Division of Microbiology, Indian Agricultural Research Institute
New Delhi-110 012, India

[2]Applied Mycology Laboratory, Department of Botany
University of Delhi, Delhi-110 007, India.

1. INTRODUCTION

Phytopathogens adversely affect crop production and account for 25-100% yield reduction (Glick and Bashan, 1997). Among the various methods available to control phytopathogens, environment friendly biological approaches includes the use of biocontrol PGPB (Dowling and O'Gara, 1994; Elsherif and Grossmann, 1994).

Free-living soil and rhizosphere bacteria promote plant growth and suppress phytopathogens (Weller, 1988). It is suggested that soil bacteria that indirectly stimulate plant growth may be referred to as biocontrol PGPB. Biochemical and genetic studies have correlated the biocontrol trait of PGPB with production of siderophores, antibiotics, hydrolytic enzymes, hydrogen cyanide and with the ability for colonising of root surface (Bull *et al.*, 1991; Hammer *et al.*, 1997; Loper and Buyer, 1991; Voisard *et al.*, 1989). There is ample avidence that induced systemic resistance (ISR) can be an alternative mechanism to antagonism for achieving biocontrol of plant diseases.

The potential benefits of biocontrol plant growth promoting bacteria (PGPR) is not fully realised because of limitations like inconsistent performance at different locations and seasons. Numerous biotic and abiotic factors contribute to this variability. Inadequate colonisation of the rhizosphere, limited tolerance to fluctuations in environmental parameters, variability in production or activation of antifungal metabolites are some of the important factors. Application of a mixture of biocontrol strains, enhance the level and consistency of biocontrol of soil borne phytopathogens (Capper and Higgins, 1993; Cook *et al.*, 1988; Duffy and Weller, 1992, 1995).

This article makes an attempt to review the research on the mechanism of action of biocontrol plant growth promoting rhizobacteria and possible approaches to increase their effectivity.

2. SIDEROPHORES

Inspite of the abundance, poor solubility of ferric iron (10^{-18} M) at pH 7.4 results in limited availability in aerated and oxidised soils (Neilands *et al.*, 1987). In rhizosphere, where there is a high competition for nutrients among the heterogenous microflora, secretion of low molecular mass (400-1000 daltons) iron binding molecules termed as siderophores permit the producers to outgrow other microflora (Neilands, 1981). Siderophores produced under iron limiting condition, chelate the ferric ions (Fe^{3+}) with a high affinity (Kd=10^{-20} to 10^{-50}) (Castignetti and Smarvelli, 1986) and transport it to microbial cell mediated by specific membrane receptors and transport systems that recognise the iron siderophore complex (Briat, 1992; Neilands, 1981; Neilands and Leong, 1986). The bacteria that produce the siderophore take up the iron-siderophore complex using a receptor which is specific for this complex and is present in the outer cell membrane of the bacterium (O'Sullivan and O'Gara, 1991). Fungal siderophores generally have a lower affinity than do the siderophores produced by bio control PGPR (Schippers *et al.*, 1987), so biocontrol PGPR are better competitors for available iron.

Siderophores produced by fluorescent pseudomonads mediate iron regulated antagonism against a variety of phytopathogens in culture (Kloepper *et al.*, 1980a; Misaghi *et al.*, 1982) and soil (Lemanceau and Alabouvette, 1993; Leong, 1986; Loper, 1990; Scher, 1986; Schippers *et al.*, 1987). The flourescent siderophores termed pyoverdines, pyochelin and pseudobactin represent only one class of siderophores produced by fluorescent *Pseudomonas* spp. (Cox, 1980; Demange *et al.*, 1987; Teintze *et al.*, 1981). Pyoverdines produced by *Pseudomonas* spps. may deplete the medium of available iron, thereby inhibiting the growth of an indicator strain, while ferric complexes have no effect. Unlike microbial phytopathogens, plants are not adversely affected by localized depletion of iron caused by PGPR because plants can grow at much lower concentration than microorganisms (O'Sullivan and O'Gara, 1991). Also certain plants use microbial Fe (III) siderophores such as ferrioxamine B (Crowley *et al.*, 1988), the ferrichromes (Orlando and Neilands, 1982), rhodotorulic acid (Crowley *et al.*, 1988; Miller *et al.*, 1985). The mechanism is that the plants transport Fe from microbial Fe(III) siderophores by reduction to Fe (II) and ligand exchange with phytosiderophores respectively or by the direct uptake of the Fe (III) siderophores (Bar-Ness *et al.*, 1991, 1992; Visca *et al.*, 1993; Wang *et al.*, 1989).

Evidence for involvement of siderophores in fungal pathogen caused disease suppression has been evaluated in studies whose absence of disease suppression was observed under iron deplete soil conditions.

2.1. Iron Concentration of Soil

Raising the soil concentration of iron to 40 μ mol Fe^{3+} /litre caused a proportionate decrease in the amount of fluorescent siderophore produced and the suppressive effect on *Gaeumannomyces graminis* var. *tritici*, a fungal pathogen of wheat, in a collection of 70 separate isolates of fluorescent pseudomonads (Elsharif and Grossmann, 1994). Similar observations were reported when iron availability in soil was enahanced by addition of chelators with low affinities for Fe (III) (Kloepper *et al.*, 1988 a,b; Scher and Bakar, 1982), $FeSO_4$ (Snch *et al.*, 1984), $FeCl_3$ (Elad and Baker, 1985) or H_2SO_4 to decrease soil pH (Elad and Baker, 1985) thereby increasing available iron concentration by about 10^3 per pH unit (Lindsay, 1979). Soil amendment with low affinity Fe III chelators resulted in failure of growth promotion of potato (Kloepper *et al.*, 1980a) or wilt disease suppression of various crops caused by *Fusarium oxysporum* (Elad and Baker, 1985; Kloepper *et al.*, 1988b; Scher and Baker, 1982; Snch *et al.*, 1984). These experimental evidences clearly show that antagonism through pyoverdine mediated iron competition is inversely related to the level of iron available to the phytopathogen in soil as in culture. However, the impact of altered iron availability on the status of other soil minerals and their cumulative effect on the production of microbial metabolites, other than pyoverdines and their role in biocontrol is still not clearly known.

2.2. Mutant Deficient in Pvd

Mutants deficient in pyoverdine production (Pvd^-) failed to promote plant growth unlike the parental strains of *P. fluorescens* and *P. putida* (Kloepper and Schroth, 1981). Similarly a derivative of *Pseudomonas aeruginosa* that was selected for its lack of siderophore production, lost the ability shown by the parental strain, to protect tomato plants against *Pythium* damping off (Buysens *et al.*, 1994). Insertion of a single Tn5 inactivates both the pyoverdine production and plant growth promotion phenotypes of *P. putida* WCS 358, suggesting a role for siderophores in biological control of minor pathogens of potato (Bakker *et al.*, 1986; Schippers *et al.*, 1987). Derivatives of *Pseudomonas putida* that over produced siderophores were more effective than the parental strain in biocontrol of a strain of *Fusarium oxysporum* pathogenic tomatoes (Vandenbergh and Gonzalez, 1984).

The general iron regulated fungistasis associated with pyoverdines was established by Misaghi *et al.*(1982), who demonstrated that pyoverdines produced by 156 strains of fluorescent pseudomonads isolated from soil or plant are antagonistic against *Geotrichum candidum*. Buyer *et al.* (1993) developed an ELISA assay to quantify the amount of pseudobactin, produced by a fluorescent pseudomonad, in response to iron limiting conditions of barley rhizosphere soil sample.

Iron regulated antagonism cannot be attributed only to pyoverdine production, however, since *Pseudomonas* spp. are also known to produce antifungal compounds (Azegami *et al.*, 1988; Gill and Warren, 1988) under iron-limiting conditions (Hamdan *et al.*, 1991; Wei *et al.*, 1996). Tn5 induced non fluorescent siderophore negative mutant of *Pseudomonas fluorescens* 3551 showed diminished suppression of damping off disease caused by *Pythium ultimum* (Loper, 1988). In another study pyoverdin negative mutants of *Pseudomonas fluorescens* 2-79 controlled take all a major root disease of wheat caused by *Gaeumannomyces graminis* var. *tritici*, as efficiently as did the parental strain that produced this siderophores (Hamdan *et al.*, 1991).

2.3. Addition of Purified Siderophores

Experiments demonstrating that addition of purified siderophores or synthetic iron chelators e.g. ethylene - diaminedi (O-hydroxy phenylacetic acid) (EDDDHA) mimics the effect of seed inoculation with *Pseudomonas* sp. strain BIO in growth promotion of potato (Kloepper *et al.*, 1980a) or suppression of take all of wheat or *Fusarium* wilt of flax (Kloepper *et al.*, 1980b), inhibition of chlamydospore germination in rhizosphere and bulk soil (Elad and Baker 1985; Snch *et al.*, 1984), provide evidence that high affinity iron chelator have suppressive activity in the soil. The difficulties in obtaining large quantities of siderophore is the limiting factor in using this approach.

Although the concentrations of pyoverdines produced in the rhizosphere by *Pseudomonas* spp. are adequate, a number of factors like crop plant, the target phytopathogen, edaphic and other environmental parameters and the affinity of the specific siderophore for iron determine its effectivity as a biocontrol agent.

3. ANTIBIOSIS

Current evidence supports the concept that antibiosis is the dominant mechanism of disease suppression by the introduced strains that produce these secondary metabolites. Results obtained by both the application of molecular technique and direct isolation have demonstrated unequivocally that these antibiotics are produced in the spermosphere and rhizosphere and play a major role in the suppression of soil borne pathogens (Georgakopoulos *et.al.*, 1994; Keel *et al.*, 1996; Kraus and Loper, 1995; Maurhofer *et al.*, 1994; Nowak–Thompson *et al.*, 1994; Pfender *et al.*, 1993; Pierson and Thomashow, 1992; Raaijmakers *et al.*, 1997; Shanahan *et al.*, 1992a; Thomashow and Weller, 1988; Thomashow *et al.*, 1996; Vincent *et al.*, 1991; Weller, 1988). Rice is associated with antagonism to several fungal pathogens *Pseudomonas* spp. produced Pyrrolnitrin. Phenazinc - 1 - carboxylic acid and pyocyanine (Rosales *et al.*, 1995). Examples of antibiotics are pyoluteorin, pyrrolnitcin implicated in control of damping -off and wilt disease of cotton (Howell and Stipanovic, 1979; 1980), vegetable crops (Homma and Suzui, 1989) caused by *Pythium ultimum, Rhizoctonia solani, Verticillium dahliae*, (Gutterson, 1990, Howie and Suslow, 1991). Phloroglucinols, the phenolic broad spectrum antibiotic is one of the major determinant for biocontrol of activity of PGPR (Thomashow and Weller, 1996; Weller and Thomashow, 1993). 2-4-diacetylphloroglucenol, involved in control of black root rot of tobacco caused by *Thielaviopsis basicola* (Defago *et al.*, 1990; Keel *et al.*, 1989), take all of wheat caused by *Gaeumannomyces graminis* var. *tritici* (Harrison *et al.*, 1993; Keel *et al.*, 1992; Vincent *et al.*, 1991) and damping-off of sugarbeet caused by *Pythium ultimum* Oomycin A(Shanahan *et al.*, 1992a) responsible for control of damping off disease of cotton caused by *Pythium ultimum* (Gutterson *et al.*, 1986; Howie and Suslow, 1991). Phenazine-1-carboxylic acid and its derivatives (2 hydroxy phenazine-1-carboxylic acid and 2 hydroxyphenazine) are implicated in the suppression of take all (Brisbane and Rovira 1988; Gurusiddaiah *et al.*, 1986; Hamdan *et al.*, 1991; Ownley *et al.*, 1992, Thomashow and Weller, 1988) and speckled blotch of wheat caused by *Septoria tritici* Roberge in Desmaz (Flaishman *et al.*, 1990). However, some phenazinc and acetyl phloroglucinols can adversely affect plant metabolism (Keel *et al.*, 1992) suggesting their potential to stimulate defense mechanisms including induced resistance, in host plants.The antibiotics pyoluteorin (Plt), pyrrolnitrin (Prn), Phenazine-1-carboxylic acid (PCA), and 2-4 diacetyl phloroglucinol (Phl) are currently a major focus of research in biological control.

The experimental approaches to confirm the involvement of antibiotics production in PGPB mediated disease suppression are:

3.1. Application of Purified Antibiotics

Application of purified antibiotic compound instead of the producing microorganism in *in vivo* conditions inhibit the same spectrum of fungal pathogens as the biocontrol strain (Carni *et al.*, 1994). Application of purified antibiotics, e.g. pyoluteorin or pyrrolnitrin, isolated from biocontrol agents, to seeds (Homma and Suzui, 1989; Howell and Stipanovic, 1979) or fruits (Janisiewicz and Marchi, 1992) resulted in suppression of disease symptoms. A correlation has been observed between antibiotic production and pathogen inhibition *in vitro* and disease suppression *in vivo* (Ahl *et al.*, 1986; Dutrecq *et al.*, 1991; Fenton *et al.*, 1992; Howell and Stipanovic, 1979, 1980; Keel *et al.*, 1996; Lambert *et al.*, 1987, Rosales *et al.*, 1995).

3.2. Antibiotic Produced by Mutants

Mutants impaired in antibiotic production either lost or had diminished capability of disease suppression compared to the wild types. *Pseudomonas fluorescens* strain CHAO which suppressed the black root rot caused by *Thielaviopsis basicola* in iron sufficient soils (Keel *et al.*, 1989; Stutz *et al.*, 1986) protects wheat against *Gaeumanomyces graminis* var. *tritici* (Defago *et al.*, 1990) and is active against a variety of root diseases (Defago *et al.*, 1990). This strain produces several antibiotics like pyoluteorin, 2-4 diacetyl phloroglucinol

(Phl) in addition to pyoverdine and hydrogen cyanide (Ahl *et al.*, 1986; Defago *et al.*, 1990; Voisard *et al.*, 1989). A Tn5 induced mutant of strain CHAO having defect in (Phl) synthesis resulted in diminished inhibition of *T. basicola* and *Gaeumanomyces graminis* var. *tritici in vitro* and partial suppression of black root rot of tobacco and take all of wheat compared with wild type CHAO (Keel *et al.*, 1992).

Complementation of the mutant with an 11-kb DNA fragment from a genomic library of wild type CHAO, restored production of the metabolite, inhibition of the fungal pathogens *in vitro* and disease suppression. Production of (Phl.) in the rhizosphere of wheat was observed by strain CHAO, the complemented mutant, but not in case of Phl mutant. In case of *Pseudomonas aureofaciens* suppression of take-all disease of wheat caused by *Gaeumanomyces graminis* var. *tritici* is through production of three antibiotics phenazine - 1- carboxylic Acid, 2 hydroxyphenazine-1- carboxylic acid and 2 hydroxyphenazine. Addition of 2.8 kilobase segment present in cosmid pLSP259 and pLSP282 from the genomic library of strain 30-84 restored phenazine production in PhZ^- mutants (Pierson and Thomashow, 1992).

Siderophore and antibiotic negative mutants of *Pseudomonas fluorescens* biovarIV showed that antibiosis was more important than competition for iron in the damping off disease control (Yuen *et al.*, 1987). Similar observations were recorded by Poplawski *et al.*, (1988) when antibiosis negative Tn5 generated mutants of *Pseudomonas fluorescens* NRRL 15133 failed to biocontrol take all disease of wheat. *Pseudomonas fluorescens* strain Hv 37a R2 and its isogenic Afu^- mutants deficient in biosynthesis of antibiotic Oomycin A, both colonized spermosphere and rhizosphere of cotton but significantly few *Phythium ultimum* propagules in the soil after root infection by parental strain was recorded in comparison to Afu^- mutant strain and nontreated control (Howie and Suslow, 1991). Afu E expression *in vivo* by an afu E - lac Z gene fusion (Gutterson *et al.*, 1988) demonstrated a positive correlation between Lac Z activity and antibiotic Oomycin A biosynthesis. Parental strain showed antibiotic Oomycin A biosynthesis in the rhizosphere within 24 h of inoculation but was not detectable in the Afu^- mutant strains

Antibiotic pyrrolnitrin negative mutants of *Pseudomonas fluorescens* BL 915were ineffective in suppressing *Rhizoctonia solani* induced damping off disease of cotton plants (Hill *et al.*, 1994) unlike the wild type strain. When a DNA 612 kb fragment, which containes a cluster of four genes required for pyrrolnitrin biosynthesis isolated from wild type bacterium, that restored pyrrolnitrin synthesis to these mutants was transferred to pyrrolnitrin non producing strains of *P. fluorescens*, the transformed strains gained both the ability to synthesize pyrrolnitrin and inhibit *Rhizoctonia solani* induced damping-off of cotton plants. Deletion mutations in any of 4 genes resulted in pyrrolnitrin nonproducing phenotype (Hammer *et al.*, 1997). Transfer of four gene cluster to *E. coli* resulted in production of pyrrolnitrin by this organism.

Thomashow and Weller (1988) reported that Tn5 insertion mutants of *Pseudomonas fluorescens* strain 2-79 defective in phenazine-1-carboxylic acid (PCA) production have a significantly reduced capacity to suppress take-all of wheat a disease induced by *Gaeumannomyces graminis* var. *tritici*. Phenazine-1-carboxylic Acid (PCA) production and disease suppression was restored in the mutants by complementation with cloned PCA biosynthetic genes.

3.3. Genes for Antifungal Toxins

Enhanced antibiotics pyoluteorin and 2,4-diacetylphloroglucinol production in mutants of *Pseudomonas fluorescens* strain CHAO resulted in corresponding improvement in suppression of *Pythium ultimum* induced disease of cucumber plants. Introduction of cosmid pME3090 carrying 22kb of CHAO DNA into strain CHAO enhanced antibiotic pyoluteorin (Plt) and 2,4-diacetyl phloroglucinol (Phl) production by 3-5 fold *in vitro* and in an artificial wheat rhizosphere (Maurhofer *et al.*, 1992). The difference between the amounts of plt and phl produced by two strain was greater in presence than in absence of *P. ultimum*. Presence of phytopathogen induces the host plant to produce more nutrients in root exudates which in turn may stimulate *Pseudomonas fluorescens* CHAO/pME 3090 to overproduce the antibiotics (plt) and phl. Differential protection offered by mutants to different plants show that bacterial antibiotic production varies in the rhizosphere of different plants due to difference in root exudates (Maurhofer *et al.*, 1995). The correlation between six to seven fold over production of antibiotic phloroglucinol *in vitro* and in wheat rhizosphere environment with a corresponding enhancement in the biocontrol activity was observed (Bonsall *et al.*, 1997; Maurhofer *et al.*, 1995).

Complementation of specific mutants defective in antifungal property have been employed for identification of gene(s) responsible for antifungal toxin biosynthesis (Bangera and Thomashow, 1996; Carruthers *et al.*, 1994; Gutterson *et al.*, 1986; Hill *et al.*, 1994; Pfender *et al.*, 1993; Pierson and Thomashow, 1992; Poplawsky *et al.*, 1988; Vincent *et al.*, 1991; Voisard *et al.*, 1989). The biosynthetic loci for pyoluteorin (Plt) pyrolnitrin (Prn), Phenazine-1-carboxylic acid (PCA) and 2,4-diacetylphloroglucinol have been cloned, and all but plt have been fully sequenced (Bangera and Thomashow, 1996; Boronin *et al.*, 1995; Hammer *et al.*, 1995; Hill *et al.*, 1994; Kraus and Loper, 1995; Pierson *et al.*, 1995). The Phl genes are conserved in all the strains within a world wide collection of Phl producers (Bangera and Thomashow, 1996; Vincent *et al.*, 1991).

Genetic analysis show that the production of 2-4 diacetylphloroglucinol by *Pseudomonas aureofaciens* Q2-87 is coded by 4.8 kb DNA fragment and in *P. fluorescens* Q2-87 by a 5kb DNA segment (Bangera and Thomashow, 1996). This 4.8 kb DNA fragment when used as hybridization probe demonstrated that Phl genes are conserved in all Phl producer strain, many of which are known biocontrol agents (Bangera and Thomashow 1996; Vincent *et al.*, 1991). A 6.2 kb region containing a cluster of four genes encode the pathway for pyrrolnitrin biosynthesis (Hammer *et al.*, 1997).

A number of antifungal metabolites produced by pseudomonads appears to be regulated by a global regulator (gac A) (Gaffney *et al.*, 1993). This leads to difficulty in identification of structural genes for individual antifungal factors by mutagenesis, since regulatory mutants affected in the global regulation of all antifungal factors are commonly found (Hill *et al.*, 1997). For example, a phloroglucinol (Prn^-) non producing *Pseudomonas* mutant (Pfender *et al.*, 1993) actually had mutation in rpoS, which regulates secondary metabolite biosynthesis genes including those for Prn (Sarniguet *et al.*, 1995). Amplification of the gene from *Pseudomonas fluorescens* CHAO by introduction of 22 kb DNA fragment of wild type enhanced antibiotic production and corresponding improved protection against *Pythium ultimum* induced damping off of cucumber (Maurhofer *et al.*, 1995; Schnider *et al.*, 1995a). Another strategy to genetically engineer *Pseudomonas* spp. with improved biocontrol activity is through increasing the amount of antibiotic produced by a given strain or by transferring biosynthetic loci into a heterologous strain (Thomashow and Weller, 1996; Weller and Thomashow, 1993). In another study, inactivation of pqq genes, responsible for biosynthesis of pyrroloquinoline quinone, a co-factor of different hydrogenases, in *Pseudomonas fluorescens* CHAO stimulated the production of antibiotic pyoluteorin (Schnider *et al.*, 1995b). It is proposed that the possible mechanism of stimulation of antibiotics production is through shunting metabolites from other metabolic pathways to pyoluteorin biosynthesis pathway.

4. INDUCED SYSTEMIC RESISTANCE

Induced systemic resistance (ISR) is based on long term activation of plant defense mechanisms induced by pathogens, non pathogens (PGPR) and chemical agents such as salicyclic acid (White, 1979), 2-6 dichloro-isonicotinic acid (INA) (Kloepper *et al.*, 1992, Metraux *et al.*, 1991). Evidence from many experimental systems indicate that PGPR induces systemic resistance (ISR) leading to plant protection against a broad spectrum of pathogens without causing any symptoms (Alstrom, 1991; Liu *et al.*, 1995a,b; Maurhofer *et al.*, 1994; Pieterse *et al.*, 1996; Tuzun and Kloepper, 1994; Van Peer *et al.*, 1992; Wei *et al.*, 1991). It has been demonstrated that PGPR *Pseudomonas* induce systemic resistance to fungal root pathogens (Van Peer and Schippers, 1989, 1992; Zhou and Paulitz, 1994; Zhou *et al.*, 1992), fungal (Wei *et al.*, 1991) bacterial leaf pathogens (Alstrom, 1991; Wei *et al.*, 1994) and viruses (Liu *et al.*, 1992, Maurhofer *et al.*, 1994) under green house (Liu *et al.*, 1995a,b, 1989; Zhou and Paulitz, 1994) and field conditions (Wei *et al.*, 1996).

Pseudomonas spp. mediated ISR was observed in (Wei *et al.*, 1991) carnation against *F. oxysporum* (Duijff *et al.*, 1993; Van Peer *et al.*, 1990 b) in cucumber against *Colletotrichum orbiculare* (Wei *et al.*, 1991), cucumber mosaic virus (Liu *et al.*, 1992) and *Pythium aphanidermatum* (Zhou and Paulitz, 1994); in beans against *P. syringae* pv. *phaseolicola* (Alstrom, 1991); in tobacco against tobacco necrosis virus (Maurhofer *et al.*, 1994) and in radish against *F. oxysporum* f. sp. *raphani* (Leeman *et al.*, 1995 a,b, 1996). Treatment of seeds or cotyledons with *P. putida* protected cucumber plants against bacterial angular leaf spot disease and also resulted in a significant decrease in the epiphytic population of causal pathogenic agent *P. syringae* pv. *lachrymans* (Liu *et al.*, 1995a). In all of these experiments the ISR inducing pseudomonad and the pathogen were applied at spatially distant location on the plant, ruling out the possibility of antibiosis and competition as mechanisms of pathogen biocontrol.

4.1. Protective Biopolymers

Formation of protective biopolymers, e.g. lignin callose and hydroxyproline rich glycoproteins (Dixon and Lamb, 1990; Hammerschmidt and Kuc, 1982; Hammerschmidt *et al.*, 1982, 1984) and the resultant alterations in the structure of plant cells walls following pathogen infection, contributes to resistance by preventing pathogen ingress, slowing down the penetration process and activation of other plant defense mechanisms. The importance of lignification in imparting disease resistance was shown in a study where inoculation of bean plants with *P. putida* resulted in an increase in the lignin content of the roots, the surface peroxidase activity and offered a high degree of protection against phytopathogenic *Fusarium solani* f. sp. *phaseoli* (Albert and Anderson, 1987; Anderson and Guerra, 1985). The significance of callose in the expression of resistance is supported by the resistant *Arabidopsis* lsd (lesion stimulating disease) mutants, which show stronger deposition of callose at the site of pathogen penetration than the susceptible wild type plants (Dietrich *et al.*, 1994).

4.2. Pathogenesis - Related Proteins (PRS)

Induced systemic resistance increases the activity of chitinases, ß-1, 3-glucanases, protease, lipases, peroxidases and other pathogenesis related proteins (PR). These PR proteins show antimicrobial activity *in vitro* (Mauch *et al.*, 1988). Some biocontrol PGPB strains have been found to produce chitinase, ß-1,3-glucanase, protease and lipase that aids pathogen cell lysis (Chet and Inbar, 1994). *Pseudomonas stutzeri* secreting extracellular chitinase and laminarinase could digest and lyse *Fusarium solani* mycelia resulting in suppression of root rot disease. Similarly suppression of diseases caused by *Rhizoctonia solani*, *Sclerotium rolfsii* and *Pythium ultimum* was mediated by ß-1, 3 glucanase producing strain of *Pseudomonas cepacia* (Fridlender *et al.*, 1993). Transfer of chitinase gene (Chi A) into a PGPR *P. fluorescens* resulted in gene expression, secretion of chitinase enzyme and biocontrol of *Rhizoctonia solani* (Koby *et al.*, 1994).

PGPR induced ISR in *Arabidopsis* occurs without induction of transcripts of PRs (Pieterse *et al.*, 1996). Similarly in radish, *Pseudomonas fluorescens* strain WCS 417 mediated ISR is not associated with accumulation of PRs (Harrison *et al.*, 1993). However, the absence of PRs following induction by *P. fluorescens* does not offer a lower level of protection (Hoffland *et al.*, 1995). In contrast, in bean and tobacco, colonization of the roots by fluorescent bacteria is correlated with induction of PRs and associated systemic resistance (Maurhofer *et al.*, 1994; Zdor and Anderson, 1992). Whether the plant production of PRs is obligatory for manifestation of ISR or not, is an unexplored area of research.

5. ROLE OF ETHYLENE

One mechanism by which plants respond to fungal phytopathogen infection is by synthesizing "stress" ethylene (Abeles *et al.*, 1992; Hyodo, 1991). Ethylene can induce some of the PR proteins e.g ß-1-3-glucanase and Chitinase (Abeles *et al.*, 1971). Structural reinforcements of cell wall such as lignification and accumulation of hydroxyproline rich cell wall proteins are also enhanced by ethylene (Boller, 1990). However, reports exist that ethylene is not the signal involved in induction of systemic resistance (SAR) because SAR gene expression in ethylene-insensitive mutant of *Arabidopsis* is similar to the wild type plants (Bleeker *et al.*, 1988; Chang *et al.*, 1993).

The determinants which signal or triggers the PGPB mediated induced systemic resistance (ISR) are not yet identified. There is still a controversy about the nature of the long distance signal (Vernooij *et al.*, 1994). Salicylic acid is considered to be involved in the pathway and is required for establishment of ISR in some cases. The most convincing demonstration of the the involvement of SA in SAR signal transduction came from the experiments with transgenic tobacco plant (NahG tobacco) harbouring a bacterial gene for salicylate hydroxylase, an enzyme which catabolizes salicyclic acid(SA) to catechol (Gaffney *et al.*, 1993). Plants expressing this gene do not accumulate SA and failed to show SAR response after inoculation with tobacco mosaic virus (TMV). Maurhofer *et al.* (1994) showed the involvement of the pyoverdine siderophore of *P. fluorescens* CHAO in the induction of disease resistance against tobacco necrosis virus in tobacco. Iron availability was found to influence induction of systemic resistance to *Fusarium* wilt of radish by *Pseudomonas fluorescens*. Leeman *et al.* (1996) proposed that Fe^{3+} (ferric ion) chelating SA, produced by selected *P. fluorescens* strains at low iron availability is involved in the induction of systemic resistance to *Fusarium* wilt of radish. This is further strengthened by the observation that *P. putida* strain WCS 358 did not induce resistance either at low or at high iron availability and did not produce salicyclic acid (SA) *in vitro*. SA is considered to serve as a siderophore for some *Pseudomonas* spp. (Van Peer *et al.*, 1990b; Visca *et al.*, 1993). *Pseudomonas fluorescence* strains WCS 374, WCS 417, mediated induced systemic resistance in radish involved lipopolysaccharide (LPS) , O-antigen (Leeman *et al.*, 1995a; Van Peer and Schippers, 1992).

6. ROOT COLONIZATION AND COMPETITION

Extensive colonization of root surface by pseudomonads is essential for an efficient biocontrol activity because unsuccessful biocontrol field experiments are often correlated with poor colonization of the root system by the biocontrol bacteria (Bull *et al.*, 1991; Schippers *et al.*, 1987; Weller, 1988). A positive correlation was observed between population size and biocontrol activity of *Pseudomonas fluorescens* strain 2-79 against *Gaeumannomyces graminis* var. *tritici* the take all pathogen of wheat (Bull *et al.*, 1991). The application of an increased population of *Pseudomonas putida* on seed pieces of potato to control the preemergent decay caused by *Erwinia carotovora* sub sp. *atroseptica* had a positive effect as reported by Xu and Gross (1986). Similar observations were reported in case of *Pythium* damping off disease of cucumber controlled by *Pseudomonas cepacia* isolate 808 (Elad and Chet, 1987).

The rhizospheric colonization of the introduced strain reportedly is a factor of initial inoculum concentration and competition from already existing natural microflora (Iswandi *et al.*, 1987; Torkewitz and Lam, 1991). During initial phases of successful plant microbe interaction complex machinery of bacterial motility, chemotaxis and attachment plays an important role (Broek and Vanderleyden, 1995). Though *Pseudomonas* strains have been shown to migrate towards soybean seeds in soil no correlation between chemotactic ability and root colonization has been reported (Scher *et al.*, 1985; 1988).

Mutational studies to elucidate the molecular basis of rhizospheric colonization show that presence of flagella (de Weger *et al.*, 1987), the presence of O- antigen of lipopolysaccharides (LPS) (de Weger *et al.*, 1989) and ability to synthesize aminoacids are important colonization traits.

Several bacterial surface molecules have been implicated in *Pseudomonas* plant root adhesion. Pili increase the attachment of *P. fluorescens* to corn roots (Vesper, 1987). Flagella are involved in colonization by *P.putida* on potato roots (de Weger *et al.*, 1987) but are not important for colonization of wheat by *P. fluorescens* (Weller, 1988). The O-antigenic part of the lipopolysaccharides was demonstrated to be essential for efficient colonization of potato roots by *P. fluorescens* and *P. putida* but is not involved in adhesion of bacteria to the root surface (de Weger *et al.*, 1989, 1991). Root present in the outer membrane of *P.fluorescens* mediated *in vitro* adsorption to seedling roots of wheat, barley, maize, bean and sunflower (De Mot and Vanderleyden, 1991).

A *P.fluorescens* (porin F protein (O pr F) mutant showed impaired adhesion at low inoculum levels, suggesting that apart from O pr F protein of *P. fluorescens* additional surface components must be responsible for *P. fluorescens* - root adsorption. The agglutination of *P. putida* cells by a bean root surface glycoprotein termed agglutinin is reported to be an event in colonization (Anderson and Josalavich, 1979; Anderson, 1983; Van Peers, *et al.*, 1990a) *P. putida* mutants deficient in agglutination phenotype (Agg^-) showed a reduced ability to adhere to plant roots, to colonize roots from a seed inoculum and to suppress *Fusarium* attack (Anderson *et al.*, 1988; Tari and Anderson, 1988). Similarly *Agrobacterium tumefaciens* mutants unable to bind to plant cells were all shown to be completely avirulent (Cangelosi *et al.*, 1987, Douglas *et al.*, 1982; Matthysses, 1987). No differences were found between Agg^- mutants and parental strains regarding the cell surface features e.g. pili, flagella and lipopolysacharides. The agglutination phenotype is sensitive to protease treatment (Buell *et al.*, 1993). A genetic locus agg A coding for 48 K Da periplasmic protein, which expressed on plant root surface was identified (Buell and Anderson, 1992, 1993). However, the exact role of periplasmic Agg A protein, in the agglutination phenotype is still not known.

Competition for nutrients and suitable ecological niches on the root surface is also a mechanism by which some biocontrol PGPB are known to protect plants against phytopathogens (Kloepper *et al.*, 1988; O'Sullivan and Gara, 1991; Park *et al.*, 1988) reported that combined application of fluorescent *P. putida* and a non pathogenic *Fusarium oxysporum* strain was effective in suppression of *Fusarium* wilt of cucumber although each was ineffective when applied separately. The bacteria inhibited chlamydospore germination of both pathogenic and nonpathogenic strains of *Fusarium* in the rhizosphere when pH was appropriate for siderophore production. Non pathogenic *Fusarium* stimulated *P. putida* population and siderophore production in the rhizosphere through enhancement in root exudation. These factors lead to stronger competition for iron which is essential for chlamydospore germination and host penetration by *Fusarium*. The ability to metabolize the constituents of sugar beet seed exudates to compound inhibitory to *Pythium ultimum* was the mechanism exerted by biocontrol Pseudomonads (Stephens *et al.*, 1993). It was also observed that there was no relationship between the ability of *Pseudomonas* spp. to inhibit a fungal pathogen under laboratory conditions that favoured production of either antibiotics or siderophores and the biocontrol activity of bacterium *in vivo*.

Production of reactive oxygen species e.g. hydroxyl ions,the superoxide anion and hydrogen per oxide is a defense mechanism employed by the producing system against pathogenic organisms (Duffy *et al.*, 1996; Klotz *et al.*, 1989; Sutherland, 1991).

Catalases protect cells against the damage caused by reactive oxygen species to cellular components. Challenge by reactive oxygen species occurs from external sources as well as from normal metabolism.Soil pseudomonads are exposed, to active oxygen species as they colonise plant roots (Katsuwon and Anderson, 1989, 1990, 1992) and certain soil fungi also produce hydrogen per oxide as a mechanism to compete with other organisms . Phytopathogens that contain higher levels of enzyme such as superoxide dismutase, catalase and peroxidase that can protect cells against active oxygen species are more versatile in overcoming these plant defense mechanisms (Klotz and Hutcheson, 1992). *Pseudomonas putida* produces three isozymes of catalase (A, B, C) in stationary phase cells. Chromosomal mutants of *P. putida* deficient in catalase 'C' were not impaired in their ability to colonize roots of bean or wheat plants grown under sterile conditions (Miller *et al.*, 1997).

7. HYDROGEN CYANIDE

Many PGPB produce diverse range of low molecular weight metabolites with antifungal activity (Dowling and O'Gara, 1994). Many Pseudomonads produce hydrogen cyanide(HCN)from oxidation of glycine, because Fe^{3+} ions and glycine enhance HCN synthesis by *Pseudomonas* spp in culture during the late exponential phase of growth (Baker and Schipper, 1987, Cactric, 1977). There is extensive evidence that HCN producing rhizospheric (Cook, 1994; Defago *et al.*, 1990; Haas *et al.*, 1991; Lam and Gaffney, 1993; Pierson and Thomashow, 1992; Voisard *et al.*, 1989; Weller, 1988) and phylloplane (May *et al.*, 1997) colonizing pseudomonads suppress soil borne and foliar phytopathogens respectively.Though there is a reduction in potato yield by HCN producing deleterious pseudomonads (Baker and Schipper, 1987), a part of antagonistic effect of *P. fluorescens* strain CHAO on *Thielaviopsis basicola* in the rhizoplane of tobacco appears to be the result of HCN production (Sutherland, 1991, Voisard *et al.*, 1989). It has been proposed that HCN may induce plant defense mechanism. An HCN deficient mutant of *P. fluorescens* derived from HCN producer strain CHAO, is less effective than the wild type in suppressing tobacco black root rot caused by *Thielaviopsis basicola* (Stutz *et al.*, 1986). This deficiency can be restored by a plasmid carrying a 5.0 kb genomic fragment from strain CHAO. The fragment contains three continuous open reading frames designated hcn ABC which encode cyanide synthase (Voisard *et al.*, 1989). Three genes Lem A, gac A and anr have shown to be involved in the regulation of cyanogenesis in *Pseudomonas* spps. The lem A and gac A genes globally control antibiotic and HCN production (Corbell and Loper, 1995; Gaffney *et al.*, 1994), anr is an anaerobic regulator of arginine deiminase pathway and cyanogenesis (Voisand *et al.*, 1994; Zimmermann *et al.*, 1991).

Pseudomonas putida strain BK 8661 which produces siderophore(s), antibiotics(s) and low levels of hydrogen cyanide (HCN) suppressed growth of *Septoria tritici* and *Puccinia recondita* f. sp. *tritici* *in vitro* and on wheat leaves. HCN over producing derivatives,the pleiotropic mutant of strain BK 8661 are deficient in siderophore and antibiotics production.HCN produced by the over producing bacterial strains resulted in a small but significant increase in the suppression of symptoms caused by *S.tritici* and *P.recondita* f. sp. *tritici* of wheat seedling leaves.

8. CONCLUSIONS

Suppression of Soil borne plant pathogens by PGPB is mediated by different mechanisms. Biocontrol PGPB may be improved by genetically engineering them to overexpress one or more of these traits so that strains with several different antiphytopathogen traits, which can act synergistically are created. Efforts to genetically engineer *Pseudomonas* spps with enhanced biocontrol activity have focussed on increasing the amount of antibiotic produced by a strain or by transferring biosynthetic loci into heterologous strains the use of antibiotic overproducing *Pseudomonas* a way to achieve disease control with a smaller bacterial inoculum. Reducing the dose of a PGPR required for disease suppression would lower the cost and increase the acceptance of the treatment., although siderophores determine the biocontrol activity of PGPB against certain. Soil borne pathogens, their consistent production and activity in the varied environment characterizing plant surfaces must be demonstrated before they can be used for disease control. Further research is needed to elucidate the precise nature and function of the determinants in PGPB mediated induced systemic resistance.

REFERENCES

Abeles, F.B., Bosshart, R.P., Forrence, L.E. and Habig, W.H. 1971, Preparation and purification of glucanase and chitinase from bean leaves, Plant Physiol. 47:129-134.

Abeles, F.B., Morgan, P.W. and Saltveit, M.E. Jr. 1992, Regulation of ethylene production by internal, environmental and stress factors, in : Ethylene in Plant Biology, II ed., Academic Press, pp.56-119.

Ahl, P., Voisard, C. and Defago, G. 1986, Iron bound-siderophores, cyanic acid and antibiotic involved in suppression of *Thielaviopsis basicola* by a *Pseudomonas fluorescens* strain, J. Phytopathol. 116:121-134.

Albert, F. and Anderson, A.J. 1987, The effect of *Pseudomonas putida* colonization on root surface peroxidase, Plant Physiol. 85:537-541.

Alstrom, S. 1991, Induction of disease resistance in common bean susceptible to halo blight bacterial pathogen after seed bacterization with rhizosphere Pseudomonads, J. Gen. Appl. Microbiol. 37:495-501.

Anderson, A.J. and Jasalavich, C. 1979, Agglutination of Pseudomonad cells by plant products, Physiol. Plant Pathol. 15:149-159.

Anderson, A.J. 1983, Isolation from root and shoot surfaces of agglutinins that show specificity for saprophytic Pseudomonads, Can. J. Bot. 61:3438-3443.

Anderson, A.J. and Guerra, D. 1985, Response of bean to root colonization with *Pseudomonas putida* in a hydroponic system, Phytopathol. 75:992-995.

Anderson, A.J., Habibzadegah, Tari, P. and Tepper, C.S. 1988, Molecular studies on the role of a root surface agglutinin in adherros and colonization by *Pseudomons putida*, Appl. Environ. Microbiol. 54 : 375-380.

Arima, K., Imanaka, H., Kousaka, M., Fukuda, A. and Tamura, G. 1964, Pyrrol- nitrin a new antibiotic substance produced by *Pseudomonas*, Agr. Biol. Chem. 28: 575-576.

Azegami, K., Nishiyama, K. and Koto, H. 1988, Effect of iron limitation on *Pseudomonas plantarii* growth and tropolone and protein production, Appl. Environ. Microbiol. 54:844-847.

Baker, K.F. 1987, Evolving concepts of biological control of plant pathogens, Ann. Rev. Phytopathology 26:67-85.

Bakker, P.A.H.M., Lamers, J.G., Bakker, A.N., Marugg, J.D., Weisbeek, P.T. and Schippers, B. 1986, The role of siderophores in potato tuber yield increase by *Pseudomonas putida* in a short rotation of potato, Neth. J. Plant Pathol. 92:249-256.

Bakker, A.W. and Schipper, B. 1987, Cyanide production in the rhizosphere in relation to potato yield reduction and *Pseudomonas* spp., Soil Biol. Biochem. 19:451-457.

Bangera, M.G. and Thomashow, L.S. 1996, Characterization of a genomic locus required for synthesis of the antibiotic 2,4-diacetylphloroglucinol by the biological agent *Pseudomonas fluorescens* Q2-87, Mol. Plant-Microbe Interact. 9:83-90.

Bar-Ness, E., Chen, Y., Hadar, Y., Marschner, H. and Romheld, V. 1991, Siderophores of *Pseudomonas putida* as an iron source for dicot and monocot plants, in : Iron Nutrition and Interactions in Plants, Y. Chen and Y. Hada, eds., Kluwer Academic Publishers, Dordrecht, The Netherlands , pp.271-281.

Bar-Ness, E., Hadar, Y., Shanzer, A. and Libman, J. 1992, Iron uptake by plants from microbial siderophores,

A study with 7-nitrobenz-2 oxa-1,3-diazole-des-ferrioxamine as fluorescent ferrioxamine B analog, Plant Physiol. 99:1329-1335.

Bell, A.A. and Wheeler, M.H. 1986, Biosynthesis and functions of melanins, Ann. Rev. Phytopathol. 24:411-451.

Bleeker, A.B., Estelle, M.A., Somerville, C., Kende, H. 1988, Insensitivity to ethylene conferred by a dominant mutation in *Arabidopsis thaliana*, Science 241:1086-89.

Boller, T. 1990, Ethylene and plant-pathogen interactions, Curr. Top. Plant Physiol. 5:138-145.

Bonsall, R.F., Weller, D.M.and Thomashow, L.S. 1997, Quantification of 2,4-Diacetylphloroglucinol produced by fluorescent *Pseudomonas* spp. *in vitro* and in the rhizosphere of wheat, Appl. Environ. Microbiol. 63:951-955.

Boronin, A.M., Movrodi, D.V., Ksenzenko, V.N., Cook, R.J.and Thomashow, L.C.1995, Characterization of genes involved in phenazine biosynthesis in plant-growth promoting *Pseudomonas fluorescens* 2-70, in : 5th International Symposium on *Pseudomonas* (Abstr.), Mo. Biol. and Biotech.

Briat, J.F. 1992, Iron assimilation and storage in Prokaryotes, J. Gen. Microbiol. 138:2475-2483.

Brisbane, P.G. and Rovira, A.D. 1988, Mechanisms of inhibition of *Gaeumannomyces graminis* var. *tritici* by fluorescent pseudomonads, Plant Pathol. 37:104-111.

Broek, A.V. and Vanderleyden, J. 1995, The role of bacterial motility, chemotaxis and attachment in bacteria plant interactions, Mol. Plant-Microbe Interact. 8:800-810.

Buell, C.R. and Anderson, A.J. 1992, Genetic analysis of the agg A locus involved in agglutination and adherence of *Pseudomonas putida*, a beneficial fluorescent Pseudomonad, Mol. Plant-Microbe Interact. 5:154-162.

Buell, C.R. and Anderson, A.J. 1993, Expression of aggA locus of *Pseudomonas putida in vitro* and in planta as reported by the reporter gene, xyl E, Mol. Plant-Microbe Interact. 6:331-340.

Buell, C.R., Whetton, R., Tari, P. and Anderson, A.J. 1993, Characterization of cell surface properties in agglutinable and non agglutinable mutants of *Pseudomonas putida*, Can. J. Microbiol. 39: 787-794.

Bull, C.T., Weller, D.M. and Thomashow, L.S. 1991, Relationship between root colonization and suppression of *Gaeumannomyces graminis* var. tritici by *Pseudomonas fluorescens* strain 2-79, Phytopathology 81:954-959.

Buyer, J.S., Kratzke, M.G. and Sikora, L.J. 1993, A method for the detection of pseudobactin, the siderophore produced by a plant growth promoting *Pseudomonas* strain, in the barley rhizosphere, Appl. Environ. Microbiol. 59:677-681.

Buysens, S., Poppe, J. and Hofte, M. 1994, Role of siderophores in plant growth stimulation and antagonism by *Pseudomonas aeruginosa* 7NSK2, in: Iimproving Plant Productivity with Rhizosphere Bacteria, M.H. Ryder, P.M. Stephens and G.D. Bowon, eds., CSIRO Adelaide pp:139-141.

Cangelosi, G.A., Hung, L., Puvanesarajah, V., Stacey, G., Ozaga, D.A., Leigh, J.A. and Nester, E.W. 1987, Common loci for *Agrobacterium tumefaciens* and *Rhizobium meliloti* exopolysaccharides synthesis and their role in plant interactions, J. Bacteriol. 169:2086-2091.

Capper, A.L. and Higgins, K.P. 1993, Application of *Pseudomonas fluorescens* isolates to wheat as potential biological control agents against take all, Plant Pathol. 42: 560-570.

Carni, R., Carmeli, S., Levy, E. and Gough, F.J. 1994, (+) - (S) - Dihydroaeruginoic acid, an inhibitor of *Septoria tritici* and other phytopathogenic fungi and bacteria, produced by *Pseudomonas fluorescens*, J. Nat. Prod. 57:1200-1205.

Carruthers, F.L., Conner, A.J. and Mahanthy, H.K. 1994, Identification of a genetic locus in *Pseudomonas aureofaciens* involved in fungal inhibition, Appl. Environ. Microbiol. 60:71-77.

Castignetti, O. and Smarrelli, J. Jr. 1986, Siderophores the iron nutrition of plants and nitrate reductase, FEBS Lett. 209:147-151.

Cactric, P. 1977, Glycine metabolism by *Pseudomonas aeruginosa* : Hydrogen cyanide biosynthesis, J. Bacteriol. 130:826-831.

Chang, C., Kwok, S.F., Blucker, A.B. and Meyerowitz, E.M. 1993, Arabidopsis ethylene - response gene ETR-1 similarity of product to two-component regulators, Science 262:539-544.

Chet, I. and Inbar, J. 1994, Biological control of fungal pathogens, Appl. Biochem. Biotechnol. 48:37-43.

Cook, R.J., Weller, D.M. and Bassett, E.N. 1988, Effect of bacterial seed treatments on growth of recropped wheat in western Washington, Biol. Cult. Tests Control Plant Dis. 3:53.

Cook, R.J. 1994, Problems and progress in the biological control of wheat take all, Plant Pathol. 43:429-437.

Corbell, N. and Loper, J.E. 1995, A global regulator of secondary metabolite production in *Pseudomonas fluorescens* pf-5, J. Bacteriol. 177:6230-6236.

Crowley, D.E., Reid, C.P.P. and Szaniszlo, P.J. 1988, Utilization of microbial siderophores in iron acquisition by oat, Plant Physiol. 87:680-685.

Cox, C.D. 1980, Iron uptake with ferripyochelin and ferric citrate by *Pseudomonas aeruginosa*, J. Bacteriol. 169:2207-2214.

Darvill, A.G. and Albersheim, P. 1984, Phytoalexins and their elicitors - A defence against microbial infection

in plants, Annu. Rev. Plant Physiol. 35:243-275.
Defago, G., Berling, C.H., Burger, U., Hass, D., Kahr, G., Keel, C., Voisard, C., Writhner, P. and Wuthrich, B. 1990, Suppression of black rot of tobacco and other root disease by strains of *Pseudomonas fluorescens*, Potential applications and mechanisms, in : Biological Control of Soil Borne Plant Pathogens, D.Hornby, ed., Oxon CAB International pp. 93-108.
Demange, P., Wendenbaum, S., Bateman, A., Dell, A. and Abdallah, M.A. 1987, Bacterial Siderophores : Structure and physicochemical properties of pyoverdins and related compounds, in : Iron Transport in Microbes, Plants and Animals, G.Winkleman, D. Vander Helm and J.B. Neilands, eds., VCH Chemie, Weinheim, Germany.
De Mot, R. and Vanderleyden, J. 1991, Purification of a root adhesive outer membrane protein of root colonizing *Pseudomonas fluorescens*, FEMS Microbiol. Lett. 81:323-328.
de Weger, L.A., Lann, B. and Luthenberg, B. 1987, Lipopolysaccharides of *Pseudomonas* spp. that stimulate plant growth : Composition and use for strain identification, J. Bacteriol. 169:1441-1446.
de Weger, L.A., Van Loosdrecht, M.C.M., Klaasen, H.E. and Lugtenberg, B. 1989, Mutational changes in physicochemical surface properties of plant growth stimulating *Pseudomonas* spp. do not influence the attachment properties of the cells, J. Bacteriol. 171:2756-2761,
de Weger, L.A., Bakker, P.A.H.M., Schippers, B., Van Looschrecht, M.C.M. and Lugtenberg, B. 1991, Mutational changes in physicochemical surface properties of plant growth stimulating *Pseudomonas* spp. do not influence the attachment properties of the cells, J. Bacteriol. 171:2756-2761.
Dietrich, R.A., Delaney, T.P., Uknes, S.J., Ward, E.R., Ryals, J.A. 1994, Arabidopsis mutants simulating disease resistance response, Cell 77:565-77.
Dixon, R.A. and Lamb, C.J. 1990, Molecular communication in interactions between plants and microbial pathogens, Ann. Rev. of Plant Physiol. Plant Mole. Biol. 41:339-367.
Doke, N. 1983, Involvement of superoxide anion generation in the response of potato tuber tissues to infection with an incompatible race of *Phytophthora infestans* and to hyphal cell wall components, Physiol. Plant Pathol. 23:345-357.
Douglas, C., Halperin, W. and Nester, E.W. 1982, *Agrobacterium tumefaciens* affected in attachment to plant cells, J. Bacteriol. 152:1265-1275.
Dowling, D.N. and O'Gara, F. 1994, Metabolites of *Pseudomonas* involved in the biocontrol of plant disease, Trends Biotech. 12:133-141.
Duffy, B.K. and Weller, D.M. 1992, Suppression of take-all by *Trichoderma koningii* used individually and in combination with fluorescent *Pseudomonas* spp. (Abstr.), Phytopathology 82:1080.
Duffy, B.K. and Weller, D.M. 1995, Use of *Gaeumannomyces graminis* var *graminis* alone and in combination with fluorescent *Pseudomonas* spp. to suppress take all of wheat, Plant Dis. 79:907-911.
Duffy, B.K., Simon, A. and Weller, D.M. 1996, Combination of *Trichoderma koningii* with fluorescent pseudomonads for control of Take-all of wheat, Phytopathology 86:188-194.
Duijff, B.J., Meijer, J.W., Bakker, P.A.H.M. and Schippers, B. 1993, Siderophore-mediated competition for iron and induced resistance in the suppression of *Fusarium* wilt of carnation by fluorescent *Pseudomonas* spp., Neth. J. Plant Pathol. 99:277-289.
Dutrecq, A.P., Debras, J., Stevaux, and Marlier, M. 1991, Activity of 2,4-diacetylphloroglucinol isolated from a strain of *Pseudomonas fluorescens* to *Gaeumannomyces graminis* var. *tritici*., in : Biotic Interactions and Soil-borne Diseases, A.B.R. Beomster, G.J. Bollen, M. Gerlagh, M.A. Ruisson, B. Schippers, and A. Tempel, eds., Elsevier, Amsterdam, The Netherlands, pp.252-257.
Elad, Y. and Baker, R. 1985, Influence of trace amounts of cations and siderophore - producing pseudomonads on chlamydospore germination of *Fusarium oxysporum*, Phytopathology 75:1047-1052.
Elad, Y. and Chet. I. 1987, Possible role of competition for nutrients in biocontrol of *Pythium* damping off by bacteria, Phytopathology 77:190-195.
Elsherif, M. and Grossmann, F. 1994, Comparative investigations on the antagonistic activity of fluorescent pseudomonads against *Gaeumannomyces graminis* var. *tritici in vitro* and *in vivo*, Microbiol. Res. 149:371-377.
Fenton, A.M., Stephens, P.M., Crowley, J., O'Callaghan, M. and O'Gara, F. 1992, Exploitation of gene(s) involved in 2,4-diacetylphloroglucinol biosynthesis confer a new biocontrol capability to a *Pseudomonas* strain, Appl. Environ. Microbiol. 58:3872-3878.
Flaishman, M., Eyal, Z., Voisard, C. and Haas, D. 1990, Suppression of *Septoria tritici* by phenazine or siderophore deficient mutants of *Pseudomonas*, Curr. Microbiol. 20:121-124.
Flaishman, M. 1992, Mechanisms associated with the suppression of the wheat pathogen *Septoria tritici* and *Puccinia recondita* by bacteria, Ph.D. thesis, Tel Aviv University, Tel Aviv, Israel.
Fridlender, M., Inbar, J. and Chet, I. 1993, Biological control of soil borne plant pahtogens by a b-1-3 gluconase-producing *Pseudomonas cepacia*, Soil. Biol. Biochem. 25:1211-1221.
Gaffney, T., Friedrich, L., Vernooij, B., Negrotto, B., Nye, G. Ukness, S., Ward, E., Kessmann, H. and Ryals, J. 1993, Requirement of salicyclic acid for the induction of systemic acquired resistance, Science

261:754-756.

Gaffney, T.D., Lam, S.T., Ligon, J., Gates, K., Frazella, A., DiMaio, J., Hill, S., Goodwin, S., Torkewitz, N., Allshouse, A.M., Kempf, H.J. and Becker, J.O. 1994, Global regulation of expresion of antifungal factors by *Pseudomonas fluorescens* biological control strain, Mol.Plant-Microbe, Interact.7:455-463.

Georgakopoulos, D.M., Hondson, N.J., Panopoulos, and Schroth, M.N. 1994, Cloning of a phenazine biosynthetic locus of *Pseudomonas aureofaciens* PGS12 and analysis of its expression *in vitro* with the ice nucleation reporter gene, Appl. Environ. Microbiol. 60:2931-2938.

Gill, P.R. and Warren, G.J. 1988, An iron antagonized fungistatic agent that is not required for iron assimilation from a flourescsent rhizosphere pseudomonads, J. Bacteriol. 170:163-170.

Glick, B.R. and Bashan, Y. 1997, Genetic manipulation of plant growth promoting bacteria to enhance biocontrol of phytopathogens, Biotechnology Adv. 15:353-378.

Gurusiddaiah, S., Weller, D.M, Sarkar, A. and Cook, R.J. 1986, Characterization of an antibiotic produced by a strain of *Pseudomonas fluorescens* inhibitory to *Gaeumannomyces graminis* var. *tritici* and *Pythium* spp. Antimicrob. Agents Chemother. 29:488-495.

Gutterson, I. 1990, Microbial fungicides : Recent approaches to elucidating mechanisms, Crit. Rev. Biotechnol. 10:69-91.

Gutterson, N.I., Layton, T.J., Ziegle,J.S. and Warren, G.J. 1986, Molecular cloning of genetic determinants for inhibition of fungal growth by a fluorescent pseudomonad, J. Bacteriol. 165:696-703

Gutterson, N., Ziegle, J.S., Warren, G.J. and Layton, T.J. 1988, Genetic determinants for catabolite induction for antibiotic biosynthesis in *Pseudomonas fluorescens* Hv 37a, J. Bacteriol. 170:380-385.

Haas, D., Keel, C., Laville, J., Maurwhofer, M., Oberhansli, J., Schnider, V., Voisard, C., Wuthrich, B. and Defago, G. 1991, Secondary metabolites of *Pseudomonas fluorescens* strain CHAO involved in the suppression of root diseases, in : Advances of Molecular Genetics of Plant-Microbe Interactions, H. Hennecke and D.P.S. Verma, eds., Kluwer Academic Publishers, Dordrecht, The Netherlands pp.450-456.

Hamdan,H., Weller, D.M. and Thomashow, L.S. 1991, Relative importance of fluorescent siderophores and other factors in biological control of *Gaeumannomyces graminis* var. *tritici* by *Pseudomonas fluorescens* 2-79 and M 4-80R, Appl. Environ. Microbiol. 57:3270-3277.

Hammer, P.E., Hill, S. and Ligon, J. 1995, Characterization of genes from *Pseudomonas fluorescens* involved in synthesis of pyrrolnitrin, Phytopathology 85:1162.

Hammer, P.E., Hill, D.S., Stephen, T.L., Van Pee, K.H. and Ligon, J.M. 1997, Four genes from *Pseudomonas fluorescens* that encode the biosynthesis of pyrrolnitrin, Appl. Environ. Microbiol. 63:2147-2153.

Hammerschmidt, R. and Kuc, J. 1982, Lignification as a mechanism for induced systemic resistance in cucumber, Physiol. Plant Pathol. 20:61-71.

Hammerschmidt, R., Nuckles, E. and Kuc, J. 1982, Association of enhanced peroxidase activity with induced systemic resistance of cucumber to *Colletotrichum lagenarium,* Physiol. Plant Pathol. 20:7382.

Hammerschmidt, R., Lamport, D.T.A. and Muldoon, E.P. 1984, Cell wall hydroxyproline enhancement and lignin deposition as an early event in the resistance of cucumber to *Cladosporium cucumerinum,* Physiol. Plant Pathol. 24:43-47.

Harrison, L.A., Letendre, L., Kovacevich, Pierson, E.A. and Weller, D.M. 1993. Purification of an antibiotic effective against *Gaeumannomyces graminis* var. *tritici* produced by a biocontrol agent *Pseudomonas aureofaciens,* Soil Biol. Biochem. 25:215-221.

Hill, D.S., Stein, J.I., Torkewitz, N.R., Morse, A.M., Howell, C.R., Pachalatko, J.P., Becker, J.O. and Ligon, J.M. 1994, Cloning of genes involved in the synthesis of pyrrolnitrin from *Pseudomonas fluorescens* and role of pyrrolnitrin in biological control of plant diseases, Appl. Environ. Microbiol. 60:78-85.

Hill, S.E., Hammer, P. and Ligon, J. 1997, The role of antifungal metabolites in biological control of plant disease, in : Technology Transfer of Plant Biotechnology, P.M.Greshoff ed., CRC Press Inc., Boca Raton, Florida.

Hoffland, E., Pieterse, C.M.J., Bik, L. and Van Pelt,J.A. 1995, Induced systemic resistance in radish is not associated with accumulation of pathogenesis - related proteins, Physiol. Mol. Plant Pathol. 46:309-20.

Homma, Y. and Suzui, T. 1989, Role of antibiotic production in suppression of raddish damping off by seed bacterization with *Pseudomonas cepacia,* Annu. Phytopathol. Soc. Japan 55:643-652.

Howell, C.R. and Stipanovic, R.D. 1979, Control of *Rhizoctonia solani* on cotton seedlings with *Pseudomonas flourescens* and with an antibiotic produced by the bacterium, Phytopathol. 69:480-482.

Howell, C.R. and Stipanovic, R.D. 1980, Suppression of *Pythium ultimum* induced damping off of cotton seedlings by *Pseudomonas fluorescens* and its antibiotic pyoluteorin, Phytopathol. 70 : 712-715.

Howie, W.J. and Suslow, T. 1991, Role of antibiotic biosynthesis in the inhibition of *Pythium ultimum* in the cotton spermosphere and rhizosphere by *Pseudomonas fluorescens,* Mol. Plant-Microbe Interact. 4 : 393-399.

Hyodo, H. 1991, Stress wound ethylene, in : The Plant Hormone Ethylene, A.K. Matto and J.C. Suttle, eds., CRC Press, Boca Raton pp. 65-80.

Iswandi, A., Bossier, P., Vandenabele, J. and Verstraete, W. 1987, Influence of inoculation density of the rhizopseudomonad strain 7 NSK 2 on the growth and composition of root microbial community of maize *(Zea mays)* and barley *(Hordeum vulgare)*, Biol. Fert. Soils 4:119-123.

Janisiewicz, W.J. and Marchi, A. 1992, Control of storage rot on various pear cultivars with a saprophytic strain of *Pseudomonas syringae*, Plant Dis. 76:555-560.

Jaselvich, C.A. and Anderson, A.J. 1981, Isolation from legume tissues of an agglutinin of saprophytic pseudomonads, Can. J. Bot. 59:264-271.

Katsuwon, J. and Anderson,A.J. 1989, Response of plant colonizing pseudomonads to hydrogen peroxide, Appl. Environ. Microbiol. 55: 2985-2989.

Katsuwon, J. and Anderson, A.J. 1990, Catalase and superoxide dismutase of root colonizing saprophytic fluorescent pseudomonads, Appl. Environ. Microbiol. 56: 3576-3582.

Katsuwon, J. and Anderson,A.J. 1992, Characterization of catalase activities in a root colonizing isolate of *Pseudomonas putida*, Can. J. Microbiol. 38:1026-1032.

Keel, C., Voisard, C., Berling, C.H., Khar, G. and Defago, G. 1989, Iron sufficiency, a prerequisite for suppression of tobacco root rot by *Pseudomonas fluorescens* strain CHAO under gnotobiotic conditions, Phytopathol. 79 :584-589.

Keel, C., Wirthner, P., Oberhansli, T.H., Voisard, C., Burger, U., Hass, D. and Defago, G. 1990, Pseudomonads as antagonists of plant pathogens in rhizosphere : Role of antibiotic 2, 4-diacetylphloroglucinol in the suppression of black rot of tobacco, Symbiosis 9:327-341.

Keel, C., Schinder, U., Maurhofer, M., Voisard, C., Laville, J., Murger, U., Wirthner, P., Hass, D. and Defago, G. 1992, Suppression of root disease by *Pseudomonas fluorescens* CHAO, Importance of bacterial secondary metabolite 2,4-Diacetyl phloroglucinol, Mol. Plant-Microbe. Interact. 5:4-13.

Keel, C., Weller, D.M., Natsch, A., Defago, G., Cook, R.J. and Thomashow, L.S. 1996, Conservation of the 2,4-diacetylphloroglucinal biosynthesis locus among fluorescent *Pseudomonas* strains from diverse geographic locations, Appl. Environ. Microbiol. 65:552-563.

Keller, H., Bonnet, P., Galiana, E., Pruvot, L., Friedrich, L., Ryals, J. and Pierse, R. 1996, Salicyclic acid mediated Elicitin-Induced systemic acquired resistance but not necrosis in tobacco, Molecular Plant-Microbe Interac. 9:696-703.

Kessmann, H., Staub, T., Hofmann, C., Maetzka, T., Herzog, J., Ward, E., Uknes, S. and Ryals, J. 1994, Induction of systemic acquired disease resistance in plants by chemicals, Annu. Rev. Phytopathol. 32:439-459.

Kloepper, J.W., Leong, J., Teintze, M. and Schroth, M.N. 1980a, Enhanced plant growth by siderophore produced by plant growth promoting rhizobacteria, Nature 286:885-886.

Kloepper, J.W., Leong, J., Teintze, M. and Schroth, M.N. 1980b, *Pseudomonas* siderophores: a mechanism explaining disesase suppressive soils, Curr. Microbiol. 4:317-320.

Kloepper, J.W. and Schroth,M.N. 1981, Relationship of *in vitro* antibiosis of plant growth promoting rhizobacteria to plant growth and the displacement of root microflora, Phytopathology 71:1020-1024.

Kloepper, J.W., Lifshitz, R. and Schroth, M.N. 1988, *Pseudomonas* inoculants to benefit plant production, ISI Atlas Sci., Animal Plant Sci. 60-64.

Kloepper, J.W., Tuzun, S. and Kuc, J.A. 1992, Proposed definitions related to induced disease resistance, Biocon Sci. Technol. 2:349-351.

Klotz, M.G., Hoffmann, R. and Novacky, A. 1989, The critical role of the hydroxyl radical in microbial infection of plants, in : Plant Membrane Transport : The Current Position, J. Dainty, M.I. Michelis, E. Marre and F.R.Coldogono, eds., Elsevier, Amsterdam pp. 657-662.

Klotz, M.G. and Hutcheson, S.W. 1992, Multiple periplasmic catalases in phytopathogenic strains of *Pseudomonas syrinage*, Appl. Environ. Microbiol. 58:2468-2473.

Koby, S., Schickler,H., Chet,I and Oppenheim, A.B. 1994, The chitinase encoding Tn7-based chi A gene endows *Pseudomonas fluorescens* with the capacity to control plant pathogens in soil, Gene 147:81-83.

Kraus, J. and Loper, J.E. 1995, Characterization of a genomic region required for production of the anti-biotic pyoluteorin by the the biological control agent *Pseudomonas fluorescens* Pf-5, Appl. Environ. Microbiol. 61:849-854.

Lam, S.T. and Gaffney, T.D. 1993, Biological activities of bacteria used in plant pathogen control, in : Biotechnology in Plant Disease Control, I. Chet, ed., Wiley Liss Inc., N.Y. pp.291-320.

Lambert, B., Leyns, F., Van Rooyen, F., Gossele, F., Papon,Y. and Swings, B. 1987, Rhizobacteria of maize and their activities, Appl. Environ. Microbiol. 53:1866-1871.

Leeman, M., Van. Pelt, J.A., Den Ouden, F.M., Heinsbroek, M., Bakker, P.A.H.M. and Schipper, B. 1995a, Induction of systemic resistance against *Fussrium* wilt of radish by lipopolysaccharides of *Pseudomonas fluorescens*, Phytopathology 85:1021-1027.

Leeman, M., Van Pelt, J.A., den Ouden, F.M., Heinsbroek, M., Bakker, P.A.H.M. and Schippers, B. 1995b, Induction of systemic resistance by *Pseudomonas fluorescens* in radish cultivars differing in susceptibility to *Fusarium* wilt, using a novel bioassay, Eur. J. Plant Pathol. 101:655-664.

Leeman, M., Den Ouden, F.M., Van Pelt, J.A., Dirkx, F.P.M., Steijl, H., Bakker, H.M. and Schippers, B. 1996, Iron availability affects induction of systemic resistance to *Fusarium* wilt of radish by *Pseudomonas fluorescens*, Phytopathology, 86:149-155.

Lemanceau, P. and Alabouvette, C. 1993, Suppression of *Fusarium* wilt by fluorescent *Pseudomonas* : Mechanisms & Application, Biocontrol Sci. and Technol. 3:219-234.

Lemanceau, P. and Alabouvette, C. 1991, Biological control of *Fusarium* diseases by fluorescent *Pseudomonas* and non pathogenic *Fusarium*, Crop. Prot. 10:279-286.

Lenanceau, P., Bakker, P.H.A.M., de Kogel, W.J., Alabouvette, C. and Schipper, B. 1992, Effect of Pseudobactin 358 production by *Pseudomonas putida* WCS358 on suppression of *Fusarium* wilt of carnation by nonpathogenic *Fusarium oxysporum* Fo47, Appl. Env. Microbiol. 58:2978-2982,

Lemanceau, P., Bakker, P.A.H.M., de Kogel, W.J., Alabouvette, C. and Schipper, B. 1993, Antagonistic effects of nonpathogenic *Fusarium oxysporum* Fo47 and Pseudobactin 358 upon pathogenic *Fusarium*, Appl. Envi. Microbiol. 59:74-82.

Leong, J. 1986, Siderophores: Their biochemistry and possible role in the biocontrol of plant pathogens, Annu. Rev. Phytopathol. 24:187-209.

Lim, H.S., Kim, S.D. 1991, *Pseudomonas stutzeri* YPL-1 genetic transformation and antifungal mechanism against *Fusarium solani* an agent of plant root rot, Appl. Environ. Microbiol. 57:510-516.

Lindsay, W.L. 1979, Chemical equilibria in soils, John Wiley & Sons, New York.

Linthorst, H.J.M., Cornelissen, B.J.C., Vankan, J.A.L., VandeRhee, M., Meuwissen, R.L.J., Gonzolez-Jaen, M.T. and Bol, J.F. 1990, Induction of plant genes by compatible and incompatible virus plant interaction, in : Recognition and Response in Plant-Virus Interactions, R.S.S. Fraser, ed., Springer-Verlag, New York, 361 pp.

Liu, L., Kloepper, J.W. and Tuzun, S. 1992, Induction of systemtic resistance against cucumber moisac virus by seed inoculation with select rhizobacterial strains (Abstr.), Phytopathology 82:1108-1109.

Liu, L., Kloepper, J.W. and Tuzun, S. 1995, Induction of systemic resistance in cucumber against *Fusarium* wilt by plant growth promoting rhizobacteria, Phytopathology 85:695-698.

Liu, L., Kloepper, J.M., Tuzun, S. 1995b, Induction of systemic aquired resistance in cucumber by plant growth promoting bacteria : duration of protection and effect of host resistance on protection and root colonization, Phytopathology 85:1064-68.

Loper, J.E. 1988, Role of fluorescent siderophore production in biological control of *Pythium ultimum* by a *Pseudomonas fluorescens* strain, Phytopathology 78:166-172.

Loper, J.E. 1990, Molecular and biochemical bases for activities of biological control agents: The role of siderophores, in : New Directions in Biological Control : Alternatives for Suppressing Agricultural Pests and Diseases, Alan R., ed., Liss. Inc. New York.

Loper, J.E. and Buyer, J.S. 1991, Siderophores in microbial interactions on plant surfaces, Molecular Plant-Microbe Interac. 4:5-13.

Matthysses, A.G. 1987, Characterization of non attaching mutants of *Agrobacterium tumefaciens*, J. Bacteriol. 169:313-323.

Mauch, F., Mauch-Mani, B. and Boller, T. 1988, Antifungal hydrolase in pea tissue, II, Inhibition of fungal growth by combinations of chitinase and B-1,3-glucanases, Plant Physiol. 88:936-942.

Maurhofer, M., Keel, C., Schnider, U., Voisard, C., Hass, D. and Defago, G, 1992, Influence of enhanced antibiotic production in *Pseudomonas fluorescens* strain CHAO on its disease suppressive capacity, Phytopathol. 82:190-195.

Maurhofer, M., Hase, C., Meuwly, P., Metraux, J.P. and Defago, G. 1994, Induction of systemic resistance of tobacco to tobacco necrosis virus by the root colonizing *Pseudomonas fluorescens* stsrain CHAO : Influence of the gac A gene and of pyoverdine production, Phytopathology 84:139-146.

Maurhofer, M., Keel, C., Haas, D. and Defago, G. 1995, Influence of plant species on disease suppression by *Pseudomonas fluorescens* strain CHAO with enhanced antibiotic production, Plant Pathol. 44:40-50.

May, R., Volksch, B. and Kampmann, G. 1997, Antagonistic activities of epiphytic bacteria from soybean leaves against *Pseudomonas syringane* pv. *glycinea in vitro* and in planta, Microb. Ecol. 34:118-124.

Metraux, J.P. Ahl-Goy, P., Staub, T., Speich, J., Steinemann, A., Ryals, J. and Ward, E. 1991, Induced systemic resistance in cucumber in responses to 2,6-dichloro-isonicotinic acid and pathogens, in : Advances in Molecular Genetics of Plant-Microbe Interactions, Vol.1, H. Hennecke and P.S. Verma, eds., Kluwer Academic Publishers, Dordrecht, Netherlands.

Meyer, J.M., Azelvandre, P. and Georges, C. 1992, Iron metabolism in *Pseudomonas* salicylic acid, a siderophore of *Pseudomonas fluorescens* CHAO, Biofector. 4:23-27.

Miller, C.D., Kim, Y.C. and Anderson, A.J. 1997, Cloning and mutational analysis of the gene for the stationary

phase inducible catalase (cat C) from *Pseudomonas putida*, J. Bacterio. 179:5241-5245.
Miller, G.W., Pushnik, J.C., Brown, J.C., Emery, T.E., Jolley, V.D. and Arnick, K.Y. 1985, Uptake and translocation of iron from ferrated rhodotorulic acid in tomato, J. Plant Nutr. 8:249-264.
Misaghi, I.J., Stowell, L.J., Grogan, R.G. and Spearman, L.C. 1982, Fungistatic activity of water-soluble fluorescent pigments of fluorescent pseudomads, Phytopathology 72:33-36.
Neilands, J.B. 1982, Microbial envelope proteins related to iron, Annu. Rev. Microbiol. 36:285-309.
Neilands, J.B. 1981, Microbial iron compounds, Ann. Rev. Biochem. 50:715-731.
Neilands, J.B., and Leong, S.A. 1986, Siderophores in relation to plant growth and disease, Annu. Rev. Plant Physiol. 37:187-208.
Neilands, J.B., Konopka, K., Schwyn, B., Coy, M., Francis, R.T., Paw, B.H. and Bagg, A. 1987, Comparative biochemistry of microbial iron assimilation, in : Iron Transport in Microbes, Plants and Animals, G. Winkelmann, D. Vander Helm and J.B. Neiland, eds., pp 3-33, Verlagsgesells chaft mbH, Weinheim.
Nowak- Thompson, B., Gould, S.J., Kraus, J. and Loper, J.E. 1994, Production of 2,4-diacetylphoroglucinol by the biocontrol agent *Pseudomonas fluorescens* pf-5, Can. J. Microbiol. 40:1064-1066.
Orlando, J.A. and Neilands, J.B. 1982, Ferrichrome compounds as a source of iron for higher plants, in : Chemistry and Biology of Hydroxamic Acids, H. Kehl and S. Karger, eds., Basel, Switzerland.
O'Sullivan, D.J. and O'Gara, F. 1991, Traits of fluorescent *Pseudomonas* spp. involved in suppression of plant root pathogens, Microbiol. Rev. 56:662-676.
Ownley, B.H., Weller, D.M. and Thomashow, L.S. 1992, Influence of *in situ* and *in vitro* pH on suppression of *Gaeumannomyces graminis* var. *tritici* by *Pseudomonas fluorescens* 2-79, Phytopathology 82:178-184.
Park, C.S., Paulitz, T.C. and Baker, R. 1988, Biocontrol of *Fusarium* wilt of cucumber resulting from interaction between *Pseudomonas putida* and non pathogenic *Fusarium oxysporum*, Phytopathology 78:190-194.
Paulitz, T.C., Ahmad, J.S. and Baker, R. 1990, Integration of *Pythium nunn* and *Trichoderma harzianum* isolate T-95 for biological control of *Pythium* damping off of cucumber, Plant Soil 121:243-250.
Pfender, W.F., Kraus, P. and Loper, J.E. 1993, A genomic region from *Pseudomonas fluorescens* PF-5 required for *Pyrrolnitrin* production reduction and inhibition of *Pyrenophore tritici-vepentis* in wheat straw, Phytopathology 83:1223-1228.
Pieterse, C.M.J., Van Wees, S.C.M., Hoffland, E., Van Pelt, J.A. and Van Loon, L.C. 1996, Systemic resistance in arabidopsis induced by biocontrol bacteria is independent of salicylic acid accumulation and pathogenesis related gene expression, Plant cell 8:1225-1237.
Pierson, E.A. and Weller, D.M. 1994, Use of mixtures of fluorescent pseudomonads to suppress take all and improve the growth of wheat, Phytopathology 84:940-947.
Pierson, L.S. and Thomashow, L.S. 1992, Cloning and heterologous expression of the phenazine biosynthetic locus from *Pseudomonas aureofaciens* 30-84, Mol. Plant-Microbe Interact. 5:330-339.
Pierson L.S., Gaffney, T., Lam, S. and Gong, F. 1995, Molecular analysis of genes encoding penazine biosynthesis in the biological control bacterium *Pseudomonas aureofaciens* 30-84, FEMS Microbiol. Lett. 134:299-307.
Poplawsky, A.R., Peng, Y.F. and Ellingboe, A.H. 1988, Genetics of antibiosis in bacterial strains suppressive to take-all, Phytopathol. 78: 423-432.
Raaijmakers, J.M., Weller, D.M. and Thomashow, L.S. 1997, Frequency of antibiotic producing *Pseudomonas* spp. in natural environments, Appl. Environ. Microbiol. 63:881-887.
Rosales, A.M., Thomashow, L., Cook, R.J. and Mew, T.W. 1995, Isolation and identification of antifungal metabolites produced by rice-associated antagonistic *Pseudomonas* spp., Phytopathology 85:1028-1032.
Sarniguet, A., Kraus, J., Henkels, M.D., Muehlchen, A.M. and Loper, J. 1995, The sigma factor affects antibiotic production and biological control activity of *Pseudomonas fluorescens* Pf-5, Proc. Natl. Acad. Sci., USA 92:12255-12259.
Scher, F.M. and Baker, R. 1982, Effect of of *Pseudomonas putida* and a synthetic iron chelotor on induction of soil suppressiveness to *Fusarium* wilt pathogens, Phytopathology 72:1567-1573.
Scher, F.M., Kloepper, J.M. and Singleton, C.A. 1985, Chemotaxis of fluorescent *Pseudomonas* spp. to soybean seed exudates *in vitro* and in soil, Can. J. Microbiol. 31:570-574.
Scher, F.M. 1986, Biological control of *Fusarium* wilts by *Pseudomonas putida* and its enhancement with EDDHA, in : Iron, Siderophores and Plant Diseases, T.R. Swinburne, ed., Plenum Press, New York pp.109-117.
Scher, F.M., Kloepper, J.W., Singleton, C.A., Zaleska, I. and Laliberte, M. 1988, Colonization of soybean roots by *Pseudomonas* and *Serratia* species : relationship to bacterial motility, chemotaxis and generation time, Phytopathol. 78:1055-1059.
Schnider, U., Keel, C., Blumer, C., Troxler, J., Defago, G. and Haas, D. 1995a, Amplification of house keeping sigma factor in *Pseudomonas fluorescens* CHAO enhances antibiotic production and improves

biocontrol abilities, J. Bacteriol. 177:5387-5392.
Schinder, U., Keel, C., Voisard, C., Defago, G. and Haas, D. 1995b, Tn5-directed cloning of pqq genes from *Pseudomonas fluorescens* CHAO:mutational inactivation of the genes results in overproduction of antibiotic pyluteorin, Appl. Environ. Microbiol. 61:3856-3864.
Schippers, B., Bakker, A.W. and Bakker, A.H.M. 1987, Interactions of deleterious and beneficial rhizosphere microorganisms and the effect of cropping practice, Ann. Rev. Phythopathol. 25:339-358.
Shanahan, P., Borro, A., O' Gara, F. and Glennon, J.D. 1992a, Isolation, trace enrichment and liquid chromatographic analysis of diacetyl phloroglucinol in culture and amperometric detection, J. Chromatogr. 606:171-177.
Shanahan, P., Glennon, J.D., Crowley, J.J., Donnelly, D.F. and O'Gara, F. 1992b, Isolation of 2,4-diacetylphloroglucinol from a fluorescent pseudomonad and investigation of physiological parameters influencing its production, Appl. Environ. Microbiol. 58:353-358.
Singh, L. 1978, *In vitro* screening of some chemicals against three phytopathogenic fungi, J. Indian Bot. Soc. 57:191-195.
Snch, B., Dupler, M., Elad, Y. and Baker, R. 1984, Chlamydospore germination of *Fusarium oxysporum* f. sp. *cucumerinum* as affected by fluorescent and lytic bacteria from *Fusarium* suppressive soil, Phytopathology 67:622-628.
Snch, B. 1998, Use of non pathogenic or hypovirulent fungal strains to protect plants against closely related fungal pathogens, Biotechnol. Adv. 16:1-32.
Stephens, P.M., Crowley, J.J. and O'Connell, C. 1993, Selection of Pseudomonad strains inhibiting *Pythium ultimum* on sugarbeet seeds in soil, Soil Biol. Biochem. 25:1283-1288.
Stutz, E.W., Defago, G. and Kern, H. 1986, Naturally occuring fluorescent pseudomonads involved in suppression of black rot of tobacco, Phytopathol. 76:181-185.
Sutherland, M.W. 1991, The generation of oxygen radicals during host plant response to infection, Mol. Physiol. Plant. Pathol. 39:79-93.
Tari, P.H. and Anderson, A.J. 1988, *Fusarium* wilt suppression and agglutinability of *Pseudomonas putida*, Appl. Environ. Microbiol. 54:2037-2041.
Teintze, M., Hossain, M.B., Barnes, C.L., Leong, J. and Vander Helm, D. 1981, Structure of ferric pseudobactin, a siderophore from a plant growth promoting *Pseudomonas*, Biochemistry 20:6446-6457.
Thomashow, L.S. and Weller, D.M. 1988, Role of a phenazine antibiotic from *Pseudomonas fluorescens* in biological control of *Gaeumannomyces graminis* var. *tritici*, J. Bacteriol. 170:3499-3508.
Thomashow, L., Weller, D.M., Bonsall, R.F. and Pierson III, L.S. 1990, Production of the antibiotic phenazine-1-carboxylic acid by fluorescent *Pseudomonas* species in the rhizosphere of wheat, Appl. Environ. Microbiol. 56:908-912.
Thomashow, L.S. 1991, Molecular basis of antibiosis mediated by rhizosphere Pseudomonads, in : Plant Growth Promoting Rhizobacteria, C. Keel, B. Koller and G. Defago, eds., WPRS Bulletin pp.109-114.
Thomashow, L.S., Bonsall, R.F. and Weller, D.M. 1996, Antibiotic production by soil and rhizosphere microbes *in situ*, in : Mannual of Environmental Microbiology, C.J. Hurst, G.R. Knudsen, M.J. McInerney, L.D. Stetzenbach and M.V. Walter, eds., A.S.M. Press, Washington, D.C. pp.493-499.
Thomashow, L.S. and Weller, D.M. 1996, Current concepts in the use of introduced bacteria for biological disease control : mechanisms and antifungal metabolites, in : Plant-Microbe Interactions, Vol. 1, G. Stacey and N. Keen, eds., Champman and Hall., New York, N.Y. pp.187-235.
Torkewitz, N.R. and Lam, S.T. 1991, The effect of population level of biocontrol bacteria on disease control efficacy, Phytopathol. 81:1179.
Tuzun, S. and Kloepper, J. 1994, Induced systemic resistance by plant growth promoting rhizobacteria, in : Improving Plant Productivity with Rhizosphere Bacteria, M.H. Ryder, P.M. Stoprenes and G.D. Bowen, eds., CSIRO, Adelaide pp. 104-109.
Vandenbergh, P.A. and Gonzalez, C.F. 1984, Method for protecting the growth of plants employing mutant siderophore producing strains of *Pseudomonas putida*, U.S. Patent #4, 479,936.
Van Peer, R. and Schippers, B. 1989, Plant growth responses to bacterization and rhizosphere microbial development in hydroponic cultures, Can. J. Microbiol. 35:456-463.
Van Peer, P., Punte, H.L.M., de Weger, L.A. and Schippers, B. 1990a, Characterization of root surface and endorhizosphere pseudomonads in relation to their colonization potential, Appl. Environ. Microbiol. 56:2462-2470.
Van Peer, R., Van Kuik, A.J., Rattink, H. and Schippers, B. 1990b, Control of *Fusarium* wilt of carnation growth on rock wool by *Pseudomonas* sp. strain WCS417r and by FeEDDHA, Neth. J. Plant Pathol. 96:119-132.
Van Peer, R., Niemann, G.J., Schippers, B. 1991, Induced resistance and phytoalexin accumulation in biological control of *Fusarium* wilt of carnation by *Pseudomonas* sp. strain WCS 417 r, Phytopathology 81:728-734.

Van Peer, R. and Schippers, B. 1992, Lipopolysaccharides of plant growth promoting *Pseudomonas* sp. strain WCS417 induce resistan in carnation of *Fusarium* wilt, Neth. J. Plant. Pathol. 98:129-139.

Vernooij, B., Ukness, S., Ward, E. and Ryals, J. 1994, Salicylic acid as a signal molecule in plant-pathogen interaction, Curr. Opinion Cell Biol. 6:275-279.

Vesper, S.J. 1987, Production of pili (fimbriae) by *Pseudomonas fluorescens* and correlation with attachment to corn roots, Appl. Environ. Microbiol. 53:1397-1405.

Voisard, C., Bull, C.T., Keel, C., Laville, J., Maurhofer, M., Schinider, U., Defago, G. and Haas, D. 1994, Biocontrol of root diseases by *Pseudomonas fluorescens* CHAO:current concepts and experimental approaches, in : Molecular Ecology of Rhizosphere Microorganisms, F. O'Gara, D.N. Dowling and B. Boesten, eds., VCH Weinhaim Germany pp.67-89.

Vincent, M.V., Harrison, L.A., Brackin, J.M., Kovacevic, P.A., Mukerji, P., Weller, D.M. and Pierson, E.A. 1991, Genetic analysis of the antifungal activity of a soilborne *Pseudomonas aeurofaciens* strain, Appl. Environ. Microbiol., 57:2928-2934.

Voisard, C., Keel, C., Hass, D. and Defago, G., 1989, Cyanide production by *Pseudomonas fluorescens* helps suppress black rot of tobacco under gnotobiotic system, EMBO J. 8:351-358.

Visca, P., Ciervo, A., Sanfilippo, V. and Orsi, N. 1993, Iron. regulated salicylate synthesis by *Pseudomonas* spp., J. Gen. Microbiol. 139:1995-2001.

Wang, J., Budde, A.D. and Leong, S.A. 1989, Analysis of ferrichrome biosynthesis in the phytopathogenic fungus *Ustilago maydis* : cloning of an ornithine-N-Oxygenase gene, J. Bacteriol.171:2811-2818.

Wang, Y., Brown,H.N., Crowley, D.E. and Szaniszlo, P.J. 1993, Evidence for direct utilization of a siderophore, ferrioxamine B, in axenically grown cucumber, Plant, Cell and Environ. 16:579-585.

Wei, G., Kloepper, J.W. and Tuzun, S. 1991, Induction of systemic resistance of cucumber to *Colletrichum orbiculare* by select strains of plant growth promoting rhizobacteria, Phytopathology 81:1508-1512.

Wei, G., Kloepper, J.W. and Tuzun, S. 1994, Induced systemic resistance to cucumber diseases and increased plant growth by plant growth promoting rhizobacteria under field conditions, in : Improving Plant Productivity with Rhizosphere Bacteria, M.H. Ryder, P.M. Stephens and G.D. Bowen, eds., CSIRO, Division of soils, Glen Osmond, Australia pp.70-71.

Wei, G., Kloepper, J.W. and Tuzun, S. 1996, Induced systemic resistance to cucumber diseases and increased plant growth by plant growth promoting rhizobacteria under field conditions, Phytopathology 86:221-224.

Weinberg, E.D. 1986, Regulation of secondary metabolism by trace metals, in : Cell Metabolism Growth and Environment, T.A.V. Subramanian, ed., Chemical Rubber Corporation Press, Boca Raton, FL.

Weller, D.M. and Cook, R.J. 1986, Suppression of root disease of wheat by flourescent pseudomonads and mechanisms of action, in : Iron, Siderophores and Plant Diseases, T.R.Swinburne, ed., Plenum Press, New York pp. 99-107.

Weller, D.M. 1988, Biological control of soil borne plant pathogens in the rhizosphere with bacteria, Ann. Rev. Phytopathol. 26:379-407.

Weller, D.M. and Thomashow, L.S. 1993, Use of rhizobacteria for biocontrol, Curr. Opin. Biotechnol. 4:306-311.

Weller, D.M. and Thomashow, L.S. 1994, Current challenges in introducing beneficial microorganisms into the rhizosphere, in : Molecular Ecology of Rhizosphere Microorganisms : Biotechnology and the Release of GMO's, F. O'Gara, D.N. Dowling and B. Boesteneds, eds., VCH, Weinheim, Germany 1-18.

White, R.F. 1979, Acetylsalicylic acid induces resistance to tobacco mosaic virus of tobacco, Virology 99:410-412.

Xu, G.W. and Gross, D.C. 1986, Selection of fluorescent pseudomonads antagonistic to *Erwinia carotovora* as suppressive of potato seed piece decay, Phytopathol. 76:414-422.

Yuen, G., Hendson, M., Rella, M. and Schroth, M.N. 1987, The roles of Fe competition and antibiosis in the inhibition of *Pythium ultimum* by *Pseudomonas fluorescens* biovar. IV, Phytopathol. 77:1758.

Zdor, R.E. and Anderson, A.J. 1992, Influence of root colonizing bacteria on the defense response of bean, Plant Soil 140:99-107.

Zhou, T., Rankin, L. and Paulitz, T.C. 1992, Induced resistance in the biological control of *Pythium aphanidermatum* by *Pseudomonas* spp. on European cucumber (Abstr), Phytopathology 82:1080.

Zhou, T. and Paulitz, T.C. 1994, Induced resistance in the biocontrol of *Pythium aphanidermatum* by *Pseudomonas* spp. on cucumber, J. Phytopathol. 142:51-63.

Zimmermann, A., Reimmann, C., Galimand, M. and Haas, D. 1991, Anaerobic growth and cyanide synthesis of *P. aeruginosa* depend on *anr*, a regulatory gene homologous with *fnr* of *E. coli*, Mol. Microbiol. 5:1483-1490.

MYCORRHIZA IN CONTROL OF PLANT PATHOGENS : MOLECULAR APPROACHES

K.G.Mukerji

Applied Mycology Laboratory, Department of Botany
University of Delhi, Delhi-110007, INDIA.

1. INTRODUCTION

A wide diversity of fungi form several morphological types of mycorrhizal associations. The present review is giving more emphasis on ecto and endomycorrhizae (Bhandari and Mukerji, 1993; Mukerji and Mandeep, 1997; Mukerji and Sharma, 1996). Researches on the biological control of plant pathogens has received much attention in recent years as a means of increasing crop production by avoiding a number of problems related to use of agrochemicals (e.g. environmental pollution, increasing pathogen resistance) and hence developing practices compatible with sustainable agriculture (Barea *et al.*, 1996; Cook and Baker, 1983; Hooker *et al.*, 1994; Linderman, 1994; Mukerji and Garg, 1988 a, b; Trotta *et al..*, 1996).

There is growing awareness regarding the negative repercussions of indiscriminate use of chemical pesticides which are not only toxic to human life but also lead to environmental as well as ecosystem pollution. The long term use of broad spectrum chemical pestices has been identified as one of the major cause of environmental pollution and deterioration of agricultural land and ecosystem as a whole. The need of the hour is the use of safer biocontrol measures that are easily degradeable, environmentally friendly with greater selectively, low dosage rates and less harmful effects on non target organisms.

Mycorrhizal fungi provide an effective alternative method of disease control especially for those pathogens which affect the below ground plant parts. In mycorrhizal fungi lies enormous potential for use as biological control agent for soil borne diseases as the root diseases are one of the most difficult to manage and lead to losses in disturbing proportions (Mukerji *et al.*, 1996).

Biotechnological Approaches in Biocontrol of Plant Pathogens
Edited by Mukerji *et al.*, Kluwer Academic / Plenum Publishers, New York, 1999

1.1. Biological Control

Biological control of any disease can be done by manipulation of the resident micro-organisms or by introducing antagonists to reduce the amount of inoculum or disease producing activity of pathogen (Baker and Cook, 1974; Chet, 1987; Cook and Baker, 1983, Deacon, 1983; Mukerji and Garg 1988 a,b; Sharma *et al.*, 1998). A number of conditions must be met for a natural pesticide (Biocontrol agent) to be effective at the operational level. (i) it should have no effect on other environmental components than its target pathogens; (ii) it should be non-phytotoxic at dosages that are effective for disease control; (iii) it should be stored and shipped without needs for special facilities; (iv) it should be efficient against a wide spectrum of root pathogens; and (v) its structure should be amenable to inexpensive chemical synthesis (Duchesne, 1994).

2. HOST PHYSIOLOGY

Mycorrhizal association results in altered host physiology. Suppression or promotion of a plant disease as affected by mycorrhizal development may ultimately be determined by several biotic and abiotic interacting forces involved in the pathogen-host-mycorrhiza system. Changes in soil environment, particularly in terms of its nutrient and water status, significantly affect the germination, growth, development and pathogenic behaviour of many soil organisms. Early establishment of mycorrhizal symbiont appears to be vital if the microbial system is to be employed affectively in the control of diseases (Kaushik and Mukerji, 1996).

Mycorrhizal plants develop extensive root system as compared to non-mycorrhizal plants, which ensures the plant with increased availability of nutrients, thereby helping the plant for better growth and development (Carling *et al.*, 1978; Smith *et al.*, 1986a, b; Young *et al.*, 1984). Mycorrhizal plants usually have more vascular bundles, hence lignification in the xylem is greater (Daft and Okasanya, 1973; Dehne and Schönbeck, 1979b). The size of the nuclei increases in host cells and accumulation of ER, cisterne and ribosomes in the vicinity of the fungus is known. Mycorrhizal roots show increased respiration (Schönbeck, 1979a). Mycorrhization strengthens the cell walls by increasing lignification and production of other polysaccharides. The growth of pathogen is restricted/inhibited because of this (Dehne and Schönbeck, 1979b). A stronger vascular system will increase the flow of nutrients, impart greater mechanical strength and diminish derogatory effects from vascular pathogens (Hussey and Roncadori, 1982).

3. ECTOMYCORRHIZA AND DISEASE CONTROL

Ectomycorrhizae are known to prevent growth of pathogenic fungi on root surface of host plants, thereby protecting plants from the harmful effect of pathogens (Mukerji *et al.*., 1997). Schisler and Linderman (1987) observed change in the soil microbial population in ectomycorrhizal Douglas Fir. This change in soil microflora was associated with the release of volatile substances by ectomycorrhiza. Lindermann (1988) also reported increased population of beneficial bacteria, fluorescent pseudomonads around ectomycorrhizal roots which are known to produce plant growth promoting susbstances.

Duchesne *et al.* (1987; 1988 a,b) observed suppression of *Fusarium oxysporum* f.sp. *pini* when inoculated along with ectomycorrhizal fungus *Paxillus involutus* on roots of *Pinus resinosa*. Mycorrhizal inoculation led to 41% increase in survival when compared to plant inoculated with *Fusarium*. Presence of *P. involutus* decreased sporulation rate of *F. oxysporum*. Increased survival ability of pines inoculated with *Fusarium* is a result of fungistatic activity of the rhizosphere microflora of *P. involutus* inoculated ectomycorrhizal roots. Ethanol extracts of the rhizosphere indicated that fungitoxic effects of *P. involutus* were present within three days of inoculation of Pine seedlings with *P. involutus* (Duchesne *et al.*. 1989). Kope *et al.*, (1991) isolated two antifungal compounds from a liquid medium in which *Pisolithus arhizus/ Pisolithus tinctorius* was growing. These were identified as hydroxy forms of benzoylformic and mandelic acid and given the names Pisolithin A and B respectively. Effects of ectomycorrhizal roots of *Pinus sylvestris* was studied on root rot fungi *Rhizoctonia solani* and wilt fungus *Fusarium moniliforme*. Two ectomycorrhizal fungi *Laccaria laccata* and *Pisolithus tinctorius* in plant roots were against harmful effects of these pathogens. The extracts of these fungi were inhibitory to the pathogens. Malajczuk and Sanfaliew (1984) found that forest litter stimulates growth of antagonistic microbes in the mycorrhizosphere of ectomycorrhizal roots which helps in reducing survival and infectivity of *Phytophthora cinnamomi*. Malajczuk and Mc Comb (1979) observed increased bacterial population in the mycorrhizosphere which were highly antagonistic towards *Phytophthora* and *Pythium*. Actinomycete population in rhizosphere of *Pinus radiata* were inhibitory to the growth of *Phytophthora cinnamomi*. A pathogen attacking an ectomycorrhizal root system is initially confronted externally with highly inter-woven network of fungal mycelium (mantle) and then internally with cortical cells whose cell walls are surrounded by fungal hyphae (Hartig's Net). Malajczuk (1984) observed that zoorpores of *P. cinnamomi* failed to germinate or penetrate ectomycorrhizal mantle.

Pisolithus tinctorius and *Thelephora terrestris* reduces the impact of the root pathogen *Phytophthora cinnamomi* on *Pinus* sp. (Marx, 1969, 1973). *Laccaria laccata* reduce the incidence of disease caused by *Fusarium oxysporum* in *Pseudotsuga menziesii* (Sylvia and Sinclair, 1983a) *Picea abies* (Sampangi and Perrin, 1985) and *Pinus sylvestris* (Chakravarty and Unestam, 1987a, b). Protection against fungal pathogen may be due to physical barrier caused by hyphal mantle (Marx, 1973) or by production of phenolic compounds volatile terpenes and sesquiterpenes in the plant tissue in response to the presence of mycorrhizal fungus (Sylvia and Sinclair, 1983 b). Some mycorrhizal fungi such as *Leucopaxillus cerealis* var. *picina* and *Pisolithus tinctorius* provide resistance against *Phytophthora cinnamomi* (Marx, 1975). Ectomycorrhizal fungi provides host plants (macro-symbionts) with growth harmones including auxins, gibberrellins, cytokinins and growth regulating 'B' vitamins (Suvercha *et al.*, 1991).

High peroxidase and laccase activity was observed in *Picea abies* (Norway spruce) and *Larix decidua* (Larch) in mycelia of the ectomycorrhizal fungus *Laccaria amethystea*. Most of the activities were derived from the fungus after mycorrhization. Peroxidase (H_2O_2 oxidoreductase) are very commonly present in most plants. They oxidise a variety of phenolic compounds and incorporate their phenolics into cell walls, lignification, wound healing and pathogen defence (Gasper *et al.*, 1991). Laccases (p-diphenol : O_2 oxidoreductase) are not commonly present in plants, but when present they also appear to play a role in lignification and pathogen defence (Mayer, 1987; O'Malley *et al.*, 1993). The specific supression of peroxidase suggests that the fungal symbiont is able to modify host defence response in mature mycorrhizae (Münzenberger *et al.*, 1997). Role of peroxidases is to prevent spreading of pathogens by rigidification of cell wall through the incorporation of phenolics (Moerschbacher, 1992). There is a transient increase in phenyl-alanine ammonia lyase activity, a key enzyme in the biosynthesis of plant phenolics, during the early stages of mycorrhiza formation of birch roots by *Paxillus involutus*, when hartig's net is initiated (Simoneau *et al.*, 1996). The induction of chitinases and peroxidases is strongly related to the aggresiveness of the fungal strain *Pisolithus* sp/in *Eucalyptus globosis* var. *bicostata* (Albrecht *et al.*, 1994). β-N-Acetylglucosaminidase activity was greatly increased in *Pinus sylvestris* roots when challenged with *Pisolithus tinctorius* and *Heterobasidion annosum* and not with *Phytophthora cinnamomi*. β-N-Acetylglucosaminidase might have an early response to chitinous pathogens (Hodge *et al.*, 1995). A few ectomycorrhizal fungi have been shown to produce antibiotics in pure culture (Garrido *et al.*, 1982; Kope and Fortin, 1989, 1990). This antibiotic production by ectomycorrhizal fungi *in vitro* are due to accidification of the growth substrate by the fungus and not due to specific antimicrobial metabolites (Rasanayagam and Jeffries, 1992).

3.1. Ectomycorrhizal Biocontrol Agents

Ectomycorrhizal fungi for large scale application should be developed with four main aims (Duchesne, 1994) : (i) the selection of ectomycorrhizal fungi with wide action ranges; (ii) use of ectomycorrhizal fungi in conjunction with other means of disease protection; (iii) the use of natural pesticides derived from ectomycorrhizal fungi; and (iv) genetic engineering of ectomycorrhizal fungi for enhanced disease suppression.

3.2. Selection of Ectomycorrhizal BCA's

Selection of ectomycorrhizal fungi should be done by taking isolates that display as many of the following features as possible (Duchesne *et al.*, 1989): (i) Shelf life: mass production should be readily and inexpensively achieved for long periods of time without loss of biocontrol potential; (ii) the organism's biology should be amenable to storage and transportation prior to field application; (iii) field application should require little manipulation; (iv) selected isolates should display their protective effect soon after inoculation; (v) they should compete efficiently with the indigenous soil microflora and colonise soils extensively under variable field conditioins; (vi) they should colonise roots at fast rate to form mycorrhiza then the pathogen can attack the roots; (vii) they must suppress most of the pathogenic species; and (viii) they should not disrupt normal beneficial soil process and do not disturb the microbial equilibrium.

3.3. Genetic Engineering of Ectomycorrhizal BCA's

Genetic engineering of ectomycorrhizal fungi, offers the potential to increase the efficiency of these microbes as biological control agents. Genomic transformation of *Laccaria laccata* was achieved using a marker gene encoding for resistance to the drug hygromycin B (Barret *et al.*, 1990).

4. VESICULAR ARBUSCULAR MYCORRHIZA (VAM) IN DISEASE CONTROL

VAM are widespread in nature and a fundamental component of agroecosystem. They are stable plant fungus associations, with mutual benefit, in which the fungus develops in two distinct functional phases: the intraradical phase with intercellular hyphae and specialised intracellular structures, "arbuscules", involved in exhange of substances betwen the two partners; the extraradical phase extending from the root into the soil, which channels water and minerals, mainly phosphorus to the plant (Harley and Smith, 1983, Srivastava *et al.*, 1996).

VAM plant-pathogen interactions range from mainly disease reduction (Bali and Mukerji 1991; Caron *et al.*, 1986 a,b; Davis and Menge, 1980, Graham and Menge, 1982, Kaye *et al.*, 1984; Krishna and Bagyaraj, 1983; Liu, 1995; Rosendahl and Rosendahl, 1990; Sharma and Mukerji, 1992; Smith *et al.*, 1990), to a neutral action (Baath and Hayman, 1983, Davis, 1980; McGraw and Schenck, 1981; Reddy *et al.*, 1989), and occasional increase in disease severity (Davis *et al.*, 1979). There contrasting results may be due to many factors, contributing positively or negatively to the development of the disease, as well as to the heterogenecity of the experimental designs (Linderman, 1994). VAM fungi improve plant health through the enhancement of plant resistance/tolerance against abiotic and biotic stress (Barea *et al.*, 1996).

VAM fungi impart resistance to host plants against root pathogens. Root pathogens are major limiting factors for plants growth and yield. Most widespread root-borne pathogens, belong to genera, *Phytophthora, Pythium, Fusarium, Verticillium* and *Rhizoctonia*. They kill roots or reduce their ability to absorb water and nutrients by penetrating root tissues and producing toxins. They share their mode of nutrition and habitat with mycorrhizal fungi. Hence, two categories of fungi strongly compete with one another for nutrition and space.

4.1. VAM AS Bio-protection Agent

When both the VAM fungi and pathogens are free living in the soil before penetration, pathogen often have competitive advantage because of their saprophytic ability, many of them are less strictly biotrophic than the mycorrhizal fungi. But situation is different when mycorrhizal fungus is already established, the mycorrhizal fungus does not have to compete for substrate and may prevent penetration by pathogens. Mycorrhizal root with *Glomus intraradices* exhibited reduction in root necrosis in tomato due to *Fusarium oxyoporum* f. sp. *radicis lycopersici* independent of the sequence of inoculation of the two fungi. The effect was more pronounced when *Glomus intraradices* was inoculated four weeks prior to *Fusarium oxysporum*. VAM fungi also reduced number of propagules of pathogenic fungi. Inoculation by *Glomus intraradices* increased plant dry mass and saved loss due to infection of *F. oxysporum* even when VAM fungus was inoculated later (Caron *et al.*, 1986 a,b).

Mycorrhizal tomato plants were more resistant to *Phytophthora nicotianae* var. *parasitica* than controls. Resistance was shown by a reduction in both root necrosis and plant growth inhibition caused by the pathogen (Trotta *et al.*, 1996). Results are similar to that reported earlier in tomato infected by *Fusarium* causing root rot (Caron *et al.*, 1986 a,b). Reduction in tomato root necrosis in mycorrhizal plants with *Glomus intraradices* may be due to induced resistance with biochemical modifications in the host e.g. production of defence related molecules or modifications in enzymatic activity of the pathogen (Caron *et al.*, 1985). The extensive network of external mycelium could also (i) supply other nutrients than P or (ii) prevent water shortage, so compensating for the loss of active roots rotted by the pathogen (Keen and Yoshikawa, 1983).

Presence of VAM fungi can reduce disease development in a non-VAM species. The presence of *Glomus intraradices* clearly reduced symptoms caused by *Fusarium oxysporum dianthi*, seedling mortality and growth reduction in the non VAM species – *Dianthus caryophyllus*. Reduction in disease severity was associated with reduced *Fusarium oxysporum dianthi* propagule in the soil/host and was not related to plant nutrition. The induction of disease resistance in *Dianthus caryophyllus* was related to mycorrhizal fungus or indirect microbial interactions in the soil (St Arnaud *et al.*, 1997). Effect of the mycorrhizal fungus *G. intraradices* on disease development by *Fusarium oxysporum* f. sp. *dianthi* in the non-mycorrhizal species *Dianthus caryophyllous* was studied by co-culture of carnation plants with mycorrhizal species *Tagetes petula*. Presence of VAmycorrhizal *T. petula* plants more than doubled the survival of *D. caryophyllous,* significantly reduced the disease symptoms, and decreased *Fusarium oxysporum dianthi* propagules by 4:1 in soil. Schönbeck and Dehne (1977) also observed that damage caused to cotton seedlings was less even when VAM fungi were inoculated after the infection with the pathogen *Thielaviopsis basicola*. Chlamydospore production by *T. basicola* was negatively correlated to percent mycorrhizal colonisation of tobacco and alfalfa (Baltruschat and Schönbeck, 1972 a,b), Chlamydospore reduction is attributed to high levels of free amino acids, arginine and citrulline. Growth of three strawberry cultivars Elsanta,Cambridge Fabourite and Rhapsody inoculated with *Glomus fasciculatum* or *Glomus etunicatum* were compared with non-inoculated plants. Roots of all inoculated plants were 55-70% colonised after 98 days. There was increased branching in AMF inoculated plants. The remaining plants were then inoculated with root pathogen *Phytophthora fragariae* and allowed to grow for 58 days more before harvest. There was reduced necrosis approximately by 60 and 30% respectively in Cambridge Favourite and Elasanta in AMF inoculated plants. Since Rhapsody was least susceptible there was no reduction in necrosis in AMF colonised plants. Reduction in necrosis was due to enhanced root system branching (Norman *et al.*, 1996).

Root colonisation by *Glomus fistulosum* BEG 31 in a population of outbreeding wild strawberries (*Fragaria vesca*) ranged from 42-80%. Such plants showed significant increase in fruit trusses, berries and leaves. Mycorrhizal plants produced significantly more primary and secondary runners and few lateral runners. One hundred randomly selected seeds were multiplied *in vitro* and 4 clonal populations were selected. Root colonisation in the clones varied from 16 to 75% (Mark and Cassells, 1996). Two susceptible mycorrhizal clones became resistant to *Phytophthora fragariae*. Much of the VAM fungi reduce soil borne diseases or the effects of diseases caused by fungal pathogens. Disease damage was reduced in 17 out of 32 cases (Dahne, 1982). Some indicated no effects of VAM fungus on diseases (Baath and Hayman, 1983,84; Davis, 1986; Zambolim and Schenck, 1983,84). Others reported increased disease severity (Davis and Menge 1980, 1981 Davis *et al.*, 1979).

Population of *Pseudomonas syringae* pv. *syringae* was lower in VAM then in non-VAM tomato plants (Garcia-Garrido and Occampo, 1989). Viral and other foliage pathogens are generally increased in VAmycorrhizal plants (Dehne, 1982; Jalali and Jalali, 1991; Schenck and Killam, 1978; Schönbeck, 1979).

4.2. Plant Defence Reactions

There is considerable evidence for the role of arbuscular mycorrhizal (AM) fungi in the control of root pathogens (Caron, 1989; Dehne, 1982; Hooker *et al.*, 1994; St. Arnaud *et al.*, 1995). The possible mechanism may be competition for colonisation sites, direct antibiosis, nutritional aspects and plant defence reactions (Azcon-Aguillar and Barea, 1996). Colonisation of roots by AM fungi induces biochemical changes within host tissues. These include stimulation of the phenylpropanoid pathway (Harison and Dixon 1993; Morandi *et al.*, 1984, 1996), changes in levels of aliphatic polyamines (El Ghachtouli *et al.*, 1995); activation of defence related genes (Franken and Gnadinger, 1994; Gianinazzi - Pearson *et al.*, 1992; Harrison and Dixon, 1993) and enhancement of certain hydrolase activities (Dumas Gaudot *et al.*, 1992; Spanu *et al.*, 1989); synthesis of protein of unknown function (Dumas - Gaudot *et al.*, 1994; Samara *et al.*, 1996). Plant chitanase and/or β-1,3 glucanase elicitation has been reported in AM roots (Blee and Anderson, 1996) and these are antifungal against pathogenic soil/root-borne fungi. Mycorrhiza induced chitinase isoforms appear to be a general phenomenon in AM symbiosis. These chitinases release oligosaccharide elicitors from the chitinous AM fungal cell walls which in turn stimulates the general defence responses of plants (Cordier *et al.*, 1996; Hahn *et al.*, 1989; Dumas - Gaudot *et al.*, 1996). Localised morphological lignification of endosperm cell walls and biochemical antifungal chitinase alterations in VAmycorrhizal roots increased resistance against wilt in tomato and cucumber (Dehne and Schönbeck, 1979; Dehne *et al.*, 1978). Increased gene activity could be associated with a decrease of DNA methylation. A high energy status of the host plant is indicated by increased respiration rate of VAmycorrhizal roots with higher metabolic activity which might enable plants to react more rapidly and more effectively against root pathogens (Baas *et al.*, 1989). AM roots shows increased ethylene production and DNA demethylation. A higher demethylation can be related to gene expression for higher resistance of plants against pathogens (Kraska, 1996; Dugasse *et al.*, 1996).

The expression of chitinases, β-1,3 gluconases and genes involved in the biosynthesis of isoflavonoid phytoalexins has been detected in symbiotic associations. Mycorrhizal root in comparision to non-mycorrhizal roots showed a transient induction of endochitinase activities followed by suppression under low and high P conditions (Lambais and Mehdy 1995). Arbuscular mycorrhizal (*Glomus mosseae*) bean (*Phaseolus vulgaris*) roots were less susceptible to the pathogen *Fusarium solani* f.sp. *phaseoli*. There was little change in the expression of the defence related genes chitinase, ß-1,3 glucanase and phenylalamine ammonia-lyase as compared to non-mycorrhizal control roots. A transient increase in chalcone synthatase transcripts was observed at later stages of mycorrhizal colonisation (Mohr *et al.*, 1998). VAM fungi can activate a part of the plants metabolic pathways associated with defence processes but they do so in an uncoordinated, weak, transient and/or very localised fasion (Table 1). The cellular and molecular responses induced in myc^{-1} mutants triggered by AM fungi confirm that the latter produce elicitors of defence - related mechanisms, and indicate that their expression is somehow controlled by specific plant genes during normal symbiotic interactions so that resistance reactions do not occur in outer root cells (Gianinazzi-Pearson *et al.*, 1996). It has been postulated that (pathogenesis-related) PR-related gene expression in epidermal tissues may contribute to the formation in roots of a defence barrier to pathogens (Mylona *et al.*, 1994). Products of symbiosis related plant genes, under the influence of a fungal activator, might directly limit the expression of defence genes (Gollotte *et al.*, 1993). Phytohormones, the concentration of which might change in mycorrhizal roots have been proposed as molecules possibly responsible for the signalling and suppression of defence related gene expression (McArthur and Knowles, 1992; Lambais and Mehdy, 1995). Another mechanism could depend on the induction by activated symbiosis related genes of a fungal suppressor of the plant defence system. Low expression of defence in later stages of arbuscule formation may be linked to a lack of active fungal elicitors at this morphological stage, the molecular composition of the fungal wall having become extremely simplified (Lemoine *et al.*, 1995). The role of microbial wall components in eliciting defence responses is currently being considered in relation to the perception of pathogens by plants (Collinge *et al.*, 1994). Low priming of defence genes by the symbiotic associations may predispose roots to respond more rapidly to a secondary infection and so contribute to their enhanced resistance to certain soil/root-borne pathogens (Gianinazzi, 1991). Polypeptide amylases have already shown that compatible interactions in mycorrhizal pea and tobacco roots are dominated by *de novo* gene expression (Dumas - Gaudot *et al.*, 1994; Sagan *et al.*, 1995; Samara *et al.*, 1996).

Table 1. Some plant resistance marker molecules investigated in fungus-root interactions in arbuscular mycorrhiza (after Gianinazzi - Pearson *et al.*, 1996)

Molecules	Modification	References
Phytoalexins	Late or transient increase in some isoflavonoids, PAL, CHS and CHI transcripts during root colonization. Localization of PAL and CHS transcripts in arbuscule-containing cells. No increase in IFR transcripts.	Morandi *et al.*, (1984), Harrison and Dixon (1993, 1994), Volpin *et al.*, (1994, 1995).
Callose	β-1,3 glucans in host wall at the base of arbuscule trunks.	Gollotte *et al.*, (1995), Gianninazzi-Pearson (1995).
Peroxidase	Increase in total and wall-bound activity in early stages of root colonization. No localization in arbuscule containing cells.	Spanu and Bonfante-Fasolo (1988), Gianianazzi and Gianinazzi-Pearson (1992), McArthur and Knowles (1992).
Chitinase	Early increases in transcripts and activity, generally followed by suppression in later stages of colonization. New isoforms.	Spanu *et al.*, (1989), Volpin *et al.*, (1994), Lambais and Mehdy (1993), Dumas Gaudot *et al.*, (1992a, b, 1994a).
β-1,3 glucanase	No detectable quantitative change in protein and decrease in transcripts in later stages of colonization	Dumas Gaudot *et al.*, (1989), Lambais and Mehdy (1993)
PR-1 protein	Slight increase in transcripts. Localization around living arbuscules.	Gianinazzi-Pearson *et al.*, (1992)

Tomato plants pre-colonised by *Glomus mosseae* showed decreased root damage by the pathogen *Phytophthora nicotianae* var *parasitica*. The amount of *P. nicotianae* v. *parasitica* hyphae growing in the root cortex was greatly reduced in mycorrhizal root system and in mycorrhizal tissue infected by the pathogen. No necrosis occurred. The mechanism of bioprotection may be due to (i) accumulation of plant derived compounds in response to mycorrhizal colonization like arginine (Baltruschat and Schönbeck, 1975); phenolics (Dehne *et al.*, 1978) and isoflavonoids (Morandi *et al.*, 1984); (ii) improvement in mineral nutrition of host plant and more precisely phosphorus uptake (Davis and Menge, 1980; Graham and Menge, 1982). Recent ultrastructural and immunocytochemical studies suggest that enhanced protection of mycorrhizal Ri T-DNA transformed carrot roots against infection by *Fusarium oxysporum* f. sp. *chrysanthemi* involves accumulation of electron dense deposits of phenolics and hydrolytic enzymes which restricts pathogen progression (Benhamou *et al.*, 1994). Necrosis did not occur in arbuscule containing cortical cells by intercellular hyphae of pathogen *P. nicotianae* var. *parasitica* with weakly accumulated phenolic components. This suggests that mycorrhizal effects on disease/pathogen are not only to reduce the spread of overall pathogen development, but also to specifically limit host tissue damage caused by the pathogen, where the mycorrhizal fungus is present (Cordier *et al.*, 1996).

Diseases caused by bacterial pathogens are decreased by VAM, while diseases caused by viruses or many foliage pathogens are often increased because they thrive in or on plants made healthier by VAM (Dehne, 1987; Schönbeck, 1979; Linderman and Pfleger, 1994). There are many reports of the incidence and severity of disease caused by plant pathogens in mycorrhizal plants. The effects are variable because of changed plant nutrition, relative density of the pathogen inoculum, mycorrhizal fungus and whether or not the plants were mycorrhizal before they faced propagules of the pathogen (Fitter and Garbaye, 1994; Graham, 1988; Harley and Smith, 1983; Linderman, 1992; Smith and Read, 1997). Mycorrhizal fungi also produce antimicrobial compounds, which directly effects pathogen development. Negative effect of VAM fungal hyphae on *Fusarium* sp. has been seen in mycorrhizal carrot roots (Benhamou *et al.*, 1994).

4.3. Root Exudate

Plant root exudate, although a small proportion of the total rhizo-deposition, are important as they influence soil, soil nutrient availability both directly and indirectly through the activity of the microbial biomass (Helal and Sauerbeck, 1984; Jones *et al.*, 1994). Carbon losses due to exudation are small compared to other processes going in the plant like respiration for growth and maintenance (Lambers, 1987). For the microbial biomass in the rhizosphere however, exudates in addition to secretions, lysates and gases released from the plant root, provide an available substrate to support growth (Bowen and Rovira, 1991; Hodge *et al.*, 1996; Lynch and Whipps, 1990). VAM fungi improve soil aggregation and inhibit some soil bacteria (Andrade *et al.*, 1995; Bethlenfalvay *et al.*, 1997). Root exudates change rhizosphere microflora in response to plant nutrient stress (Jakobson and Rosendahl, 1990); fluctuations in soil aggregation with bacterial populations (Roldan *et al.*, 1994) and nutrient allocation to above and below ground plant organs (Kothari *et al.*, 1990). Soil bacteria and VAM fungi may form preferential associations (Schreiner *et al.*, 1997).

The concept of mycorrhizosphere implies that mycorrhiza significantly influence the microflora of the rhizosphere by altering the physiology and exudation of the roots (Bansal and Mukerji, 1995; Garbaye, 1991; Paulitz and Linderman, 1991). The net effect is that other microorganisms in the soil are affected by the change, resulting in a new rhizosphere balance.

4.3.1. Change in Microflora

VAM fungi affect the composition of bacterial communities and fungi in the mycorrhizosphere which elicits such plant responses as resistance to stress and disease (Paulitz and Linderman, 1991; Linderman, 1992; Bethlenfelway and Schüepp, 1994). Because of the changed nature of exudates, VAM fungi on hyphosphere causes increase in composition and proliferation of rhizobacteria (Andrade *et al.*, 1997). In managing rhizosphere populations to result in biological control of plant diseases, compatible VAM fungi and effective antagonists should be delivered to the production system to guarantee their dominance (Linderman, 1986, 1988, 1991).

4.3.2. VAM in Control of Nematodes

Changes in root exudation by VAmycorrhizal plants may change the attractiveness of roots to nematode pathogens (Mittal *et al.*, 1991; Mukerji *et al.*, 1996). VAM improves host plant vigor and reduces yield losses caused by nematode infection, especially in soils with low phosphorus. Physiological change in VAmycorrhizal roots could change resistance to nemtodes by increased production of inhibitory substances (Suresh *et al.*, 1985) or by changes in root exudation which could alter mycorrhizosphere population and also affect nematode population and survival. Plant growth response as a result of mycorrhiza assisted nutrition improves plant health, confor protection, prophylactic effect increase the host tolerance to nematode infection (Francl, 1993; Pinochet *et al.*, 1996). Growth of mycorrhizal seedlings of rough lemon became significantly greater than growth of non-mycorrhizal seedlings inoculated with citrus burrowing nematode *Radopholus citrophilus*, when seedlings were inoculated both with nematode and VAM fungus *Glomus etunicatum* suppression of seedling growth caused by nematode was alleviated by *Glomus etunicatum* inoculated seedlings as compared to non-mycorrhizal plants (O'Bannon and Nemec, 1979). Inoculation with two isolates of *Glomus mosseae* increased growth of banana cultivar Musa AAA plants in comparision to non-mycorrhizal plants. *Glomus mosseae* increased growth of banana plant by enhancing plast nutrition uptake (N,P,K,Ca and Mg) and suppressing nematodes (*Meloidogyne incognita)* reproduction and galling during the early stages of plant development (Jaizme-Vega *et al.*, 1997).

4.4. Phytoalexins

Mycorrhiza formation leads to a symbiotic relationship between the plants and the fungus, it has some similarities with certain pathological interactions. Also the presence of Myc$^-$ phenotype on Nod$^-$ mutants of *Pisum sativum* (Duc *et al.*, 1989) and *Medicago truncatula* (Sagan *et al.*, 1995) suggests a common mechanism shared by mycorrhizal and nodulation interactions.

Many studies indicate a beneficial effect of mycorrhiza in improving plant protection or tolerance to soil borne pathogens (Calvet *et al.*, 1995; Hwang *et al.*, 1992; Paulitz and Linderman, 1991). Increased phenolic metabolism in plant roots has been suggested as part of the mechanisms involved in biocontrol. The phenolics in relation to ectomycorrhizal infection, in particular, phytoalexins and flavonoids/isoflavonoids has attracted lot of attention (Table 2).

Increased levels of phytoalexins in mature mycorrhizal soybean roots could be elicited by the release of fungal cell wall molecules in senescent arbuscules. Several steps of mycorrhizal development can therefore be explained in relation to phenolic compounds (Morandi, 1987, 1996) :

(i) Primary contact of the mycorrhizal fungus with the roots that induces a specific defence mechanisms;
(ii) a specific suppression of phytoalexin biosynthesis;
(iii) in aging arbuscules, an elicitation of phytoalexins due to the disintegration of the arbuscule and the release of fungal cell wall components.

Table 2. Occurrence of phytoalexion and phenolic compounds in mycorrhizal compared to non-mycorrhizal roots of different plants (after Morandi 1996)

Analysed phenolics	Studied interaction	Concentration in Mycorrhizal (M) Non-mycorrhizal (NM) roots	Root fraction analysed	Reference
Lignin	*Cucumis sativus/Glomus mosseae*	+	Total	Dehne and Schönbeck, 1979
Total soluble phenols	*Lycopersicon lycopersicum/G. mosseae*	+	Total	Dehne and Schonbeck, 1979
	Arachis hypogaea/G. fasciculatum	+	Total	Krishna and Bagyaraj, 1984
	Pisum sativum/G. mosseae	+	Total	El Ghachtouli, 1995
Total soluble and bound phenolics	*Allium porrum/G. versiforme*	--	Cell wall	Cadignola *et al.*, 1989
	Ginkgo biloba/G. versiforme	--	Cell wall	Cadignola *et al.*, 1989
Bound tyrosol	*Allium porrum/G. versiforme*	--	Cell wall	Codignola *et al.*, 1989
Bound syringic acid	*Allium porrum/G. versiforme*	+ (slight)	Cell wall	Codignola *et al.*, 1989
Bound ferulic acid	*Allium porrum/G. versiforme*	+ (slight)	Cell wall	Codignola *et al.*, 1989
Bound ferulic acid	*Ginkgo biloba/G. versiforme*	+ (slight)	Cell wall	Codignola *et al.*, 1979
	Allium cepa/G. intraradices	+	Cell wall	Grandmaison *et al.*, 1993
Bound ferulic acid	*Allium cepa/G. versiforme*	+	Cell wall	Grandmaison *et al.*, 1993
Bound p-coumaric acid	*Allium cepa/G. intraradices*	+	Cell wall	Grandmaison *et al.*, 1993
	Allium cepa/G. versiforme	+	Cell wall	Gandmaison *et al*, 1993
	Ginkgo biloba/G. versiforme	--	Cell wall	Codignola *et al.*, 1989

Bound N-feruloyltyramine	*Allium cepa/G. intraradices*	+	Cell wall	Grandmainson *et al.*, 1993
	Allium cepa/G. versiforme	+	Cell wall	Grandmaison *et al.*, 1993
Soluble N-feruloyltyramine	*Allium cepalG. intraradices*	-	Total	Grandmaison *et al.*, 1993
	Allium cepa/G. versiforme	-	Total	Grandmaison *et al.*, 1993
Blumenin	*Hordeum vulgae/G. intraradices*	+++	Total	Maier *et al.*, 1995
	Triticum aestivum/G. intraradices	+++	Total	Maier *et al.*, 1995
	Secale cereale/G. intraradices	+++	Total	Maier *et al.*, 1995
	Avena sativa/G. intraradices	+++	Total	Maier *et al.*, 1995
Isoflavnoid	*Glycine max/G. fasciculatum*	+	Total	Morandi *et al.*, 1984
	Glycine max/G. mosseae	+	Total	Morandi *et al.*, 1984
	Glycine max/G. mosseae+G. fasciculatum	+	Total	Morandi *et al.*, 1984
	Glycine max/G. fasciculatum	--untill 6 weeks, + later	Total	Morandi and Gianinzzi Pearson, 1986
	Glycine max/G. intraradices	+	Total	Morandi *et al.*, 1991
Glyceollin I+II+III	*Glycine max/G. mosseae*	+	Total	Morandi, 1989
	Glycine max/G. fasciculatum	+	Total	Morandi, 1989
	Glycine max/G. intraradices	+	Total	Morandi, 1989
	Glycine max/G. mosseae	--	Total	Wyss *et al.*, 1989
	Glycine max/G. mosseae	--(low and high P levels)	Total	Wyss *et al.*, 1989
Coumestrol	*Glycine max/G. fasciculatum*	+	Total	Morandi *et al.*, 1984
	Glycine max/G. mosseae	+	Total	Morandi *et al.*, 1984
	Glycine max/G. mosseae+G. fasciculatum	+	Total	Morandi *et al.*, 1984
	Glycine max/G.	+ untill 6 weeks,	Total	Morandi and

	fasciculatum	-- later		Gianniazzi Pearson 1986
	Glycine max/G. intraradices	+	Total	Morandi *et al.*, 1991
	Glycine max/G. fasciculatum	+ at 9 weeks	Total	Morandi, 1989
	Glycine max/G. mosseae	-- at 10 weeks	Total	Morandi, 1989
	Glycine max/G. intraradices	+ at 7 weeks, - at 10 weeks	Total	Morandi, 1989
	Medicago truncatula/G. vesiforme	+	Total	Harrison and Dixon, 1993
Coumestrol	*Medicago sativa/G. versiforme*	+	Total	Harrison and Dixon, 1993
Daidzein	*Glycine max/G. fasciculatum*	+	Total	Morandi *et al.*, 1984
	Glycine max/G. mosseae	+	Total	Morandi *et al.*, 1984
	Glycine max/G. mosseae+G. fasciculatum	+	Total	Morandi *et al.*, 1984
	Glycine max/G. fasciculatum	+ at 9 weeks	Total	Morandi, 1989
	Glycine max/G. fasciculatum	-- at 10 weeks	Total	Morandi, 1989
	Glycine max/G. intraradices	-- at 7 weeks, - at 10 weeks	Total	Morandi, 1989
	Glycine max/G. intraradices	--	Total	Morandi, and Le Quere, 1991
	Medicago truncatula/G. versiforme	+	Total	Harrison and Dixon, 1993
	Medicago sativa/G. versiforme	+	Total	Harrison and Dixon, 1993
	Vigna unguiculata/G. fasciculatum	+	Total	Sundaresan *et al.*, 1993
Isosojagol	*Glycine max/G. intraradices*	++	Total	Morandi and Le Quere, 1991
Wyerone	*Vicia faba/Glomus* mixture*	+ (8 weeks)	Total	Kape *et al.*, 1992
Medicarpin	*Medicago truncatula/G. versiforme*	++	Total	Harrison and Dixon, 1993

Medicarpin	*Medicago sativa/G. versiforme*	+	Total	Harrison and Dixon, 1993
Formononetin	*Medicago truncatula/G. versiforme*	++	Total	Harrison and Dixon, 1993
	Medicago sativa/G. versiforme	+	Total	Harrison and Dixon, 1993
	Medicago sativa/G. intraradices	+	Total	Volpin *et al.*, 1994, 1995
Formononetin malonyl glucoside	*Medicago truncatula/G. versiforme*	+	Total	Harrison and Dixon, 1993
	Medicago sativa/G. versiforme	-	Total	Harrison and Dixon, 1993
Formononetin-7-O-glucoside	*Medicago sativa/G. intraradices*	--	Total	Volpin *et al.*, 1994
	Medicago sativa/G. intraradices	-- (untill 20 days), - later	Total	Volpin *et al.*, 1995
Medicarpin malonyl glucoside	*Medicago truncatula/G. versiforme*	+	Total	Harrison and Dixon, 1993
	Medicago sativa/G. versiforme	+	Total	Harrison and Dixon, 1993
Medicarpin-3-0-glucoside 4', 7-dihydroxyflavone	*Medicago sativa/G. intraradices*	-- (untill 20 days), - later	Total	Volpin *et al.*, 1995
	Medicago truncatula/G. versiforme	++	Total	Harrison and Dixon, 1993
	Medicago truncatula/G. versiforme	++	Total	Harrison and Dixon, 1993
Medicarpin-3-0-glucoside 4', 7-dihydroxyflavone	*Medicago sativa/G. versiforme*	++	Total	Harrison and Dixon, 1993

* *Glomus mosseae* + *G. macrocarpum* + *G. aggregatum*
M : + low ++ medium +++ high; NM : - Medium -- low

4.5. VAM: Effective Biocontrol Agents

Biocontrol of plant diseases may be strongly influenced by VAM fungi by one or more mechanisms (Linderman, 1994, Sharma *et al.*, 1998) : (i) enhanced nutrition; (ii) competition for host photosynthate and infection/colonisation sites; (iii) morphological change in roots and root tissues; (iv) changes in bio-chemical constituents of plant related to plant resistance response; (v) reduction of abiotic stresses and (vi) microbial changes in the mycorrhizosphere, (vii) alleviation of physical stresses.

The potency of VAM in biocontrol depends on (i) timing and extent of VAM formation/ colonisation; (ii) inoculum level of the pathogen-inoculum potential; (iii) variation in VAM fungi, host genotype and chemical and microbial composition of soil.

5. CONCLUSIONS

Different types of mycorrhizae may differ in their ability to control plant diseases. For useful results, suitable host symbiont combination has to be found so as to produce desirable effect of improving plant growth and inducing resistance to pathogens in given environmental conditions. Since pre-inoculation of plants with VAM fungi has been found effective in controlling severity of disease, efforts should be directed to find out suitable techniques for large scale inoculum production which could be commercially exploited. The use of mycorrhizal fungi provides an effective alternative method of disease control particularly in soil and root borne diseases. In the biological control of diseases with help of mycorrhizal fungi, the results being host mediated provide some hope that in future mycorrhizal fungi can be used as a potential biocontrol agent.

REFERENCES

Albrecht, C., Burgess, T., Dell, B. and Fapeyrie, F. 1994, Chitinase and peroxidase activities are induced in eucalyptus roots according to aggressiveness of Australian ectomycorrhizal strains of *Pisolithus* sp. New Phytol. 127 : 217-222.

Andrade, G., Azcon, R. and Bethlenfalvay, G.J. 1995, A rhizobacterium modifies plant and soil responses to the mycorrhizal fungus *Glomus mosseae*, Appl. Soil. Ecol. 2 : 195-202.

Andrade, G., Mihara, K.L., Linderman, R.G. and Bethlenfalway, G.J. 1997, Bacteria from rhizosphere and hyphosphere soils of different arbuscular-mycorrhizal fungi, Plant and Soil 192 : 71-79.

Azcon - Aguilar, C. and Barea, J.M. (eds.) 1996, Mycorrhizas in integrated system from genes to plant development, Kluwer Academic Publishers, Dordrecht, The Netherlands.

Baas, R., vander Werf, A. and Lambers, H, 1989, Root respiration and growth in *Plantago major* as affected by vesicular-arbuscular mycorrhizal infection, Plant Physiol. 91 : 227-232.

Baath, E. and Hayman, D.S. 1983, Plant growth response to vesicular arbuscular mycorrhiza, XIV, Interactions with *Verticillium* wilt on tomato plants, New Phytol. 95 : 419-426.

Baath, E. and Hayman, D.S. 1984, No effect of VA mycorrhiza on red core disease of strawberry, Trans. Br. Mycol. Soc. 82 : 534-536.

Baker, K.F. and Cook, R.J. 1974, Biological Control of plant pathogens, W.H. Freeman, San Francisco, C.A.

Bali, M. and Mukerji, K.G. 1991, Interactions between VA mycorrhizal fungi and root microflora, in : Plant Roots and their Environment, B.L. McMichael and H. Persson, eds., Elsivier, Amsterdam, pp. 396-401.

Baltruschat, H. and Schönbeck, F. 1972a, Influence of endotrophic mycorrhiza on chlamydospore production of *Thielaviopsis basicola,* Phytopathol. Z. 74 : 172-183.

Baltruschat, H. and Schönbeck, F. 1972b, The influence of endotrophic mycorrhiza on the infestation of tobacco by *Thielaviopsis basicola*, Phytopathol., Z. 74 : 184-191.

Baltruschat, H. and Schönbeck, F. 1975, The influence of endotrophic mycorrhiza on the infestation of tobacco by *Thieaviopsis basicola*, Phytopathol. Z. 84 : 172-188.

Bansal, M. and Mukerji, K.G. 1995, Positive correlation between VAM induced changes in root exudation and mycorrhizosphere mycoflora, Mycorrhiza 5 : 39-44.

Barea, J.M., Calvet, C., Estaun, V. and Camprubi, A. 1996, Biological control as a key component in sustainable agriculture, Plant and Soil 185 : 171-172.
Barret, V., Dixon, R.K. and Lemke, P.A. 1990, Genetic transformation of a mycorrhizal fungus, Appl. Microbial, Biotechnol. 33 : 313-316.
Benhamou, N., Fortin, J.A., Hamel, C., St. Arnaud, M. and Shatilla, A. 1994, Resistance responses of mycorrhizal Ri T-DNA transformed carrot roots to infection by *Fusarium oxysporm* f. sp. *chrysanthemi*. Phytopatholgy 84 : 958-968.
Bethlenfalvay, G.J. and Schüepp, H. 1994, Arbuscular mycorrhizas in agrosystem stability, in : Impact of Arbuscular Mycorrhizas on Sustainable Agriculture and Natural Ecosystems, S. Ginianazzi and H. Schüepp, eds., pp.117-131, Birkhauser Verlag, Basel.
Bethlenfalvay, G.J., Andrade, G., and Azcon-Aguilar, C. 1997, Plant and soil responses to mycorrhizal fungi and rhizobacteria in nodulated or nitrate-fetilized peas, Biol. Fert. Soils (in press).
Bhandari, N.N. and Mukerji, K.G. 1993. The Haustorium, Research Studies Press Ltd., U.K. and John Wiley and Sons Inc., New York, USA, p. 308.
Blee, K.A. and Anderson, A.J. 1996, Defence related transcript accumulation in *Phaseolus vulgaris* L. colonized by the arbuscular mycorrhizal fungus *Glomus intraradices* Schenck and Smith, Plant Physiol. 110 : 675-688.
Bowen, G.D. and Rovria, A.D. 1991, The Rhizosphere : the hidden half of the hidden half, in : Plant Roots, the Hidden Half, Y. Waizel, A. Eshel and U. Kaf Kafi, eds., pp. 641-661, Marcel Dekker Inc., New York, U.S.A.
Calvet, C., Pinochet, J., Camprubi, A. and Fernandez, C. 1995, Increased tolerance to the root lesion nematode *Pratylenchus vulnus* in mycorhizal micropropagated BA-29 quince rootstock, Mycorrhiza 5 : 253-258.
Carling, D.E., Riehle, W.G., Brown, M.F. and Johnson, D.R. 1978, Effects of a vesicular-arbuscular mycorrhizal fungus on nitrate reductase and nitrogenase activities in nodulating and non-nodulating legumes, Phytopathology 68 : 1590-1597.
Caron, M. 1989, Potential use of mycorrhizae in control of soil borne diseases, Can. J. Plant Pathol. 11 : 177-179.
Caron, M., Fortin, J.A. and Richard, C. 1985, Influence of substrate on the interaction of *Glomus intraradices* and *Fusarium oxysporum* f. sp. *radicis-lycorpersici* on tomatoes, Plant and Soil 87 : 233-239.
Caron, M., Fortin, J.A. and Richard, C. 1986a, Effect of *Glomus intraradices* on infection by *Fusarium oxysporum* f. sp. *radicis-lycopersici* in tomato over a 12-week period, Can. J. Bot., 64 : 552-556.
Caron, M., Fortin, J.A. and Richard, C. 1986b, Effect of phosphorus concentration and *Glomus intraradices* on *Fusarium* crown and root rot of tomatoes, Phytopathology 76 : 942-946.
Chakravarty, P. and Unestam, T. 1987a, Mycorrhizal fungi prevent disease in stressed pine seedlings, Jour. of Phytopathology 118 : 335-340.
Chakravarty, P. and Unestam, T. 1987b, Differential influence of ectomycorrhizae on plant growth and disease resistence of *Pinus sylvestris* seedlings, Jour. of Phytopathology 120 : 104-120.
Chet, I. (ed.) 1987, Innovative approaches to plant disease control, John Wiley and Sons, N.Y., USA.
Collinge, D.B., Gregersen, P.L. and Thordal-Christensen, H. 1994, The introduction of gene expression in response to pathogenic microbes, in : Mechanisms of Plant Growth and Improved Productivity : Modern Approaches and Perspectives, A.S. Basra, ed., Marcel Dekker Inc., N.Y., USA, pp. 391-433.
Cordier, C, Gianinazzi, S. and Gianinazzi - Pearson, V. 1996, Colonization patterns of root tissues by *Phytophthora nicotianae* var. *parasitica* related to reduced disease in mycorrhizal tomato, Plant and Soil, 185 : 223-232.
Codignola, A., Verrota, L., Spanu, P., Maffei, M. Scannerini, S. and Bonfante-Fasolo, P. 1989, Cell wall bound phenols in roots of vesicular-arbuscular mycorrhizal plants, New Phytol. 112:221-228.
Cook, R.J. and Baker, K.F. 1983, The nature and practice of biological control of plant pathogens, APS Press, St. Paul, USA, 539p.
Daft, M.J. and Okasanya, B.O. 1973, Effect of endogone mycorrhiza on plant growth, V, Influence of infection on the multiplication of viruses in tomato, petunia and straw-berry, New Phytol. 72 : 975-983.
Davis, R.M. 1980, Influence of *Glomus fasciculatum* on *Thielaviopsis basicola* root rot of citrus, Plant Dis. 64 : 839-840.
Davis, R.M. and Menge, J.A. 1980, Influence of *Glomus fasciculatum* and soil phosphorus on *Phytophthora* root rot of citrus, Phytopathology 70 : 447-452.
Davis, R.M. and Menge, J.A. 1981, *Phytophthora parasitica* inoculation and intensity of vesicular-arbuscular mycorrhizae in *Citrus*, New Phytol. 87 : 705-715.
Davis, R.M., Menge, J.A. and Erwin, D.C. 1979, Influence of *Glomus fasiculatum* and soil phosphorus on *Verticillium* wilt of cotton, Phytopathology 69 : 453-456.
Deacon, J.W., 1983, Microbial control of plant pests and diseases, Am. Soc. Microbiol., Washington, D.C., USA.

Dehne, H.W. 1982, Interactions between vesicular-arbuscular mycorrhizal fungi and plant pathogens, Phytopathology 72 : 1115-1119.

Dehne, H.W. 1987, Zur Bedeutung der vesikulär-arbuskularen (VA) Mykorrhiza für die Pflanzengesundheit, Habil. Univ. Hannnover, Germany.

Dehne, W.H. and Schönbeck, F. 1978, The influence of endotrophic mycorrhiza on plant diseases, 3, Chitinase activity and Ornithine cycle, Z. Pflanzenkr. Pflanzenschutz, 82 : 666-678.

Dehne, H.W. and Schönbeck, F. 1979a, Untersuchungen Zum Einflu-β-der endotrophen Mykorrhiza auf Pflanzenkrankheiten, II, Phenolstoffwechsel und Lignitizierung, Phytopathol. Z. 5 : 210-216.

Dehne, H.W. and Schönbeck, F. 1979b, The influence of endotrophic mycorrhizae on plant disease, I, Colonization of tomato plants by *Fusarium oxysporum* f. sp. *lycopersici*, Phytopathol. Z. 95:105-109.

Dehne, H.W., Schönbeck, F. and Baltruchat, H. 1978, Untersuchungen zum Einfluß-der endotrophen Mykorrhiza auf Pflanzenkrankheiten, III, Chitinase artivität und ornithinzyklus, Z. Pflakrankh. Pflschutz. 85 : 666-678.

Duchesne, L.C. 1994, Role of ectomycorrhizal fungi in Biocontrol, in : Mycorrhizae and Plant Health, F.L. Pfleger and R.G. Linderman, eds., APS Press, St. Paul, USA, pp.27-45.

Duchesne, L.C., Peterson, R.L. and Ellis, B.E. 1987a, The accumulation of plant-produced antimicrobial components in response to ectomycorrhizal fungi : a review, Phytoprotection 68 : 17-27.

Duchesne, L.C., Peterson, R.L. and Ellis, B.E. 1987b, Pine root exudates stimulate antibiosis by *Paxillus involutus* against *Fusarium oxysporum*, in : Proc. 7th NACOM, D.M. Sylvia, L.L. Hung and J.H. Graham, eds., pp. 193, Gainesville, U.S.A.

Duchesne, L.C., Peterson, R.L. and Ellis, B.E. 1988a, Interaction beteween the ectomycorrhizal fungus *Paxillus involutus* and *Pinus resinosa* induces resistance to *Fusarium oxysporum*, Can. J. Bot., 66 : 558-562.

Duchesne, L.C., Peterson, R.L. and Ellis, B.E. 1988b, Pine root exudate stimulates antibiotic synthesis by the ectomycorrhizal fungus *Paxillus involutus*, New Phytol. 108 : 471-476.

Duchesne, L.C., Peterson, R.L. and Ellis, B.E. 1989a, The future of ectomycorrhizal fungi as biological control agents, Phytoprotection 70 : 51-58.

Duchesne, L.C., Peterson, R.L. and Ellis, B.E. 1989b, The time-course of disease suppression and antibiosis by the ectomycorrhizal fungus *Paxillus involutus*, New Phytol. 111 : 693-698.

Duc, G., Trouvelot, A., Gianinazzi-Pearson, V. and Silvia, G. 1989, First report of non-mycorrhizal plant mutants (Myc^-) obtained in pea (*Pisum sativum* L.) and fababean (*Vicia faba* L.), Pl. Sci. 60 : 215-222.

Dugassa, G.D., von Alten, H and Schönbeck F., 1996, Effect of arbuscular mycorrhiza (AM) on health of *Linum usitatissimum* L. infected by fungal pathogens, Plant and Soil 185 : 173-182.

Dumas-Gardot, E., Furlan, V., Grenier, J. and Asselin, A. 1992 a, New acidic chitinase isoforms induced in tobacco roots by vesicular-arbuscular mycorrhizal fungi, Mycorrhiza 1:133-136.

Dumas-Gaudot, E., Grenier, J., Furlan, V. and Asselin, A. 1992b, Chitinase, Chitosanase and β-1,3-glucanase activities in *Allium* and *Pisum* roots colonised by *Glomus* species, Pl. Sci. 84 : 17-24.

Dumas, E., Gianianazzi-Pearson, V. and Gianinazzi, S. 1989, Production of new soluble proteins during VA endomycorrhizae formation, Agri. Eco. Environ. 29 : 111-114.

Dumas-Gaudot, E., Asselin, A., Gianinazzi - Pearson, V., Gollotte, A. and Gianinazzi, S. 1994a, Chitinase isoforms in roots of various pea genotypes infected with arbuscular mycorrhizal fungi, Pl. Sci. 99 : 27-37.

Dumas-Gaudot, E., Guillaume, P., Tahiri-Alaoui, A., Gianinazzi-Pearson, V. and Gianinazzi, S. 1994b., Changes in polypeptide patterns in tobacco roots colonised by two *Glomus* species, Mycorrhiza 4 : 215-221.

Dumas-Gaudot, E., Slezack, S., Dassi, B., Pozo, M.J., Gianinazzi-Pearson V. and Gianinazzi, S. 1996, Plant hydrolytic enzymes (Chitnases and β-1,3-glucanases) in root reactions to pathogenic and symbiotic micro-organisms, Plant and Soil 185 : 211-221.

El Ghachtouli, N. 1995, Polyamines and mycorrhiza development in mycorrhiza-susceptible (Myc^+) and resistant (Myc^-) pea genotypes, Doctoral Thesis, Burgundy University, Dijon, France, p. 222.

El Ghachtouli, N., Paynot, M., Morandi, D., Martin-Tanguy, J. and Gianinazzi, S. 1995, The effect of polyamines on endomycorrhizal infection of wild type *Pisum sativum* cv. Frisson (nod^+ myc^+) and two mutants (nod^- myc^+ and nod^- myc^-), Mycorrhiza 5 : 189-192.

Fitter, A.H. and Garbaye, J. 1994, Interaction between mycorrhizal fungi and other soil organisms, Plant and Soil 159 : 123-132.

Francl, L.J. 1993, Interactions of nematodes with mycorrhizae and mycorrhizal fungi, in : Nematode Interactions, M.W. Khan, eds., pp. 203-216, Chapman and Hall, London.

Franken, P. and Gnadinger, F. 1994, Analysis of parsley arbuscular endomycorrhiza : Infection developent and m RNA levels of defence related genes, Mol. Plant-Microbe Interaction 7 : 612-620.

Garbaye, J. 1991, Biological interactions in the mycorrhizosphere, Experientia 47 : 370-375.

Garcia-Garrido, J.M. and Ocampo, J.A. 1989, Effect of VA mycorrhizal infection of tomato on damage caused

by *Pseudomonas syringae*, Soil Biol. Biochem. 21 : 165-167.
Garrido, N., Becerra, J., Manticorea, C., Oehrens, E., Sylvia, M. and Horak, E. 1982, Antibiotic properties of ectomycorrhizae and saprophytic fungi growing on *Pinus radiata* D. Don. I, Mycopathologia 77 : 93-98.
Gasper, T., Penel, C., Hagege, D. and Greppin, H. 1991, Peroxidases in plant growth, differentiation and development processes, in : Biochemical, Molecular and Physiological Aspects of Plant Peroxidases, J. Labarzweski, H. Greppin, C. Penel and T. Gasper, eds., University Press, Geneva, Switzerland, pp. 251-280.
Gianinazzi, S. 1991, Vesicular-arbuscular (endo) mycorrhizas : cellular, biochemical and genetic aspects, Agri., Eco. and Environ. 35 : 105-109.
Gianinazzi, S. and Gianinazzi-Pearson, V. 1992, Influence of intergeneric grafts between host and non-host legumes on vesicular arbuscular mycorrhiza formation, New Phytol. 120 : 505-508.
Gianinazzi Pearson, V. 1995, Morphological compatibility in interactions between roots and arbuscular endomycorrhizal fungi : molecular mechanisms, genes and gene expression, in : Pathogenesis and Host Specificity in Plant Disease, K. Kohomoto, R.P.Singh and R.S. Singh, eds., Oxford : Pergamon Press, Elsevier Science, Vol.II : 251-263.
Gianinazzi-Pearson, V., Dumas-Gaudot, E., Gollote, A., Tahiri-Alaoui, A. and Gianinazzi, S. 1996, Cellular and molecular defence related root responses to invasion by arbuscular mycorrhizal fungi, New Phytol. 133 : 45-57.
Gianinazzi-Pearson, V., Tahiri-Alaoui, A., Antoniw, J.F., Gianinazzi, S. and Dumas-Gaudot, E. 1992, Weak expression of the pathogenesis related PR-b 1 gene and localization of related protein during symbiotic endomycorrhizal interactions in tobacco roots, Endocytobiosis Cell Res. 8 : 177-185.
Gallotte, A., Gianinazzi-Pearson, V. and Gianinazi, S. 1995, Etude immunocytochinique des interfaces planta-champignon endomycorrhizien a arbuscular chez des pois isoginiques myc$^+$ on resistant 'a l' endomycorhization (Myc$^-$), Acta Botonica Gallica 141:449-454.
Gallotte, A., Gianinazzi-Pearson, V., Giovanneti, M., Sbrana, C., Avio, L. and Gianinazzi, S. 1993, Cellular localization and cytochemical probing of resistance reactions to arbuscular mycorrhizal fungi in a 'locus A" myc mutant of *Pisum sativum* (L.), Planta 191 : 112-122.
Graham, J.H. 1988, Interactions of mycorrhizal fungi with soil borne plant pathogens and other organisms : an introduction, Phytopathology 78 : 365-366.
Graham, J.H. and Menge, J.A. 1982, Influence of vesicular arbuscular mycorrhizae and soil phosphorus on take-all disease of wheat, Phytopathology 72 : 95-98.
Grandmaison, J., Olah, G.M., Vancalsteren, M.R. and Furlan, V. 1993, Characterisation and localization of plant phenolics likely involved in the pathogen resistance expressed by endomycorrhizal roots, Mycorrhiza, 3 : 155-164.
Hahn, M.G., Bucheli, P., Cervone, F., Doares, S.H., O'Neil, R.A., Darvill, A. and Albersheim, P. 1989, Roles of cell wall constituents in plant-pathogen interactions, in : Plant-Microbe Interactions : Molecular and Genetic Perspectives, T. Kosuge and E.W. Nester, eds., pp. 131-181, McGraw Hill, N.Y., U.S.A.
Harley, J.L. and Smith, S.E. 1983, Mycorrhizal symbiosis, Academic Press, London, U.K. 483 p.
Harrison, M.J. and Dixon, R.A. 1993, Isoflavonoid accumulation and expression of defence gene transcripts during the establishment of vesicular - arbuscular mycorrhizal associations in roots of *Medicago truncatula*, Mol. Plant-Microbe Interaction 6 : 643-654.
Harrison, M.J. and Dixon, R.A. 1994, Spatial patterns of expression of flavonoid/isoflavonoid pathway genes during interactions between *Medicago truncatula* and the mycorrhizal fungus *Glomus vesiforme*, Plant J. 6 : 9-20.
Helal, H.M. and Sauerbeck, D.R. 1984, Influence of plant roots on C and P metabolism in soil, Plant and Soil 76 : 175-182.
Hodge, A., Alexander, I.J. and Gooday, G.W. 1995, Chitinolytic activities of *Eucalyptus pilularis* and *Pinus sylvestris* root system challenged with mycorrhizal and pathogenic fungi, New Phyol. 131 : 255-261.
Hodge, A., Grayston, S.J. and Ord, B.G. 1996, A novel method for characterisation and quantification of plant root exudates, Plant and Soil 184 : 97-104.
Hooker, J.E., Jaizme-Vega, M. and Atkinson, D. 1994, Biocontrol of plant pathogens using arbuscular mycorrhizal fungi, in : Impact of Arbuscular Mycorrhizas on Sustainable Agriculture and Natural Ecosystems, S. Gianianazzi and H. Schüepp, eds., pp. 191-200, Birkhauser Verleg, Basel, Switzerland.
Hussey, R.S. and Rancadori, R.W. 1982, Vesicular-arbuscular mycorrhizae may limit nematode activity and improve plant growth, Plant Dis. 66 : 9-13.
Hwang, S.F., Chang, K.F. and Chakravarty, P. 1992, Effect of vesicular arbuscular mycorrhizal fungi on the development of *Verticillium* and *Fusarium* wilts of alfalfa, Plant Dis. 76 : 239-243.
Jaizme-Vega, M.C., Tenoury, P., Pinochet, J. and Jaumot, M. 1997, Interactions between the root knot nematode *Meloidogyne incognita* and *Glomus mosseae* in banana, Plant and Soil 196 : 27-35.
Jakobsen, I. and Rosendahl, L. 1990, Carbon flow into soil and external hyphae from roots of mycorrhizal

cucumber plants, New Phytol. 115 : 77-83.

Jalali, B.L. and Jalali, I. 1991, Mycorrhiza in plant disease control, in : Handbook of Applied Mycology, D.K. Arora, B. Rai, K.G. Mukerji and G.K. Knudsen, eds., Vol. I : 131-154, Marcel Dekker, N.Y., U.S.A.

Jones, D.L., Edwards, A.C., Donachie, K. and Darrah, P.R. 1994, Role of proteinaceous amino acids released in root exudates in nutrient acquisition from the rhizosphere, Plant and Soil 158 : 183-192.

Kape, R., Wex, K., Parniske, M., George, E., Wetzel, A. and Warner, D. 1992, Legume root metebolites and VA mycorrhiza development, J. Plant Physiol. 141: 54-60.

Kaushik, A. and Mukerji, K.G. 1996, Mycorrhiza in control of Diseases, in : Disease Scenario in Crop Plants, Vol. II, Cereals, Pulses, Oil Seeds and Cash Crops, V.P.Agnihotri, O. Prakash, R. Kishun. and A.K Mishra, eds., International Books and Periodical Supply Service, Delhi, India, pp. 217-231.

Kaye, J.W., Pfleger, F.L. and Stewart, E.L. 1984, Interaction of *Glomus fasciculatum* and *Pythium ultimum* on house grown *Poinsetia*, Can. J. Bot. 62 : 1575-1579.

Keen, N.T. and Yoshikawa, M. 1983, Physiology of disease and the nature of resistance to *Phytophthora*, in : *Phytophthora* its Biology, Taxonomy, Ecology and Pathology, D.C. Erwin, S. Bartnicki-Garcia and P.H. Tsao, eds., pp. 279-287, APS Press, St. Paul, U.S.A.

Kope, H.H. and Fortin, J.A. 1989, Inhibition of phytopathogenic fungi *in vitro* by cell free culture media of ectomycorrhizal fungi, New Phytologist 113 : 57-63.

Kope, H.H. and Fortin, J.A. 1990, Antifungal activity in culture filtrates of the ectomycorrhizal fungus *Pisolithus tinctorius*, Can. J. Bot. 68 : 1254-1259.

Kope, H.H., Tsantrizos, Y.S., Fortin, J.A. and Ogilvie, K.K. 1991, p-Hydroxybenzoylformic acid and (R)- (-) p-hydroxymandelic acid, two antifungal compounds isolated from the liquid culture of the ectomycorrhizal fungus *Pisolithus arhizus*, Can. J. Microbiol. 37 : 258-264.

Kothari, S.K., Marschner, H. and George, E. 1990, Effect of VAmycorrhizal fungi and rhizosphere microorganisms on root and shoot morphology, growth and water relations in maize, New Phytol. 116 : 303-311.

Kraska, T. 1996, Vergleichende Untersuchungen von Resistenzinduktoren, deren Wirkungsweisen und dem Einfluβ auf die DNA-Methylierung, Dissertation Univ. Hannover, Germany.

Krishna, K.R. and Bagyaraj, D. 1983, Interaction between *Glomus fasciculatum* and *Sclerotium rolfi* in peanut, Can. J. Bot. 61 : 2349-2351.

Krishna, K.R. and Bagyaraj, D.J. 1984, Phenols in mycorrhizal roots of *Arachis hypogaea*, Experientia 40 : 85-86.

Lambais, M.R. and Mehdy, M.C. 1993, Suppression of endochitinase, β-1,3-endoglucanase and chalcone isomerase expression in bean vesicular-arbuscular mycorrhizal roots under different soil phosphate conditions, Mol. Plant-Microbe Interactions 6 : 75-83.

Lambais, M.R. and Mehdy, M.C. 1995, Differential expression of defence related genes in arbuscular mycorrhiza, Can. J. Bot. 73 : 553-540.

Lambers, H. 1987, Growth, respiration, exudation and symbiotic associations, The fate of carbon translocated to the roots, in : Root Development and Function, P.J. Gregory, J.V. Lake and D.A. Rose, eds., pp. 125-145, Cambridge University Press, Cambridge, U.K.

Lemoine, M.C., Gollotte, A. and Gianinazzi-Pearson, V. 1995, Localization of β(1-3) glucan in walls of the endomycorrhizal fungi *Glomus mosseae* (Nicol. and Gerd.) Gerd, and Trappe and *Acaulospora laevis* Gerd. and Trappe during colonization of host roots, New Phytol. 129:97-105.

Linderman, R.G. 1986, Managing rhizosphere microorganisms in the production of horticultural crops, Hort. Sci. 21 : 1299-1302.

Linderman, R.G. 1988a, Mycorrhizal interactions with the rhizosphere microflora: The mycorrhizosphere effect, Phytopathology 78 : 366-371.

Linderman, R.G. 1988b, VA mycorrhizal symbiosis, ISI Atlas of Science, Animal and Plant Sciences, Section 1:183-188.

Linderman, R.G. 1991, Mycorrhizal interactions in the rhizosphere, in : Rhizosphere and Plant Growth, D.L. Keister and P.B. Gregan, eds., Kluwer Academic Publishers, Dordrecht, The Netherlands, pp. 343-348.

Linderman, R.G. 1992, Vesicular-arbuscular mycorrhizae and soil microbial interactions, in : Mycorrhiza in Sustainable Agriculture, G.J. Bethlenfalvay and R.G. Linderman, eds., pp. 45-70. Am., Soc. Agron. (Sp. Publ.) No. 54. Am Soc. Agron. Madison, Wis., U.S.A.

Linderman, R.G. 1994, Role of VAM fungi in biocontrol, in : Mycorrhizae and Plant Health, E.L. Pfleger and R.G. Linderman, eds., pp.1-25, APS Press, St. Paul, U.S.A.

Linderman, R.G. and Paulitz, T.C. 1994, Mycorrhizae and plant health, APS Press, St. Paul, USA, pp. 337-344.

Liu, R.J. 1995, Effect of vesicular arbuscular mycorrhizal fungi on *Verticillium* wilt of cotton, Mycorrhiza 5 : 293-297.

Lynch, J.M. and Whipps, J.M. 1990, Substrate flow in the rhizosphere, Plant and Soil 129 : 1-10.

Maier, W., Peipp, H., Schmidt, J., Wray, V. and Strack, D. 1995, Levels of a terpenoid glycocide (blumenin) and cell wall-bound phenolics in some cereal mycorrhizas, Plant Physiol. 109 : 465-470.
Malajczuk, N. 1984, Perspective on ectomycorrhizae-pathogen interaction, Proc. 6th NACOM R. Molina, ed., Oregon, USA, pp. 123.
Malajczuk, N. and Mc Comb, A.J. 1979, The microflora of unsuberized roots of *Eucalyptus calophylla* R. Br. and *Eucalyptus marginata* Donn. ex. Sm. seedlings grown in soils suppressive and conducive to *Phytophthora cinnamomi* Rands, I, Rhizoplane bacteria, actinomycetes and fungi, II, Mycorrhizal roots and associated microflora 27 : 235.
Malajczuk, N. and Sanfelieu, C.L. 1984, Survival of *Phytophthora cinnamomi* zoospores on non-mycorrhizal and ectomycorrhizal roots of jarrah, Proc, 6th NACOM, R. Molina, ed., Oregon, USA pp. 336.
Mark, G.L. and Cassells, A.C. 1996, Genotype dependence in the interaction between *Glomus fistulosum, Phytophthora fragariae* and the wild strawberry (*Fragaria vesca*), Plant and Soil, 185 : 233-239.
Marx, D.H. 1969, The influence of ectotrophic ectomycorrhizal fungi on the resistance of pine roots to pathogenic infections, I, Antagonism of mycorrhizal fungi to pathogenic fungi and soil bacteria, Phytopathology 59 : 153-163.
Marx, D.H. 1973, Mycorrhizae and feeder root diseases, in : Ectomycorrhizae : Their Ecology and Physiology, G.C. Marks and T.T. Kozlowski, eds., Academic Press, N.Y. USA, pp. 351-382.
Marx, D.H. 1975, Role of ectomycorrhizae in the protection of pine from root infection by *Phytophthora cinnamomi*, in : Biology and Control of Soil-borne Plant Pathogens, J.W. Bruehl, ed., APS Press, St. Paul, USA, pp. 77-83.
Mayer, A. 1987, Polyphenol oxidases in plants-recent progress, Phytochemistry 26 : 11-20.
Mc Arthur, D.A.J. and Knowles, N.R. 1992, Resistance responses of potato to vesicular arbuscular mycorrhizal fungi under varying abiotic phosphorus levels, Plant Physiol. 100 : 341-351.
McGraw, A.C. and Schenck,N.C. 1981, Effect of two species of vesicular arbuscular mycorrhizal fungi on the development of fusarium wilt of tomato (Abst), Phytopathology, 71 : 894.
Mittal, N. Sharma, M., Saxena, G. and Mukerji, K.G. 1991, Effect of VA mycorrhiza on gall formation in tomato roots, Incompatibility News Letter, 23 : 39-43.
Moerschbacher, B.M. 1992, Plant peroxidases : involvement in response to pathogens, in : Plant Peroxidases 1980-1990, Topics and Detailed Literature on Molecular, Biochemical and Physiological Aspects, C. Penel., T. Gasper, and H. Greppin eds., University Press, Geneva, Switzerland, pp. 91-99.
Mohr, U., Lange, J., Boller, T., Wiemken, A. and Vogeli - Lange, R. 1998, Plant defense genes and induced in the pathogenic interaction between been roots and *Fusarium solani*, but not in the symbiotic interaction with the arbuscular mycorrhizal fungus *Glomus mosseae*, New Phytol. 138 : 589-598.
Molina, R. and Trappe, J.M. 1982, Patterns of ectomycorrhizal host specificity and potential among Pacific North West conifers and fungi, For. Sci. 28 : 423-458.
Morandi, D. 1989, Effect of xenobiotics on endomycorrhizal infection and isoflavonoid accumulation in soybean roots, Plant Physio., Biochem. 27 : 697-701.
Morandi, D. 1996, Occurrence of phytoalexins and phenolic compounds in endomycorrhizal interactions and their potential role in biological control, Plant and Soil 185 : 241-251.
Morandi, D., Bailey, J.A. and Gianinazzi - Pearson, V. 1984, Isoflavonoid accumulation in soybean roots infected with vesicular-arbuscular mycorrhizal fungi, Physiol. Plant Pathol. 24:357-364.
Morandi, D., Branzanti, B. and Gianinazzi-Pearson, V. 1992, Effect of some plant flavonoids on *in vitro* behaviour of an arbuscular mycorrhizal fungus, Agronomie 12 : 811-816.
Morandi, D. and Gianinazzi-Pearson, V. 1986, Influence of mycorrhizal infection and phosphate nutrition on secondary metabolite contents of soybean roots, in : Physiological and Genetical Aspects of Mycorrhizae, V. Gianinazzi-Pearson and S. Gianinazzi, eds., pp. 787-791, INRA Press, Paris.
Morandi, D. and Le Quere, J.L. 1991, Influence of nitrogen on accumulation of isojagol, a newly detected coumestan in soybean and associated isoflavonoids in root and nodules of mycorrhizal and non-mycorrhizal soybean, New Phytol. 117 : 75-79.
Mukerji, K.G., Chamola, B.P. and Sharma, M. 1997, Mycorrhiza in control of plant pathogens, in : Management of Threatening Plant Diseases of National Importance, V.P. Agnihotri, A.K. Sarbhoy, and D.V. Singh, eds., Malhotra Publishing House, New Delhi, pp. 298-314.
Mukerji, K.G. and Garg, K.L. (eds.) 1988a, Biocontrol of Plant Diseases, Vol. I, CRC Press Inc., Florida, USA.
Mukerji, K.G. and Garg K.L. (eds.) 1988b, Biocontrol of Plant Diseases, Vol. II, CRC Press Inc. Florida, USA.
Mukerji, K.G. and Mandeep 1997, Mycorrhizal relationships of wetland and rivers associated plants, in : Ecology of Wetland and Associated Systems, S.K. Mazumdar, ed., The Pensylvania Acad. of Science, pp. 91-108.
Mukerji, K.G. and Sharma, M. 1996, Mycorrhizal relationships in Forest Ecosystems, in : Forests - A Global Perspective, S.K. Mazumdar, E.W. Muller and Fred. J. Brenner, eds., The Pensylvania Acad. of

Sciences, pp. 95-125.
Mukerji, K.G., Upadhyaya, R.K. and Kaushik, A. 1996, Mycorrhiza and Integrated Disease Management, in : IPM System in Agriculture, Vol. II, Biocontrol in Emerging Biotechnology, Rajeev, K. Upadhyay, K.G. Mukerji and R.L. Rajak, eds., Aditya Books Pvt. Ltd., New Delhi, pp. 423-452.
Münzenberger, B., Olter, T., Wüstrich, D. and Polle, A. 1997, Peroxidase and laccase activities in mycorrhizal and non-mycorrhizal fine roots of Norway spruce (*Picea abies*) and larch (*Larix decidua*), Can. J. Bot. 75 : 932-938.
Mylona, P., Moerman, M., Yang, W.C., Gloudemans, T., van de Kerckhove, J., van Kammen, A., Bisseling, T. and Franssen, H.J. 1994, The root epidermis-specific pea gene RH_2 is homologous to a pathogenesis-related gene, Plant Mol. Biol. 26 : 39-50.
Norman, J.R., Atkinson, D. and Hooker, J.E., 1996, Arbuscular mycorrhizal fungal-induced alteration to root architecture in strawberry and induced resistance to the root pathogen *Phytophthora fragariae*, Plant and Soil 185-191-198.
O' Bannon, J.H. and Nemec, S. 1979, The response of *Citrus lemon* seedlings to a symbiont, *Glomus etunicatus* and a pathogen, *Radophalus similis*, J. Nematol. 11 : 270-275.
O'Malley, D.M., Whetten, R., Bao, W., Chen, C.L. and Sederoff, R.R. 1993, The role of laccase in lignification, Plant J. 4 : 751-757.
Paulitz, T.C. and Linderman, R.G. 1991, Mycorrhizal interactions with soil organisms, in : Handbook of Applied Mycology, Vol.I, Plant and Soil, D.K. Arora, B.Rai, K.G. Mukerji and G.R. Knudsen, eds., p. 77-129, Marcel Dekker Inc., New York, USA.
Pinochet, J., Calvet, C., Camprubi, A. and Fernandez, C. 1996, Interactions between migratory endoparasitic nematodes and arbuscular mycorrhizal fungi in perennial crops, A review, Plant and Soil 185 : 183-190.
Rasanayagam, S. and Jeffries, P. 1992, Production of acid is responsible for antibiosis by some ectomycorrhizal fungi, Mycol. Res. 6 (11) : 971-976.
Reddy, M.V., Rao, J.N. and Krishna, K.R. 1989, Influence of vesicular arbuscular mycorrhizae on *Fusarium* wilt of pigeon pea, Int. Pig. Newsl. 9 : 23.
Roldan, A., Garcia-Orenes, F. and Albaladejo, J. 1994, Microbial populations in the rhizosphere of *Brachypodium retusum* and their relationship with stable aggregates in a semiarid soil of south eastern Spain, Arid Soil Res. Rehab. 8 : 105-114.
Rosendahl, C.N. and Rosendahl, S. 1990, The role of vesicular arbuscular mycorrhiza in controlling damping off and growth reduction in cucumber caused by *Pythium ultimum,* Symbiosis 9 : 363-366.
Sagan, M., Morandi, D., Tareughi, E. and Duc, G. 1995, Selection of nodulation and mycorrhizal mutants in the model plant, *Medicago truncatula* (Gaertn) after gamma ray mutagenesis, Plant Sci. 111 : 63-71.
Samara, A., Dumas-Gaudot, E., Gianinazzi-Pearson, V. and Gianinazzi, S. 1996a, Studies of *in vivo* polypeptide synthesis in non-mycorrhizal and arbuscular-mycorrhizal (*Glomus mosseae*) pea roots, in : Mycorrhizas in Integrated Systems from Genes to Plant Development, C. Azcon-Aguilar and J.M.Barea, eds., pp. 263-266, Kluwer Academic Publisher, Dordrecht, The Netherlands.
Samara, A., Dumas-Gaudot, E, Gianinazzi-Pearson, V. and Gianinazzi, S. 1996b, Changes in polypeptide profiles of two pea genotypes inoculated with the arbuscular mycorrhizal fungus *Glomus mosseae*, in : Integrated Systems: From Genes to Plant Development, C.Azcon-Aguilar, J.M. Barea and J. Ocampo, eds., COST, Brussels.
Sampangi, R. and Perrin, R. 1985, Attempts to ellucidate the mechanisms involved in the protective effect of *Laccaria laccata* against *Fusarium oxysporum*, in : Physiological and Genetical Aspects of Mycorrhizae, V. Gianinazzi-Pearson and S. Gianinazzi, eds., INRA, Paris, France, pp. 807-810.
Schenck, N.C. and Kellam, M.K. 1978, The influence of vesicular arbuscular mycorrhiza on disease development, Fla. Agric. Exp. Stn. Tech. Bull, 798, Gainesville, Florida.
Schisler, D.A. and Linderman, R.G. 1987, The influence of volatiles purged from soil around Douglas fir ectrmycorrhizae on soil microbial populations, in : Proc. 7th NACOM, D.M. Sylvia, H. Hung and J.H. Graham, eds., pp. 218.
Schönbeck, F. 1979, Endomycorrhiza in relation to plant diseases, in : Soil-borne Plant Pathogens, B. Schippers and W. Gams, eds., Academic Press, New York, USA pp. 271-280.
Schönbeck, F. and Dehne, H.W. 1977, Damage to mycorrhizal and non-mycorrhizal cotton seedlings by *Thielaviopsis basicola*, Plant Dis. Reptr. 61:266-269.
Schreiner, R.P., Mihara, K.L., McDaniel, H. and Bethlenfalvay, G.J. 1997, Mycorrhizal fungi influence plant and soil functions and interactions, Plant and Soil 188 : 199-209.
Sharma, M. and Mukerji, K.G. 1992, Mycorohiza-Tool for biocontrol, in : Recent Developments in Biocontrol of Plant Diseases, K.G.Mukerji, J.P. Tewari, D.K.Arora and G. Saxena, eds., Aditya Books Pvt. Ltd., New Delhi, pp. 52-80.
Sharma, M., Mittal, N. and Mukerji, K.G. 1998, Fungi : Tool for plant disease management, in : Microbes for Health, Wealth and Sustainable Environment, A. Varma, ed., Malhotra Publishing House, New Delhi,

pp. 101-154.
Simoneau, P., Feugey, L., Viémont, J.D., Swoboda, I., Heberte Bors, E. and Strullu, D.G. 1996, Induction of phenylalanine ammonia-lyase in birch challenged with ectomycorrhizal fungi, in : Mycorrhiza in Integrated Systems from Genes to Plant Development, Proc. 4th European Symp. on Mycorrhiza (ESM), Granada, 11-14th July, 1994, Azcon-Aguilar and J.M. Barea, eds., European Commission, Brussels, p.203.
Smith, G.S., Hussey, R.S. and Roncadori, R.W. 1986a, Penetration and post infection development of *Meloidogyne incognita* as affected by *Glomus intraradices* and phosphorus, J. Nematol. 18 : 429-437.
Smith, G.S., Roncadori, R.W. and Hussey, R.S. 1986b, Interaction of endomycorrhizal fungi, superphosphate and *Meloidogyne incognita* on cotton in microplot and field studies, J. Nematol. 18 : 208-214.
Smith, S.E. and Read, D.J. 1997, Mycorrhizal Symbiosis, Academic Press, London, U.K.
Smith, V.L., Wilcox, W.F. and Harmann, G.E. 1990, Potential for biological control of *Phytophthora* root and crown rots of apple by *Trichoderma* and *Gliocladium* spp. Phytopathology 80 : 880-885.
Spanu, P., Boller, T., Ludwig, A., Wiemken, A., Faccio, A. and Bonfante Fasolo, P. 1989, Chitinase in roots of mycorrhizal *Allium porrum* : regulation and localization, Planta 177 : 447-455.
Spanu, P. and Bonfante-Fasolo, P. 1988, Cell-Wall bound peroxidase activity in roots of mycorrhizal *Allium cepa*, New Phytologist 109 : 119-124.
Srivastava, D., Kapoor, R., Srivastava, S.K. and Mukerji, K.G. 1996, Vesicular-arbuscular mycorrhiza - an overview, in : Concepts in Mycorrhizal Research, K.G. Mukerji, ed., Kluwer Academic Publishers, Dordrecht, Netherlands, pp. 1-39.
St. Arnaud, M., Hamel, C., Caron, M. and Fortin, J.A. 1995, Inhibition of *Pythium ultimum* in roots and growth substrate of mycorrhizal *Tagetes petula* colonised with *Glomus intraradices*, Can. J. Plant Path. 16 : 187-194.
St. Arnaud, M., Hamel, C., Vimard, B., Caron, M. and Fortin, J.A. 1997, Inhibition of *Fusarium oxysporum* f. sp. *dianthi* in the non-VAM species *Dianthus caryophyllus* by co-culture with *Tagetes petula* companion plants colonised by *Glomus intraradices*, Can. J. Bot. 75 : 998-1005.
Sundaresan, P., Raja, N.U. and Gunasekaran, P. 1993, Induction and accumulation of phytoalexins in cowpea roots infested with a mycorrhizal fungus *Glomus fassiculatum* and their resistance to *Fusarium* wilt disease, J. Biosciences 18 : 291-301.
Suresh, A.K., Bagyaraj, D.J. and Reddy D.D.R. 1985, Effect of vesicular-arbuscular mycorrhiza on surival, penetration and development of root-knot nematode in tomato, Plant and Soil. 87 : 305-308.
Suvercha, Mukerji, K.G. and Arora, D.K. 1991, Ectomycorrhiza, in : Handbook of Applied Mycology, Vol. I, D.K. Arora, B. Rai, K.G. Mukerji and G.R. Knudson, eds., Marcel Dekker Inc., New York, USA, pp. 187-215.
Sylvia, D.M. and Sinclair, W.A. 1983a, Suppressive influence of *Laccaria laccata* on *Fusarium oxysporum* and on Douglas fir seedlings, Phytopathology 73 : 384-389.
Sylvia, D.M. and Sinclair, W.A. 1983 b, Phenolic compounds and resistance to fungal pathogens induced in primary roots of Douglas fir seedlings by the ectomycorrhizal fungus *Laccaria laccata*, Phytopathology 73 : 390-397.
Trotta, A., Varese, G.C., Gnavi, E., Fusconi, A., Sampo, S. and Berta, G. 1996, Interactions between the soilborne root pathogen *Phytophthora nicotianae* var. *parasitica* and the arbuscular mycorrhizal fungus *Glomus mosseae* in tomato plants, Plant and Soil 185 : 199-209.
Volpin, H., Elkind, Y., Okon, Y. and Kapulnik, Y. 1994, A vesicular arbuscular mycorrhizal fungus *Glomus intraradices* induces a defence response in alfalfa roots, Plant Physiol. 104 : 683-689.
Volpin, H., Phillips, D.A., Okon, Y. and Kapulink, Y. 1995, Suppression of an isoflavonoid phytoalexin defense response in mycorrhizal alfalfa roots, Plant Physiol. 108 : 1449-1454.
Wyss, P., Boller, T. and Wiemken, A. 1989, Glyceollin production in soybean during the process of infection by *Glomus mosseae* and *Rhizoctonia solani*, Agric. Ecos. Environ. 29 : 451-456.
Young, C.C., Juang, T.C. and Guo, H.Y. 1984, Effects of VA mycorrhizal inoculation on the soybean growth, yield, N-fixation and absorption of mineral phosphate, J. Agric. Assoc., China 128:29-40.
Zambolim, L. and Schenck, N.C. 1983, Reduction of the effects of pathogenic root infecting fungi on soybean by the mycorrhizal fungus, *Glomus mosseae*, Phytopathology 73 : 1402-1405.
Zombolim, L. and Schenck, N.C. 1984, Effect of *Macrophomina, Rhizoctonia, Fusarium* and the mycorrhizal fungus *Glomus mosseae* on nodulated and non-nondulated soybeans, Fitopathologia Brasiteiva 9 : 129-138.

BIOLOGICAL CONTROL OF BACTERIAL PLANT DISEASES

Surinder Kaur and K.G. Mukerji

Applied Mycology Laboratory
Department of Botany
University of Delhi
Delhi-110007, INDIA.

1. INTRODUCTION

All microorganisms occupy various ecological niches in nature, often on first-come-first-served basis. In order for a plant disease to develop, a pathogen must compete with other microorganisms to secure the necessary niche on the susceptible plant. An effective biological control agent excludes or displaces a pathogen from an available niche by preemption or by direct competition through differential depletion of nutrients or suppression of pathogen growth through the production of antibiotics or bacteriocins.

Biological control of plant pathogens has been considered as a potential control strategy in recent years. The use of biocontrol strategies offer several advantages over the chemical control, since it is economical, self perpetuating and usually free from residual side effects. However, in reality it will not immediately nor totally replace chemicals but the use of biocontrol agents can and significantly enhance quality of life, the environment and agricultural productivity. Genetic manipulation of biocontrol agents offers a possible approach for improving their potential for plant disease control. Molecular techniques allow modification of these potential biocontrol agents by the directed removal or addition of one or more genes. Some of these techniques and their uses in the biocontrol of various plant bacterial diseases are discussed in the following pages.

2. BIOCONTROL OF PLANT DISEASES CAUSED BY BACTERIA

Bacterial diseases of plants are notoriously difficult to control. A wide range of fungicides are available to combat the many, economically important, fungal diseases of plants. The continuing dearth of suitable and effective bactericides for use against the fewer, but sometimes nonetheless economically important, bacterial diseases of crop plants

Biotechnological Approaches in Biocontrol of Plant Pathogens
Edited by Mukerji *et al.*, Kluwer Academic / Plenum Publishers, New York, 1999

has stimulated a search for alternative methods of control. Although biological control is not a new agricultural technique, it is only in comparatively recent times that serious efforts have been made to exploit the interactions and mutual antagonisms that undoubtedly occur among the complex populations of microorganisms in the relatively stable soil environment. Such interactions have probably been overlooked in the past because there is rarely any notable evidence of their occurrence. However, it is just one such natural interaction that has been successfully explained in the biological control of crown gall.

Biological control of bacterial plant diseases in nature seems to be the rule rather than the exception. In fact, early in this century biological control was the only means to manage plant diseases. However, during the last few decades, biological control has been replaced with chemical control because of the wide spectrum of chemicals available in the market, the ease of use and the efficacy of pest control. Although biological control recently has replaced chemical control due to increased environmental awareness. Unfortunately, implementation of biological control is currently hindered by the lack of effective biological control agents and strategies. It is clear that wild-type agents taken from nature are seldom adequate for biological control. Molecular techniques allow modification of these potential bio control agents by the directed removal or addition of one or more genes. For example, the antagonistic properties of a bio control organism towards a target pathagen could be increased by expression of noval antimicrobials.

Because of the inability of the biocontrol agent to compete with pathogen, under conditions, which favour the growth of the pathogen, these biocontrol strategies have met with little success in the field. One record approach to this problem is to minimize differences in the bio control agent and the target pathogen by deriving the bio control agent from the pathogen. For example, description of the *hrp* genes of *Pseudomonas solanacearum* does not affect the ability of the mutant to penetrate host roots and multiply within host tissues, but disruption of the *hrp* genes does not affect the ability of the pathogen to cause disease (Trigalet and Trigalet-Demery 1990; Trigalet *et al*.. 1994). Thus, these avirulent derivatives are adapted to the same ecological niche as the parental strain and can be used to preempt or displace the pathogen from the susceptible host.

2.1. Crown Gall

Crown gall is caused by soil bacterium *Agrobacterium tumefaciens* (Smith and Townsend) Conn (Zambryski *et al*., 1989; Jones and Kerr, 1989; Winans, 1992). The bacterium penetrates plants at wound sites or root lenticels. These wounds may be natural, originating during seed germination, emergence of lateral roots, from freezing damage, or may be punctures made by feeding of soil fauna. Many wounds are made during normal horticultural practice e.g. root trimming or breakage preplanting or from accidental damage during cultivation. Thus, there are many sites available for ingress of the pathogen. Following infection the plant cells transform into autonomously proliferating tumor cells. The resultant unregulated cell division gives rise to clearly visible and unsightly galls. *A. tumefaciens* has a uniquely vast host range among dicotyledonous plants - over 600 species in 93 families are reported to be susceptible (De Cleene and Ley, 1976). The bacterium contains a tumor inducing (Ti) plasmid (Van Larebeke *et al*., 1975, Watson *et al*., 1995) that directs the transfer of a segment of its own DNA (T-DNA) into plant cells (Chilton *et al*., 1977). The T-DNA is integrated into the plant DNA (Chilton *et al*., 1980; Willmitzer, 1980) where it directs the synthesis of phytohormones, which cause tumerous cell growth (Akiyoshi, *et al*., 1984; Inze *et al*., 1984; Pelegerm and Sands, 1996; Schroder *et al*., 1984.) and opines, which act as noval nutrient sources catabolized almost exclusively by the inciting bacterium (Guyon *et al*., 1980). One class of Ti plasmid direct the plant to synthesize, and the bacterium to catabolize, the opines nopaline and agrocinopine; strains of *A. tumefaciens* carrying this type of Ti plasmid are subject to biological control by *Agrobacterium radiobacter* (Beijerinck and Van Delden) Conn Strain K84 (Kerr and Htay, 1974, Kerr and Tate, 1984) whose synthesis is encoded by a 47Kb plasmid, PAgK84, carried by strain K84 (Ellis *et al.*, 1979; Farrand *et al.*, 1985). Agrocin 84, a di-substituted fraudulent analog of adenosine nucleoside (Fig. 1), specifically exhibits pathogenic strains of *A. tumefaciens* with a nopaline type Ti plasmid. Agrocinopines are phosphorylated opines that are synthesized in plant cells which have been transformed with nopaline type T-DNA. The enzymes for the transport and catabolism of agrocinopines by the *Agrobacterium* are included on the Ti plasmid. Agrocin 84 is taken up by a Ti-plasmid-encoded agrocinapine permease (Ellis and Murphy, 1981) and acts by terminating DNA synthesis in the recipient bacterium (Das *et al.*, 1978; Kerr and Tate, 1984; McCardell and Pootjer, 1976; Murphy and Roberts, 1979).

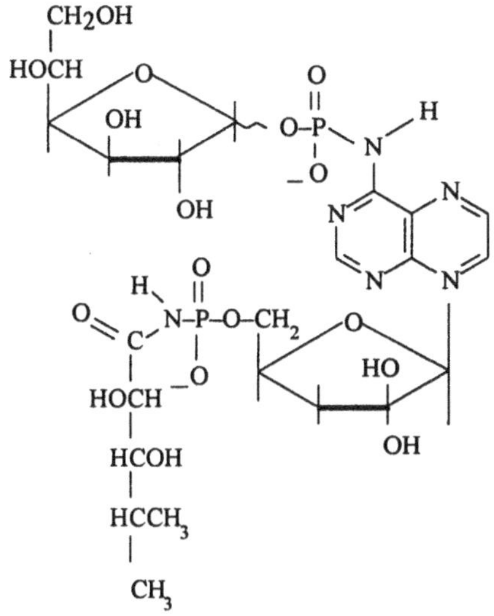

Figure 1. Molecular structure of agrocin 84.

Strain K84 has been used successfully throughout the world to control crown gall but its continued success was threatened by the ability of PAgK84 to transfer conjugatively to pathogenic agrobacteria (Ellis and Kerr, 1979; Farrand *et al.*, 1985, Panagopoulos *et al.*, 1979) which then become insensitive to agrocin 84 (Ellis *et al.*, 1979; Ryder *et al.*, 1987).

Agrocin-plasmid transfer is controlled by a defined region (Tra) on PAgK84 (Farrand *et al.*, 1985). Recombinant DNA-techniques have been used to construct a new biological control strain, K1026, that is identical to K84 apart from a 5.9 Kb deletion overlapping the Tra region of PAgK84 (Chilton *et al.*, 1980). K1026 is unable to transfer its mutant agrocin-84 plasmid designated PAgK1026, to other agrobacteria but remains inhibitory *in vitro* to strains sensitive to agrocin-84. Jones and Kerr (1989) found that new strain K1026 is as efficient as K84 in controlling crown gall. They also observed that relative to K84, K1026 has a similar ability to colonise and survive on roots as well as to colonize galls and displace inciting agrobacteria. Thus, K1026 appears to retain the ecological competence of its progenitor, K84.

Jones and Kerr (1989) found an apparent biomodial polarization in the population ratios of K84 or K1026 to K27 in the few galls on K84 or K1026 treated plants. K84 and K1026 both use the opines nopaline and agrocinopine synthesised by K27-induced galls (Ellis *et al.*, 1982; Ryder *et al.*, 1987). So, consistent with the opine concept (Guyon *et al.*, 1980), those galls with an excess of K84 or K1026 presumably reflect

K84 or K1026 colonisation of these galls with the concomitant displacement of K27 which in the presence of agrocinopine, becomes more sensitive to agrocin84 (Ellis and Murphy, 1981). The galls in which there was an excess of K27 presumably reflect opportunities for colonization that were missed.

Since K1026 is clearly as efficient as K84 in controlling crown gall and because it is unable to transfer its agrocin-84 plasmid, it should be used as a replacement for K84 to prolong the effective biological control of crown gall (Jones and Kerr, 1989).

The efficacy of a crown gall biological control agent is not solely dependent upon the production of agrocin-84. Additional mechanisms such as niche competition or production of other agrocins play a role in the suppression of the disease.

A mutant of *A. radiobacter* (Agr-) that was cured of plasmid PAgK84 did not prevent crown gall of tomato stems when coinoculated with a pathogenic strain of *A. tumefaciens* i.e. there is a good correlation between agrocin 84 sensitivity in *A. tumefaciens* and biological control by K84 when wounded tomato stems are coinoculated with the pathogen and antagonist (Kerr and Htay, 1974). However, when K84 is applied to tomato stem 24 hr before the pathogen,infection by agrocin-resistant strains can be prevented (Cooksey and Moore, 1980). In addition, agrocin resistant strains have been controlled by K84 in field tests with *Prunus* seedlings (Cooksey and Moore, 1980; Moore, 1977; Schroth and Moller, 1976). These data suggest that in addition to the production of agrocin-84, other mechanisms are involved in biological control by K84.

Likewise, pre-inoculation of plum with strain K84 reduced the incidence of control crown gall on plants which were subsequently inoculated with either agrocin-84 sensitive strains or agrocin-84 resistant strains of *A. tumefaciens* (Lopez *et al.*, 1989). The effectiveness of strains' K84 was greater against the agrocin-84 sensitive pathogen than against the resistant pathogen (Lopez *et al.*, 1989). A mutant of strain K84 which did not produce agrocin 84 also restricted the development of crown gall caused both agrocin 84-sensitive and resistant *A. tumefaciens.* A wild type strain of K84 was more effective than the Agr- strain controlling crown gall caused by agrocin-84 sensitive strains, but the levels of control against agrocin 84 resistant strains of *A tumefaciens* by the wild type strain K84 and the mutant strain were similar (Lopez *et al.*, 1989).

Shim *et al.*(1987) carried out experiments to find out the effectiveness of transfer deficient (Tra^-) strains as biocontrol agents and also determined if the chromosomal background for the plasmid is important.

Strain K907, a K84 derivative harboring a Tra^- Tn5 insertion of PAgK84 effectively controlled crown gall, whereas strain K823 a C58 derivative harboring the same plasmid was much less efficient. The latter strain appears to be ecologically inefficient. Differences in the effectiveness of crown gall control were presumably due to differences in colonization. Increased production of agrocin 84 did not result in increased level of biocontrol. *A. radiobacter* strains carrying a high copy number mutant of PAgK84 overproduced agrocin 84 *in vitro* but were not more effective than the wild type in controlling crown gall (Shim *et al.*,1987).

The biocontrol of crown gall disease by *A. radiobacter* K84 is primarily mediated through the production of agrocin-84. The results indicate that the production of agrocin 84, greatly enhances the effectiveness of the biological control agents, but competition for colonization sites or nutrients also contributes to the efficacy of control (Cooksey and Moore, 1982; Lopez *et al.*, 1989; Shim *et al.*, 1987).

In addition to agrocin-84, *A. radiobacter* strains K84 and K1026 also produce agrocin 434, a di-substituted cystidine nucleoside which inhibits a broader range of *Agrobacterium* than agrocin 84 (Donner *et al.*, 1993). Production of agrocin 434 by strain K84 and its derivatives may be an additional factor contributing to the biocontrol of crown gall.

The range of *A. tumefaciens* controlled by *A. radiobacter* could be expanded by further increasing the number of agrocins secreted by strains K84 or by creating a biocontrol strain which produces more effective agrocins.

2.2. Bacterial Blights

2.2.1. *Pseudomonas syringae*

Pseudomonas syringae, are opportunistic pathogens of a wide variety of plant species (Schroth *et al.*, 1991). Individual strains or pathovars (pv) cause a foliar disease in a limited subset of host plants. During pathogenesis, the bacteria invade the plant tissue through natural openings or wounds. In a susceptible host, the bacteria multiply and disperse throughout the plant tissue and cause water-soaked lesions on leaves and other plant parts. In a resistant reaction on non-host plants or resistant host plants, the bacteria rapidly elicit the hyper sensitive response (HR) (Klement, 1982). The HR of higher plants is characterised by the rapid, localized death of plant cells at the site of pathogen invasions, inhibiting the multiplication and spread of the pathogen within the host tissue (Pelgeram and Sand, 1996). Five structurally distinct classes of toxins that cause either chlorotic or necrotic symptoms in infected plant tissues are produced by pathovars of *P. syringae*. The chlorosis inducing toxins are the most common among pathovars of *P. syringae* and are classified broadly as either tabtoxin, phaseolotoxin, coronative or tagetitoxin. In contrast, the necrosis-inducing toxins are restricted to pathovar *syringae* and encompass a family of structurally related lipopeptides called the syringomycins (Gross, 1991). The molecular structure of these toxins are shown in figure 2.

Lindgren *et al.* (1986) characterized eight insertion mutants of *P. syringae* pv *phaseolicola* the causative agent of halo blight of bean, which had lost the ability to elicit the HR on non host plants such as tobacco and the ability to colonise or cause disease on host plants. The mutations did not alter the growth of the mutants on minimal media. A single recombinant plasmid from a wild type genomic library was isolated that restored the wild type phenotype to seven of the insertion mutants, suggesting that the genes affected by the insertions were clustered. The cluster of genes was designated as *hrp* (hypersensitive reaction and pathogenicity) (Lindgren *et al.*, 1986, 1989; Rahme *et al.*, 1991) *hrp* genes as *hrp*-like mutations have been described in *P. s.* pv *phaseolicola* (Lindgren, 1986), *P. s.* pv *syringae*, (Huang *et al.*, 1988; Mukhopadhyay, 1988), *P. s.* pv *tomato* (Cuppels, 1986), *P. s.* pv *morsprunorum* (Liang and Jones, 1995) *P. s.* pv *pisi* (Malik *et al.*, 1987), *P. s.* pv *tabaci* (Lindgren *et al.*, 1988), *P. s.* pv *glycinea* (Lindgren *et al.*, 1988) as well as in several other phytopathogenic bacteria including *Erwinia stewartii* (Coplin *et al.*, 1992b), *Erwinia chrysanthemi* (Laby *et al.*, 1992) *Erwinia amylovora* (Bauer and Beer, 1992). *Pseudomonas solanacearum* (Boucher *et al.*, 1985, 1987) and some pathovars of *Xanthomonas compestris* (Boucher et al., 1987, Bonas *et al.*, 1991).

The use of avirulent Hrp-Mutants of *P. syringae* or *X. compestris* as biocontrol agents has not been reported despite the effectiveness of biological control using Hrp-mutants of *P. solanacearum* and *E. chrysanthemi*. Avirulent mutants of *P. syringae* and *X. compestris* have also been characterized with inactivated avirulence genes (avr) (Lorang *et al.*, 1994).

Bacterial speck of tomato has been controlled using copper compounds and/or antibiotics. The effectiveness of the treatment is dependent upon environmental conditions and the development of resistance in the pathogen population. Bacterial speck of tomato (*Lycopersicon esculentum* Mull) has been controlled by copper compounds and antibiotics (Conlin and McCarter, 1983, MacNab, 1979, 1980, Yunis *et al.*, 1980). However, the effect of these treatments depend upon the environmental conditions (Jardine and Stephens, 1987). Reduced effectiveness of copper compounds has also been associated with copper resistance in *P. s.* pv *tomato* which was widespread in Southern California isolates of *P. s.* pv *tomato* (Bender and Cooksey, 1986; Cooksey, 1987).

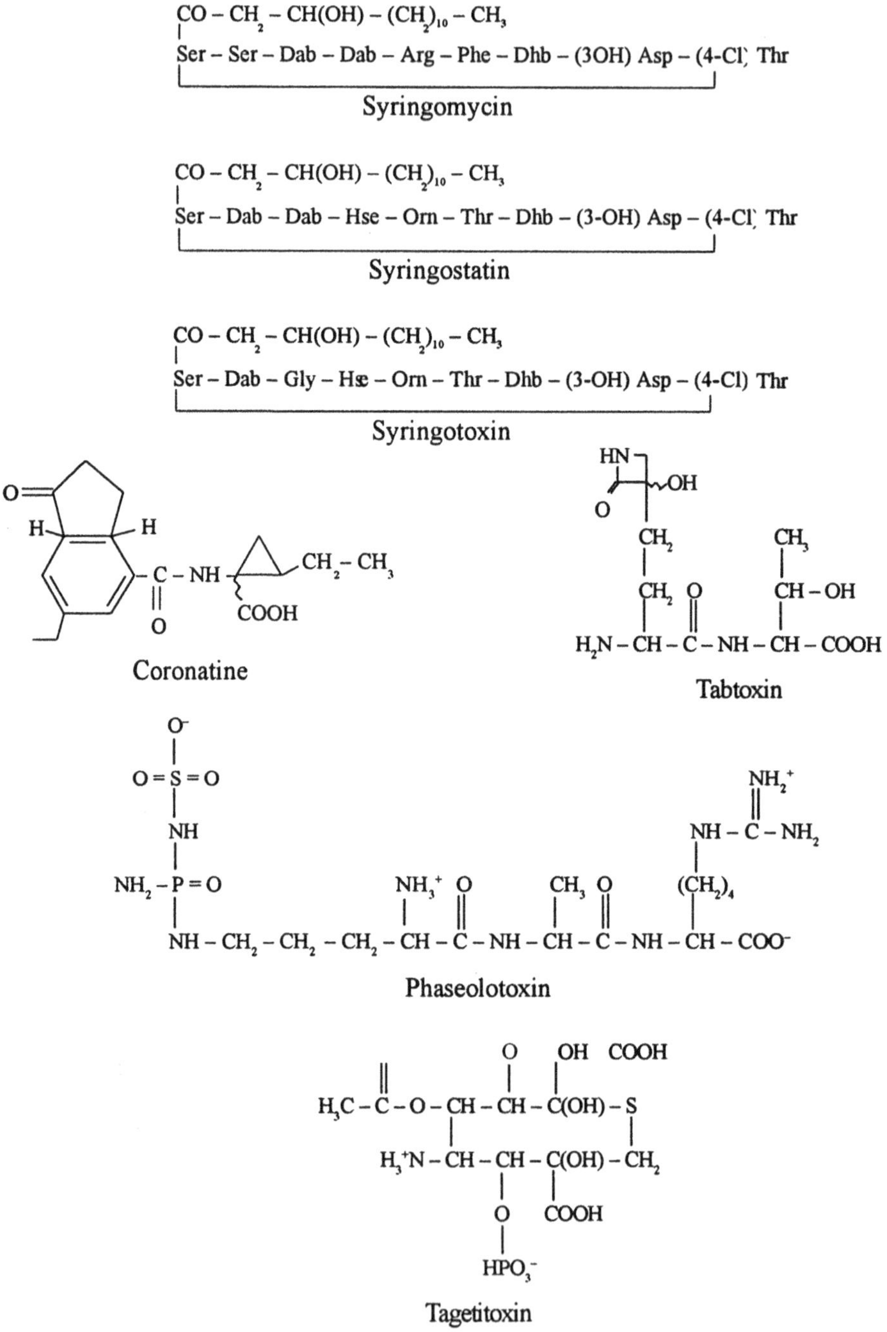

Figure 2. Toxin structures. Abbreviations : Arg, arginine; (3-OH) Asp, 3-hydroxyaspartic acid; (4-Cl) Thr, 4-chlorothreonine; Dab, 2,4-diaminobutryic acid; Dhb, 2,3-dehydro-2-aminobutyric acid; Gly, glycine; Hse, homoserine; Orn, ornithine; Phe, phenylalanine; Ser, serine; Thr, threonine. (From Gross, 1991).

Cooksey (1988) found a non pathogenic Tn 5 mutant of *Pseudomonas syringae* pv *tomato* retained its ability to survive epiphytically on tomato leaves, and coinoculations of the mutant with a pathogenic strain significantly reduced the incidence of bacterial speck in green house. When a derivative of the nonpathogen with resistance to high levels of copper was coinoculated with a copper-sensitive pathogenic strain to plants treated with a copper bactericide, significantly greater control was achieved than with either the nonpathogen or the copper treatment alone.

Pseudomonas syringae pv *savastanoi* (Smith) Young *et al.* causes the olive knot disease in olive (*Olea europaea* L.) and is characterized by the formation of galls on stems and, less frequently, on other parts of the plant. A similar disease occurs on oleander (*Nerium oleander* L.) (Wilson and Magie, 1963) and privet (*Ligustrum japonicum* Thunb) (Botalico and Ercolani, 1971).

Pathovar savastanoi synthesizes Indole-3-acetic acid (IAA) (Beltra, 1958, 1961) and substances with cytokinin like activity (Surico *et al.*, 1975). Smidt and Kosug (1978) demonstrated that IAA deficient (Iaa^-) mutants of pv *savastanoi* from oleander could not induce gall formation on oleander. Surico *et al.* (1984) compared the pathogenicity on olive and oleander plants of wild type and IAA deficient (Iaa^-) mutant strains of pv *savastanoi* from olive, oleander and privet. All wild-type strains from olive, oleander and privet induced galls on olive. Only wild type isolates from oleander were tumorigenic on oleander. An exception was observed with isolate EW 1017 from olive which also caused galls on oleander, however, this strain resembled strains from oleander in that IAA genes were plasmid borne. However, it differs from typical isolates from olive in two major traits, it induces knots on oleander plants and bears the genes for IAA production on a 70kb plasmid. Further, a fragment of PI IAA containing the monooxygenese gene hybridizes with the 70kb plasmid, but not with chromosomal DNA of strain EW 1017. Isolates of EW 1017 cured of the plasmid are Iaa and avirulent on oleander. Iaa^- mutants of the strains from olive and privet did not induce galls on olive and oleander. Iaa^- mutants of strains from oleander produced no symptoms on oleander plants, but induced a typical galls on olive plants. The role of bacterial production of IAA in gall formation on the olive and oleander knot diseases was thus confirmed (Surico *et al.*, 1984).

2.2.2. Fire Blight

Fire blight, a necrotic disease of rosaceous plants caused by *Erwinia amylovora* is especially destructive to apple and pear trees (Schroth *et al.*, 1974). In fact, this is the disease, which has generated two important concepts for Science (Baker, 1971). First, that bacteria may cause diseases of plants, and that insects provide an efficient means of dissemination of bacterial pathogens.

The disease is currently managed using sprays of streptomycin during the bloom period. Streptomycin resistant *E. amylovora* have been isolated from Orchards in California, Washington, Oregon and Michigan (Wodzinski *et al.*, 1994).

The role of saprophytic bacteria in reducing populations of *E. amylovora* and infections is a purely speculative subject. Rosen in 1928 stated that a yellow bacterium apparently limited *E. amylovora* in the amount of host tissue it could invade, but uncharacteristically, he provided no supportive data. Since then various reports (Farabee and Lockwood, 1952; Goodman, 1966, 1967; Smith and Powell, 1966) have suggested that certain yellow bacteria have a beneficial effect in reducing disease, although data to support this also are limited. Later on yellow bacteria was found to belong to the *E. herbicola* group (Dye, 1969; Kamagata, *et al.*, 1968). *E. herbicola* is a non pathogenic bacterium and it occurs as a common epiphyte on apple and pear trees and produces an antibiotic which is inhibitory to *E. amylovora* and can reduce fire blight incidence when sprayed onto apple blossoms before inoculation with *E. amylovora*. The role of the inhibitory substance was demonstrated by comparing the ability of mutant of *E. herbicola* lacking antibiotic production (Ant-) to control fireblight with the ability of wild type strains to control fire blight (Vanneste *et al.*, 1992). The Ant- were not as effective as the wild type in controlling plant disease, suggesting that antibiotic production is one of the mechanisms involved in the biological control of fireblight. Although it is not the only mechanism by which *E. herbicola* suppresses fire blight, as the Ant- mutants retained some ability to reduce fire blight incidence on ornamental trees (Vanneste *et al.*, 1992) and the disease symptoms caused by an antibiotic resistant strain of *E. amylovora* were reduced when the pear plants were pre-inoculated with the antibiotic producing *E. herbicola* (Wodzinski *et al.*, 1994).

One drawback of *E. herbicola* as a biological control agent of fire blight is that the bacterium does not multiply outside of the area of the fruit where it has been introduced (Beer *et al.*, 1984). Therefore, the growth of *E. amylovora* is only restricted in wounds where both bacteria are present.

Vanneste *et al.* (1990) have identified three classes of mutants of *E. amylovora* affected in pathogenicity using transposon mutagenesis. The first class of mutants were non pathogenic on all plant species tested and unable to reduce the HR on tobacco (*hrp* mutant). Class two mutants were nonpathogenic on apple calli but retained the ability to induce a hypersensitive response on tobacco (*dsp* mutant) disease specificity. Class three mutants were impaired in EPS (Extracellular Polysaccharide) production.

3. BACTERIAL WILT

Pseudomonas solanacearum is a soilborne plant pathogen, which causes wilt of solaneceous crops and of many other plants (Buddenhagen and Kelman, 1964; Hayward, 1991). Over 200 plant species are known hosts of the bacterium and new host pathogen combinations are continually being described. Cultivars of several commercial crops have been developed which are resistant to strains of *P. solanacearum* within a geographical area, but the cultivars may or may not be resistant to strains from other areas.

Various attempts have been made to utilize bacteria which are antagonistic towards *P. solanacearum* or avirulent strains of *P. solanacearum* to control bacterial wilt (Chen and Echandi, 1984; El-Abyad *et al.*, 1993; Trigalet and Trigalet Demery, 1990; Trigalet *et al.*, 1994).

Although promising results have been obtained under controlled conditions none of the biocontrol agents have proven reliable under field conditions.

Many attempts have been made to identify the biochemical determinants of pathogenicity of *P. solanacearum* and to study their mode of action in the host plant (Dudman, 1959; Husain and Kelman, 1958; Kelman and Cowling, 1965; Phelps and Sequeira, 1957; Whatley *et al.*, 1980). Most of these studies have been done with collections of wild type strains or with avirulent pleiotropic mutants. Two classes of such mutants of *P. solanacearum* have been described. The first class corresponds to spontaneous avirulent mutants, easily recognizable by their red non-slimy colonies on a tetrazolium chloride/glucose rich medium whereas on the same medium the wild-type strain gives rise to white or pink slimy colonies (Kelman, 1954). Mutants of this type ('red' mutants) have a rough type lipopolysaccharide and are defective for exopolysaccharide production (Hendrick and Sequeira, 1984). The second class corresponds to avirulent mutant resistant to acridine orange (Message *et al.*, 1978). They have the wild type colony morphology but exhibit noticeable differences in the composition of their lipopolysaccharide and exopolysaccharide (Drigues *et al.*, 1985).

Induced mutagenesis should increase the number of avirulent mutants available. Transposons are tools of choice to induce mutagenesis since they generate polar mutations and should further facilitate genetic and physical analysis of the mutants. In addition they allow easy cloning of genes of interest. Transposon mutagenesis has already been used extensively in the study of plant - bacteria interactions with the pathogenic *Pseudomonas syringae* (Anderson and Mills, 1984), with agrobacteria and with rhizobia (de Bruijn and Lupski, 1984). Boucher *et al.* (1981) showed that Tn5, Tn7 and Tn10 can be used for mutagenesis of *P. solanacearum*. Staskawicz *et al.* (1983) obtained *P. solanacearum* avirulent mutants, using Tn5, which are similar in their colony morphology to the spontaneous red mutants.

Boucher *et al.* (1986) reported transposon mutagenesis in tomato isolate of *Pseudomonas solanacearum* using Tn5 and Tn7 inserted in suicide conjugative plasmids. Tn7 integrates at high frequency in a particular site of the genome and Tn5 appears to transpose much more randomly, allowing isolation of auxotrophic mutants with a frequency of 0.35%. The mutants showed a wide range of nutritional requirements. Following Tn mutagensis, screening of 8250 clones on axenic tomato seedlings led to isolation of 12 avirulent mutants. For avirulent mutants, insertion of Tn 5 occurred in at least 10 different EcoRI restriction fragments. Additional indendent insertion of 1*S50* were also detected in four of these mutants.

The invasiveness of 11 Tn-5 induced avirulent mutants of *P. solanacearum* was compared to the invasiveness of their wild-type parent (Trigalet and Demery, 1986). Invasiveness was estimated by the ability of bacterial strains to penetrate into and multiply within the root system of a susceptible host. Of the 11 avirulent mutants, 10 were invasive, but showed reduced colonisation and multiplication relative to the virulent wild type. When susceptible plants were inoculated with mixtures of the avirulent and virulant strains of *P. solanacearum*, the virulent strains were no longer able to efficiently colonize the host tissue (Trigalet and Trigalet, Demery, 1990). Four of the mutants capable of excluding the virulent wild type have insertions mapping to *hrp* genes and one had an insertion mapping to a *dsp* (disease specificity) gene.

The limited invasiveness of *hrp* mutants of *P. solanacearum* and their ability to exclude pathogenic strains from a susceptible host suggest that they could be used as endophytic biocontrol agents (Frey *et al.*, 1994). Three pathogenic strains of *P. solanacearum* were rendered nonpathogenic by insertion of the nontransposable Ω-km interposon (Fellay *et al.*, 1987) within an *hrp* gene (Frey *et al.*, 1994). The ability of the Hrp-mutants to colonize the host tissues was reduced compared with that of a pathogenic strain, but they persisted at low levels in the plant. Upon challenge with pathogenic *P. solanacearum* there was a reduced incidence of disease in plants pre-inoculated with Hrp-strains when compared with the controls.

Erwinia stewartii causes a vascular wilt of sweet corn and a leaf blight of field corn (Pepper, 1967). Water soaking (wts) genes are necessary for pathogenicity and the production of water soaked lesions (Coplin *et al.*, 1992a; Federick *et al.*, 1993). The *wts* genes were required for lesion formation and wilting in corn but were not required for growth within host plant tissue (Coplin and Majerczak, 1990; Coplin *et al.*, 1992a; Dolph *et al.*, 1988). The *wts* genes hybridized with *hrp* genes from *Erwinia* and *Pseudomonas* (Coplin *et al.*, 1992a, Federick *et al.*, 1993).

P. solanacearum produces a variety of virulence factors in addition to the extra cellular polysaccharide (EPS) which increase the severity of wilting in susceptible hosts (Denney et al., 1990). Mutants defective in the export of major extra cellular proteins and Enzymes (ccp) failed to infect susceptible hosts via the roots and disseminates throughout plant tissues (Kang *et al.*, 1994).

P. solanacearum sometimes spontaneously loses pathogenicity under certain conditions. This loss is most frequently due to disruption of the *phc* A gene (Brumbley *et al*., 1993) but has also been linked to the increased expression of a *trans*-acting factor, espR109 or the presence of a small plasmid pJTPS1 (Negishi *et al*., 1993).

4. BACTERIAL SOFT ROT

Bacterial soft rot is the leading cause of decay of many vegetables after harvest. It accounts for over 80% of disorder of vegetables at transit or in the markets (Lund, 1983). *Erwinia carotovora* is generally thought to be the principal cause of the problem (Bartz and Eckert, 1987). In Columbia basin in the State of Washington, *E. carotovora* causes preemergence seed piece decay, blackleg, soft stem rot, and soft rot of daughter tubers. Contaminated seed tubers, soil and irrigation water serve as sources of inocula of *E. carotovora* (Apple and Powelson, 1983; De Boer and Kelman, 1975; Powelson and Apple, 1984). The pathogen colonizes potato root and tuber surfaces and under favourable environmental conditions, populations commonly exceed 10^6 colony-forming units (CFU) per gm of root by mid season. Because soil is an important source of *E. carotovora* primarily, *E.c.* subsp. *corotovora* (Powelson and Apple, 1984), disease control based on planting certified pathogen free potato seed has not been successful in the Columbia basin. Therefore, biological control is a potential method to control soft rot disease. Potatoes could be treated with bacteria, originally selected from potato surfaces, that are inhibitory to the growth of *E. carotovora* Fluorescent pseudomonads are attractive candidates for biological control of *E. carotovora* because they colonize the potato rhizoplane and rhizosphere and produce high populations (Azad *et al*., 1985; Kloepper and Schroth, 1981; Kloepper *et al*., 1980; Loper *et al*., 1984) and they produce a variety of secondary metabolites and substances that can alter the composition of the rhizosphere microflora (Burr and Caesar, 1984; Suslow, 1982). Kloepper (1983) reported that some fluorescent pseudomonas strains when applied to potato seed pieces, reduce the populations of *E. carotovora* on root and daughter tubers by 95-100% and 22-95% respectively, compared to untreated plants. These bacteria are known as plant growth promoting rhizobacteria (PGPR) because of their ability to improve plant growth through suppression of "deleterious" root colonising microorganisms (Suslow and Schroth, 1982). Suppression was attributed to production of fluorescent siderophores that were essential for uptake of iron by the pseudomonads (Kloepper *et al*., 1980). Because the fluorescent pseudomonads produce siderophores with high affinities for iron, deleterious microorganisms which could not obtain iron from these siderophores or produce siderophores with lower affinities for iron were deprived of the iron necessary for growth. Strain B10 of *Pseudomonas fluorescens* (Trevisan) Migula-P *putida* (Trevisan) Migula, for example was originally isolated from potato, it produces the fluorescent siderophore, pseudobactin, which inhibited *E. carotovora* both *in vitro* and *in vivo* (Kloepper *et al*., 1980). Strain B10 suppressed growth of *E. carotovora* and its colonization of potato root and tuber surfaces (Kloepper, 1983). Colyer and Mount (1984) reported the use of fluorescent pseudomonads to control postharvest soft rot of potatoes. Preplant treatments of potato seed pieces with antibiotic producing strain M17 of *P. putida* caused a 6.8 to 18.2% reduction in rotted tissues of the harvested daughter tubers. A mutant of strain M17 unable to produce an antibiotic was about 50% less effective than the parental strain. Burr and Caesar (1984) also noted a decreased incidence of soft rot caused by *Erwinia* spp. in potato tuber disks inoculated with a plant growth-promoting fluorescent pseudomonad.

Numerous other examples are there for successful use of fluorescent pseudomonads (usually *P. fluocrecence* or *P. putida*) for biological control (Geels and Schipper, 1983a, b;

Howie and Echandi, 1983; Howell and Stipanovic, 1980; Scher and Baker, 1982; Sneh *et al.*, 1984; Vrany *et al.*, 1981; Weller and Cook, 1983). For example, strain Pf 5 of *P. fluorescence*, which was isolated from a, cotton rhizosphere, suppressed seedling diseases caused in cotton by *Rhizoctonia solani* and *Pythium ultimum*. Strains Pf 5 produced *in vitro* the antibiotics pyrrolnitrin and pyoluteorin, which were inhibitory to the seedling pathogens. Fluorescent pseudomonad siderophores also have been reported to have leading roles in the biocontrol of several important fungal pathogens (Sneh *et al.*, 1984; Vandenbergh *et al.*, 1983; Xu and Gross, 1986a,b). However, pseudomonads may not be ideally suited as biological control agents under all conditions favouring tuber soft rot because environmental conditions maximising population of *Pseudomonas* spp. are not identical to those conductive for disease development. In certain environments, an avirulent strain of *Erwinia* which would proliferate under the same conditions as the target pathogen might be a superior biocontrol agent (Costa and Loper, 1994; Wilson and Lindow, 1993).

Studies with mutants of *E. chrysanthemi* (Andro *et al.*, 1984; Beaulieu and Gijsegem, 1992; Beaulieu *et al.*, 1993; Hugouviux-Cotte-Pattat *et al.*, 1986; Kelemu and Collmer, 1993; Thurn and Chatterjee, 1985; Payne *et al.*, 1987) and *E. carotovora* (Costa and Loper, 1994; Murata *et al.*, 1990) and *Pseudomonas viridiflava* (Liao *et al.*, 1988) have demonstrated that pectinolytic enzyme production and export are tightly linked with the ability of these bacteria to elicit soft rot disease *Erwinia carotovora* subsp. *carotovora* mutants defective in the production of pectate lyase have been used in the biological control of soft rot (Stromber *et al.*, 1990). However, sequential disruption of structural genes on coding pectolytic enzyme is a laborious process because of the complexity of pectate lyase expression.

All pathogenic strains of *E. chrysanthemi* and *E. carotovora* produce complex sets of pectate-inducible pectate lyase (*pel*) isozymes (Collmer and Keen, 1986; Reverchon *et al.*, 1986). Genetic manipulations of Pel production in *E. chrysanthemi* and pel+ *Escherichia coli* strains have confirmed that a single pel isozyme is sufficient to cause extensive maceration in potato tubers (Boccara et al., 1988; Payne *et al.*, 1987).

Mutants of *E. chrysanthemi* with directed deletions in all of their known pel genes (Δ pel) were still able to macerate plant tissues, although at reduced rates (Beaulieu *et al.*, 1993; Ried and Collmer, 1988). Additional mutations in either the exopolygalacturonate lyase gene (peh x) (He and Collmer, 1990) or the exogalacturonosidase gene (*pelx*) (Brooks *et al.*, 1990) did not diminish the maceration capacity of the pectate lyase-deficient mutant of *E. chrysanthemi*. *E. chrysanthemi* mutant was constructed with deletions or insertions in all of its known genes coding for pectic enzymes by Kelemu and Collmer (1993). The mutant still retained some ability to macerate chrysanthemum tissues although it did not cause pitting on pectate semisolid medium used to detect pectolytic activity in bacteria. It was observed that *E. chrysanthemi* produces a set of plant-inducible pectate lyase isozymes as well as a set of pectate-inducible pectate lyase isozymes. However, only the expression of plant-inducible pectic enzymes is not sufficient to cause systemic disease in whole plants (Beaulieu *et al.*, 1993).

Erwinia carotovora subsp. *betavasculorum* produces and secrete pectolytic enzymes, causing vascular nacrosis and root rot of sugar beet (Thomson *et al.*, 1977). The pathogen can also colonize and macerate potato tubers. *Erwinia carotovora* subsp. *betavasculorum* strain Ecb 168 produces an antimicrobial substance(s) that suppresses the growth of the related bacterium *E. c.* subsp. *carotovora* in culture and in wounds of potato tubers. Therefore, an avirulent mutant of *E. c.* subsp. *betavasculorum* may be an effective agent for the biological control of tuber soft rot caused by *E. c* subsp. *carotovora* (Costa and Lopez, 1994).

The effectiveness of *Erwinia* biocontrol agents might be further enhanced by additional manipulations such as modification of *hrp* genes. The ability of the soft rot pathogens *E. chrysanthemi* and *E. carotovora* to elicit the HR on non-host plants has been uncertain due to the destructive activity of extracellular enzymes on both host and non-host tissues. A Pel ABCE-out-mutant of *E. chrysanthemi* caused rapid necrosis in tobacco leaves that was typical of that elicited by narrow host range pathogens (Bauer *et al.*, 1994). The *hrp* genes of *E. chrysanthemi* were identified using an *hrp* gene cluster from *E. amylovora* (Laby and Beer, 1992) and mutated using transposon mutagenesis (Bauer *et al.*, 1994). The *hrp* constructs were transformed into the out-mutant, and two transformants were identified which did not elicit the HR in tobacco leaves. The mutated *hrp*-containing DNA corresponding to the two mutants was then introduced into a virulent wild-type strain of *E. chrysanthemi*. Introduction of the *hrp* mutation resulted in decreased virulence of both the wild-type and the out-strains on witloof chicory leaves. *Erwinia chrysanthemi* therefore can produce two independent types of necrosis in plant tissues, a macerative necrosis dependent on the secretion of pectic enzymes and a hypersensitive necrosis dependent on the Hrp system.

Bacterial cell surface components such as Lipopolysaccharide (LPS) or proteins might account for plant-bacterium recognition processes. Two mutants of *E. chrysanthemi* which were resistant to EC2, or *Erwinia* tranducing bacteriophage (Resibois *et al.*, 1989) had modifications of the LPS structure of the outer membrane and were avirulent on *Saintpaulia ionantha* (Schoonejnans et al., 1987) and of the avirulent mutants induced systemic resistance in inoculated plants against infection with pathogenic wild-type strains.

Epiphytic bacteria antagonistic to soft rot pathogens of harvested fruits and vegetables are commonly present on the surface of plants. Twenty seven strains antagonistic to *E.c.* subsp. *carotovora* on King's medium B (KB) were isolated and characterized by Liao (1989), of which 24 were identified as either *P. fluorescens* or *P. putida*. Fluorescent pseudomonads usually coexist with other microorganisms in diverse environments, including soil, water, and biomaterials. The prevalence of this group in nature may be due to theircatabolic verstality and their ability to produce a broad spectrum of secondary metabolites that may be toxic to other microorganisms (Leisinger and Margraff, 1979). Two species *P. putida* and *P. fluorescens*, are frequently isolated from plant tissue or soil and reported as potential biocontrol agent of phytopathogens (Colyer and Mount, 1984; Xu and Gross, 1986a,b). Antibacterial activity of *P. putida* strain PP22 may have been associated with the production of both fluorescent siderophores and antibiotics. In addition to ability to produce an antibacterial compound, strain PP22 possesses several other properties that meet the requirements of a biocontrol agent. It colonised the tubers and roots of potato plants and reduced the development of soft rot on potato tubers. Several strains of *P. putida* have been previously isolated and shown to be potential biocontrol agents of soft rot bacterium *E. carotovora* (Burr and Caesar 1984; Colyer and Mount, 1984; Schroth *et al.*, 1984; Xu and Gross, 1986a,b). Strain PP22 has been shown to be antagonistic, at least *in vitro*, to an extremely broad spectrum of bacteria. This strain is therefore potentially useful for control of not only diverse groups of soft rot bacteria but also other field pathogens such as *P. syringae* and *X. compestris* (Liao, 1989).

5. FROST INJURY

Frost injury of sensitive plants may be incited by ice nucleation active (Ice +) strains of *P. syringae*, *P. viridiflora*, *P. fluorescens*, *X. compestris* pv *translucens* and *E. herbicola* which commonly inhibit the leaf surfaces of healthy plants (Hu *et al.*, 1992).

Most frost-sensitive plants have the capacity to super cool to -5 or -6°C before water within the tissue freezes. However, ice+ bacteria catalyze ice formation at temperatures as warm as -2°C, and leaf surface populations of these bacteria initiate damaging ice formation at temperatures of -1 to -5°C. Frost injury is related to the number of bacterial ice nuclei on the plant at the time of freezing. Treatments that reduce the numbers of ice + bacteria on plants reduce the severity of frost injury (Pilgeram and Sands, 1996).

Suppression of ice nucleation-active bacteria can be accomplished in several ways. One way is *via* direct chemical suppression with bacteriocides and the other is *via* antagonism between ice nucleation-active and inactive epiphytic bacteria presumably competing for the same leaf surface niches. Lindow and his coworkers (1983) have demonstrated the effectiveness of both approaches.

In theory, the best competition for the leaf surface niche of an ice nucleation-active bacterium should be the same bacterium lacking the capacity for ice nucleation. The genes coding for ice nucleation activity have been identified using techniques of modern molecular biology (Lindow, 1983) and engineered bacterial strains which combine high competitiveness with a lack of ice nucleation activity have been developed.

Ice nucleation deficient (INA^-) mutants of *Pseudomonas* strains derived by site-directed mutagenesis were tested for their efficacy as biological control agents of frost injury on blossoms of greenhouse-grown strawberry plants (Lindemann and Suslow, 1987). Inhibition of one bacterial strain by its near-isogenic counterpart was dose dependent rather than strain dependent. The INA deletion mutants of *P. syringae* and *P. fluorescens* bio var. I inoculated on 10^7 cfu per blossom inhibited growth of their ice nucleation active (INA^+) potential strains inoculated at 10^2 cfu per blossom. The INA^+ parental strains inhibited INA^+ derivatives when inoculum doses were reversed. Inhibition was incomplete unless doses differed by about 10^4 fold. No inhibition occurred when two strains were inoculated simultaneously at equal doses. *P. syringae* strain protected blossoms against freezing by other *P. syringae* strains but did not inhibit or protect against INA^+ *P. fluorescens*. The INA^- *P. fluorescens* strain was much more effective as an exhibitor of *P. syringae* strains than of other *P. fluorescens* strains (Lindemann and Suslow, 1987).

6. CONCLUSIONS

Biological control of plant disease involve the use of antagonistic microorganisms that interact with the parasite to reduce its effectiveness. It from a theoretical point of view, has a lot of appeal because it does not add chemicals to the environment. The major challenge has been to make it an effective control method, and at the same time, maintain the balance so the biological control agent does not have to be added each time.

The development of molecular biology techniques to obtain and handle DNA has provided the tools for transferring and evaluating genetic characteristics from many different organisms. This has resulted in progress in the development of a new generation of biological control agents whose traits have been modified in a concerted effort to improve the wild type. Important achievements include the creation of biological strains of *A. radiobacter* which control grown gall but are unable to transfer genes encoding bacteriocin resistance to target pathogens. The second achievement is the disruption of pathogenecity genes in *Erwinia* and *Pseudomonas*, resulting in biocontrol strains capable of growing within plant tissue and excluding the parental pathogen but incapable of eliciting plant disease in non compromised hosts. The third accomplishment is the derivation of Ice-strains of epiphytic bacteria which may reduce the severity of frost injury caused by Ice + strains.

REFERENCES

Akiyoshi, D.E., Klee, H., Amasino, R.M., Nestes, E.W., Gardon, M.P. 1984, T-DNA of *Agrobacterium tumefaciens* encodes an enzgme of cytokinin biosynthesis, Proc. Natl. Acad. Sci. 81 : 5994-5998.

Anderson, D.M. and Mills, D. 1984, The use of transposon mutagenesis in the isolation of nutritional and virulence mutants in two pathovars of *Pseudomonas syringae*, Phytopath. 75: 104-108.

Andro, T., Chambost, J.P., Kotoujansky, A., Cattaneo, J., Bertheau, Y., Barras, F., Van Gijsegem, F. and Coleno, A. 1984, Mutants of *Erwinia chrysanthemi* defective in secretion of pectinase and cellulase, J. Bacteriol. 160: 1199-203.

Apple, J.D. and Powelson, M.L. 1983, Potato blackleg in progeny plantings from diseased and symptomless parent plants (Abst), Phytopath. 73: 806.

Azad, H.R., Davis, J.R., Schnathorst, W.C. and Kado, C.I. 1985, Relationship between rhizoplane and rhizosphere bacteria and verticillium wilt disease in potato, Arch. Microbiol. 140: 347-351.

Baker, K.F. 1971, Fire blight of pome fruits : the genesis of the concept that bacteria can be pathogenic to plants, Hilgardia 40 : 603-633.

Bartz, J.A. and Eckert, J.W. 1987, Bacterial diseases of vegetable crops after harvest, in: Postharvest Physiology of Vegetables, J. Weichamann, ed., Marcel Dekker Inc. New York, pp. 351-375.

Bauer, D.W. and Beer, S.V. 1987, Cloning of gene from *Erwinia amylovora* involved in induction of hypersensitivity and pathogenicity, in : Plant Pathogenic Bacteria, E.L. Civerolo; A. Collmer; R.E. Davis and A.G. Gillaspie, eds., Martinus Nijhott, Boston, 425.

Bauer, D.W., Bogdanove, A.J., Beer, S.V. and Collmer, A. 1994, *Erwinia chrysanthemi hrp* genes and their involvement in soft rot pathogenesis and elicitation of the hypersensitive response, Mol. Plant-Microbe Interact. 7: 573.

Beaulieu, C. and Van Gijsegem, F. 1992, Pathogenic behaviour of several mini-mu-induced mutants of *Erwinia chrysanthemi* on different plants, Mol. Plant Microbe Interact 5: 340.

Beaulieu, C., Boccora, M. and Van Gijsegem, F. 1993, Pathogenic behavior of pectinase defective *Erwinia chrysanthemi* mutants on different plants, Mol. Plant-Microbe Interact 6: 197.

Beer, S.V., Rundle, J.R. and Norelli, J.L. 1981, Biological control of fireblight by *Erwinia herbicola*, in: Proc. Int. Conf. Plant Pathog. Bac. 5th Coli, Columbia, pp. 596-597.

Beer, S.V., Rundle, J.R. and Norelli, J.L. 1984, Recent progress in the development of biological control for fire blight - a review, Acta Hortic. 151 : 195.

Beltra, R. 1958, Relacion entre la concentration de triptofano en los tallos do olivo Y la localizacion de los tumorer bacterianos, Microbiol. Espan. 11: 401-410.

Beltra, R. 1961, Effecto morfogenetico observado en los extractos hormonales de los tumores del olivo, Microbiol. Espan. 14: 177-187.

Bender, C.L. and Cooksey, D.A. 1986, Indigenons plasmids in *Pseudomonas syringae* pv. *tomato* : Conjugative transfer and role in copper resistance, J. Bacteriol. 165: 534-541.

Boccara, M., Diolez, A., Rouve, M. and Kotoujansky, A. 1988, The role of individual pectate lyases of *Erwinia chrysanthemi* strain 393 in pathogenicity on *Saintpaulia* plants, Physiol. Mol. Plant Pathol. 33: 95.

Bonas, U., Schutle, R., Fenselau, S., Minsavage, G.V., Starkawicz, B.L. and Stall, R.E. 1991, Isolation of a gene cluster from *Xanthomonos compestris* pv. *vesicatoria* that determines pathogenicity and the hypersensitive response on pepper and tomato, Mol. Plant-Microbe Interact. 4: 81.

Botalico, A. and Ercolani, G.L. 1971, *Pseudomonas savastanoi* (F.F. Smith) Stevens Su Ligustro giappunese in Pugila, Phytopath, Mediterr. 10: 132-135.

Boucher, C.A., Barberis, P.A., Trigalet, A.P. and Demery, D.A. 1985, Transposon mutagenesis of *Pseudomonas solanacearum* : Isolation of Tn5 induced avirulent mutants, J. Gen. Microb. 131: 2449-2457.

Boucher, C., Message, B., Dedieu, D. and Zischek, C. 1981, Uses of P-1 incompatability group plasmids to introduce transposons into *Pseudomonas solanacearum*, Phytopath. 71: 639-642.

Boucher, C.A., Van Gijsegem, F., Barberis, P.A., Arlat, M. and Zischek, C. 1987, *Pseudomonas solanacearum* genes controlling both pathogenecity on tomato and hypersensitivity on tobacco are clustered, J. Bacteriol. 169: 5626.

Brooks, A.P., He, S.Y., Gold, S., Keen, N.T., Collmer, A. and Hutcheson, S.W. 1990, Molecular cloning of the structural gene for exopolygalacturonate lyase from *Erwinia chrysanthemi* EC 16 and characterization of the enzyme product, J. Bacteriol. 172: 6950.

Brumbley, S.M., Corney, B.F. and Denny, T.P. 1993, Phenotype conversion in *Pseudomonas solanacearum* due to spontaneous inactivation of *Phc* A, a putative Lys R transcriptional regulator, J. Bacteriol. 175: 5477.

Buddenhagen, I. and Kelman, A. 1964, Biological and physiological aspects of bacterial wilt caused by *Pseudomonas solanacearum*, Ann. Rev. Phytopath. 2: 203-230.

Burr, T.J. and Caesar, A. 1984, Bacterization of potatoes with *Pseudomonas putida* and its influence on postharvest soft rot diseases, Plant Dis. 68: 703-706.

Chen, W. and Echandi, E. 1984, Effects of avirulent bacteriocin-producing strains of *Pseudomonas solanacearum* on the control of bacterial wilt of tobacco, Plant Pathol. 33: 245.

Chilton, M.D., Drummond, M.H., Merlo, D.J., Sciasky, D., Montoya, A.L., Gordon, M.P. and Nester, E.W. 1977, Stable incorporation of plasmid DNA into higher plant cells, The molecular basis of crown gall tumorigenesis, Cell 11: 263-271.

Chilton, M.D., Saiki, R.K., Yadav, N., Gordon, M.P. and Quetier, F. 1980, T-DNA from *Agrobacterium* Ti plasmid is in the nuclear DNA fraction of crown gall tumor cells, Proc. Natl. Acad. Sci., USA 77 : 4060-4064.

Collmer, A. and Keen, N.T. 1986, The role of pectic enzymes in plant pathogenesis, Ann. Rev. Phytopath. 24: 383-409.

Colyer, P.D. and Mount, M.S. 1984, Bacterization of potatoes with *Pseudomonas putida* and its influence on postharvest soft rot diseases, Plant. Dis. 68: 703-706.

Conlin, K.C. and McCarter, S.M. 1983, Effectiveness of selected chemicals in inhibiting *Pseudomonas syringae* pv *tomato* in vitro and in controlling bacterial speck, Plant Dis. 67: 639-644.

Cooksey, D.A. 1987, Characterization of a copper resistance plasmid conserved in copper-resistance strains of *Pseudomonas syringae* pv. *tomato*, Appl. Environ. Microbiol. 53 : 454-456.

Cooksey, D.A. 1988, Reduction of infection by *Pseudomonas syringae* pv. *tomato* using a nonpathogenic, copper resistant strain combined with a copper bactericide, Phytopath. 78: 601.

Cooksey, D.A. and Moore, L.W. 1980, Biological control of crown gall with fungal and bacterial antagonists, Phytopath. 70: 506-509.

Cooksey, D.A. and Moore, L.W. 1982, Biological control of crown gall with an agrocin mutant of *Agrobacterium radiobacter*, Phytopath. 72: 919.

Coplin, D.L. and Majerczak, D.R. 1990, Extracellular polysacchride genes in *Erwinia stewartii* directed mutagenesis and complementation analysis, Mol. Plant-Microbe Interact. 3: 286.

Coplin, D.L., Frederick, R.D. and Majerczak, D.R. 1992a, New pathogenicity loci in *Erwinia stewartii* identified by random Tn5 mutagenesis and molecular cloning, Mol. Plant Microbe Interact. 5: 266.

Coplin, D.L., Frederick, R.D., Majerczak, D.R. and Tuttle, L.D. 1992b, Characterization of a gene cluster that specified pathogenicity in *Erwinia stewartii*, Mol. Plant-Microbe Interact. 5: 81.

Costa, J.M. and Loper, J.E. 1994, Derivation of mutants of *Erwinia carotovora* subsp. *betavasculorum* deficient in export of pectolytic enzymes with potential for biological control of potato soft rot, App. Environ. Microbiol. 60: 2278.

Cuppels, D.A. 1986, Generation and characterization of Tn5 insertion mutations in *Pseudomomas syringae* pv. *tomato*, App. Environ. Microbiol. 51 : 323.

Das, P.K., Basy, M. and Chatterjee, G.C. 1978, Studies on the mode of action of agrocin 84, J. Antibiot. 31: 490-492.

De Boer, S.H. and Kelman, A. 1975, Evaluation of procedure for detection of pectolytic *Erwinia* spp. on potato tubers, Am. Potato. J. 52: 117-123.

de Brauijn, F.S. and Lupski, J.R. 1984, The use of transposon Tn5 mutagenesis in the rapid generation of correlated physical and genetic maps of DNA segments cloned into multicopy plasmids - a review, Gene 27: 131-149.

De Cleene, M. and De Ley, J. 1976, The host range of crown gall, Bot. Rev 42: 389.

Denny, T.P., Corney, B.F. and Schell, M.A. 1990, Inactivation of multiple virulence genes reduces the ability of *Pseudomonas solanacearum* to cause wilt symptoms, Mol. Plant-Microbe Interact. 3: 293.

Dolph, P.J., Majerczak, D.R. and Coplin, D.L. 1988, Characterization of a gene cluster for exopolyssacharide biosynthesis and virulence in *Erwinia stewartii*, J. Bacteriol. 170: 865.

Donner, S.C., Jones, D.A., McClure, N.C., Rosenwarne, G.M., Tate, M.E., Kerr, A., Fajardo, N.N. and Clare, B.G. 1993, Agrocin 434, a new plasmid encoded agrocin from the biocontrol *Agrobacterium* strains K84 and K1026, which inhibits biovar 2 agrobacteria, Physiol. Mol. Plant Pathol. 42: 185.

Drigues, P., Demery-Lafforgue, D., Trigalet, A., Dupin, P., Samain, D. and Asselineau, J. 1985, Comparative studies of lipopolysaccharide and exopolysaccharide from a virulent strain of *Pseudomonas solanacearum* and from three avirulent mutants, J. Bacteriol. 162: 504-509.

Duddman, W.F. 1959, Comparison of stem from tomato and banana strains of *Pseudomonas solanacearum*, Nature 184: 1969-1970.

Dye, D.W. 1969, A taxonomic study of the genus *Erwinia* 3, The "herbicola" group N.Z.J. Sci. 12: 223-236.

El-Abyad, M.S., El-Sayed, M.A., El-Stanshoury, A.R., El-Batanouny, N.H. 1993, Inhibitory effects of UV mutants of *Streptomyces corchorusii* and *Streptomyces spiroverticillatus* on bean and banana wilt pathogens, Can. J. Bot. 71: 1080.

Ellis, J.G., Keer, A., Petit, A. and Tempe, J. 1982, Conjugal transfer of nopaline and agropin Ti-plasmids - The role of agrocinopines, Mol. and Gen. Genet. 186 : 269-274.

Ellis, J.G., Kerr, A., Van Montagu, M., Schell, J. 1979, *Agrobacterium*: genetic studies on agrocin 84 production and the biological control of crown gall, Physiol. Plant Pathol. 15: 311-319.

Ellis, J.G. and Murphy, P.J. 1981, Four new opines from crown gall tumors - their detection and properties, Mol. Gen. Genet. 181: 36.

Farabee, G.J. and Lockwood, J.L. 1958, Inhibition of *Erwinia amylovora* by *Bacterium* sp. isolated from fire blight cankers, Phytopath. 48 : 209-211.

Farrand, S.K., Slota, J.E., Shim, J.S. and Kerr, A. 1985, Tn5 insertions in the agrocin 84 plasmids: the conjugal nature of PAGK84 and the locations of determinants for transfer and agrocin 84 production, Plasmid 13: 106.

Fellay, R., Frey, J. and Krish, H. 1987, Interposon mutagenesis of soil and water bacteria: a family of DNA fragments desgined for in vitro insertional mutagenesis of gram-negative bacteria, Gene 52: 147.

Frederick, R.D., Majerzak, D.R. and Coplin, D.L. 1993, *Erwinia stewartii* WtsA, a positive regulator of pathogenicity gene expression is similar to *Pseudomonas syringae* pv. *phasiolicola* HrpS, Mol. Microbiol. 9: 477.

Frey, P., Prior, P., Marie, C., Kotoujansy, A., Trigalet-Demery, D. and Trigalet, A. 1994, Hrp-mutants of *Pseudomonas solanacearum* as potential biocontrol agents or tomato bacterial wilt, Appl. Environ. Microbiol. 60: 3175.

Geels, F.P. and Schippers, B. 1983a, Selection of antagonistic *Pseudomonas* spp. and their root colonization and persistence following treatment of seed potatoes, Phytopath. Z. 108: 193-206.

Geels, F.P. and Schippers, B. 1983b, Reduction of yield depressions in high frequency potato cropping soil after seed treatments with antagonistic fluorescent *Pseudomonas* spp., Phytopath. Z. 108 : 207-214.

Goodman, R.N. 1966, The protection of apple tissue against infection by *Erwinia amylovora* afforded by avirulent strains of *E. amylovora*, a saprophytic bacterium and other bacterial plant pathogens, Rev. Roum. Biol. Ser. Bot. 11: 75-79.

Goodman, R.N. 1967, Protection of apple stem tissue against *Erwinia amylovora* infection by avirulent strains and three other bacterial species, Phytopath. 57: 22-24.

Gross, P. 1991, Molecular and genetic analysis of toxin production by pathovars of *Pseudomonas syringae*, Ann. Rev. Phytopath. 29: 247-248.

Guyon, P., Chilton, M.D., Petit, A. and Tempe, J. 1980, Agrocine in "null-type" crown gall tumours : Evidence for generality of the opine concept, Proc. Natl. Acad. Sci. 77: 2693-2697.

Hayward, A.C. 1991, Biology and epidemiology of bacterial wilt caused by *Pseudomonas solanacearum*, Ann. Rev. Phytopathol. 29: 65.

He, S.Y. and Collmer, A. 1990, A molecular cloning, nucleotide sequence and marker exchange mutagenesis of the exo-poly-̇-D-galacturonosidase-encoding *peh X* gene of *Erwinia chrysanthemi* EC16, J. Bacteriol. 172: 4988.

Hendrick, C.A. and Sequiera, L. 1984, Lipopolysaccharide defective mutants of the wilt pathogen *Pseudomonas solanacearum*, Appl. Envir. Micro. 48: 94-101.

Howell, C.R. and Stipanovic, R.D. 1980, Suppression of *Pythium ultimum* induced damping off of cotton seedlings by *Pseudomonas fluorescens* and its antibiotic, pyoluteorin, Phytopath. 70: 712-715.

Howie, W.J. and Echandi, E. 1983, Rhizobacteria: Influence of cultivar and soil type on plant growth and yield of potato, Soil Biol. Biochem. 15: 127-132.

Hu, N.T., Hung, M.N., Chiou, S.J., Tang, F., Chiang, D.G., Huang, H.Y. and Wu, C.Y. 1992, Cloning and characterization of a gene required for the secretion of extracellular enzymes across the outer membrane by *Xanthomonas compestris* pv. *compestris*, J. Bacteriol. 174: 2679.

Huang, H.C., Schuurink, R., Denny, T.P., Atkinson, M.M., Baker, C.J., Yucel, I., Hutchinson, S.W. and Collmer, A. 1988, Molecular cloning of a *Pseudomonas syringae* pv *syringae* gene cluster that enables *Pseudomonas fluorescens* to elicit the hypersensitive response in tobacco plants, J. Bacterol. 170: 4748.

Hugouvieux-Cotte-Pattat, N., Reverchon, S., Condemina, G. and Robert-Baudouy, J. 1986, Regulatory mutations affecting the synthesis of pectate lyase in *Erwinia chrysanthemi*, J. Gen. Microbiol. 132: 2099.

Hussain, A. and Kelman, A. 1958, The role of pectic and cellulolytic enzymes in pathogenesis by *Pseudomonas solanacearum*, Phytopath. 48: 377-386.

Inze, D., Follin, A. Van Lijsebettens, M., Simoens, C., Genetello, C., Van Montagu, M. and Schell, J. 1984, Genetic analysis of the individual T-DNA genes of *Agrobacterium tumefaciens*, Further evidence that two genes are involved in indole-3-acetic acid synthesis, Mol. Gen. Genet. 194 : 265-274.

Jardine, D.J. and Stephens, C.T. 1987, Influence of timing of application and chemical on control of bacterial speck of tomato, Plant Dis. 71: 405-408.

Jones, D.A. and Kerr, A. 1989, *Agrobacterium radiobacter* strain K1026, a genetically engineered derivative of strain K84 for biological control of crown gall, Plant Dis. 73: 15.

Kamagata, K., Tamagawa, Y. and Kocur, M. 1968, Differntiation of *Erwinia amylovora, Erwinia carotovora* and *Erwinia herbicola*, J., Gen. Appl. Microbiol. Tokyo 14 : 39-45.

Kang, Y., Huang, J. Mao, G., He, L. and Schell, M.A. 1994, Dramatically reduced virulence of mutants of *Pseudomonas solanaceorum* defective in export of extra cellular proteins across the outer membrane, Mol. Plant-Microbe Interact. 7 : 370.

Kelemu, S. and Collmer, A. 1993, *Erwinia chrysanthemi* EC 16 produces a second set of plant inducible pectate lyase isozymes, App. Environ. Microbiol. 59: 1756.

Kelman, A. 1954, The relationship of pathogenicity of *Pseudomonas solanacearum* to colony appearance on a tetrazolium medium, Phytopath. 44: 693-695.

Kelman, A. and Cowling, E.B. 1965, Cellulose of *Pseudomonas solanacearum* in relation to pathogenesis, Phytopath. 55: 148-155.

Kerr, A. and Htay, K. 1974, Biological control of crown gall through bacteriocin production, Physiol. Plant Pathol. 4: 37-44.

Kerr, A. and Tate, M.E. 1984, Agrocine and the biological control of crown gall, Microbiol. Sci. 1 : 1-4.

Klement, Z. 1982, Hypersensitivity, in : Phytopathogenic Prakaryotes, Vol. 2, M.S. Mount and G.H. Lacy, eds., Academic Press, New York, 149.

Kloepper, J.W. 1983, Effect of seed piece inoculation with plant-growth promoting rhizobacteria can populations of *Erwinia carotovora* on potato roots and daughter tubers, Phytopath. 713: 217-219.

Kloepper, J.W. and Schroth, M.N. 1981, Relationship of *in vitro* antibiosis of plant growth-promoting rhizobacteria to plant growth and the displacement of root microflora, Phytopath. 71: 1020-1024.

Kloepper, J.W., Schroth, M.N. and Miller, T.D. 1980, Effect of rhizosphere colonization by plant growth - promoting rhizobacteria on potato plant development and yield, Phytopath. 70 : 1078-1082.

Laby, R.J. and Beer, S.V. 1992, Hybridization and functional complementation of the *hrp* gene cluster from *Erwinia amylovora* strain Ea 321 with DNA of other bacteria, Mol. Plant-Microbe Interact. 5: 412.

Leisinger, T. and Margraff, R. 1979, Secondary metabolities of the fluorescent pseudomonad, Microb. Rev. 43: 422-442.

Liang, L.S. and Jones, A.L. 1995, Organization of the *hrp* gene cluster and nucleotide sequence of the *hrp* L gene from *Pseudomonas syringae* pv. *morsprunorum*, Phytopath. 85: 118.

Liao, C.H. 1989, Antagonism of *Pseudomonas putida* strain PP22 to phytopathogenic bacteria and its potential use as a biocontrol agent, Plant Dis. 73: 223.

Linderman, J. and Suslow, T.V. 1987, Competition between ice nucleation active wild type and ice-nucleation deficient deletion mutant strains of *Pseudomonas syringae* and *P. fluorescens* biovar. I and biological control of frost injury on strawberry blossoms, Phytopath. 77: 882-886.

Lindgren, P.B., Fredrick, R., Govindarajan, A.G., Panopoulos, N.J., Staskawicz, B.J. and Lindow, S.E. 1989, An ice nucleation reporter gene system: identification of inducible pathogenicity genes in *Pseudomonas syringae* pv. *phasiolicola*, EMBO J. 8: 1291.

Lindgren, P.B., Panopoulos, N.J., Starkawicz, B.J. and Dahlbeck, D. 1988, Genes required for pathogenicity and hypersensitivity are conserved and interchangeable among pathovars of *Pseudomonas syringae*, Mol. Gen. Genet. 211: 499.

Lindren, P.B., Peet, R.C. and Panapoulos, N.J. 1986, Gene cluster of *Pseudomonas syringae* pv. *phasiolicola* controls pathogenicity of bean plants and hypersensitivity on non host plants, J. Bacteriol. 168: 512.

lindow, S.E. 1983, Methods of preserving frost injury caused by epiphytic ice-nucleation active bacteria, Plant Dis. 67; 327.

Lio, C.H., Hung, H.Y. and Chatterjee, A.K. 1988, An extracellular pectate lyase is the pathogenicity factor of the soft-rotting bacterium *Pseudomonas viridiflora*, Mol. Plant-Microbe. Interact. 1: 99.

Loper, J.E., Suslow, T.V. and Schroth, M.N. 1984, Lognormal distribution of bacterial populations in the rhizosphere, Phytopath. 74 : 1454-1460.

Lopez, M.M., Gorris, M.T., Salcedo, C.I., Montajo, A.M. and Miro, M. 1989, Evidence of biological control of *Agrobacterium tumefaciens* strains sensitive and resistant to agrocin 84 by different *Agrobacterium radiobacter* strains on some fruit trees, Appl. Environ. Microbiol. 55: 741.

Lorang, J.M., Shen, H., Kobayashi, D., Cooksey, D. and Keen, N.T. 1994, avrA and avrE in *Pseudomonas syringae* pv. *tomato* PT 23 play a role in virulence on tomato plants, Mol. Plant-Microbe Interac. 7: 508.

Mac Nab, A.A. 1979, Tomato bacterial speck control with fungicides 1978, Fungic Nematic Tests 34 : 84.

Mac Nab, A.A. 1980, Tomato bacterial speck and early blight control with fungicides, Fungic Nematic Tests 36: 161.

Malik, A.N., Vivian, A. and Taylor, J.D. 1987, Isolation and partial characterization of three classes of mutants in *Pseudomonas syringae* pathovar *pisi* with altered behavior towards their host *Pisum sativum*, J. Gen. Microbiol. 133: 2393.

McCardell, B.A. and Pootjes C.F. 1976, Chemical nature of agrocin 84 and its effect on a virulent strain of *Agrobacterium tumefaciens*, Antimicrob. Agents Chomother. 10 : 498-502.

Message, B., Boistard, P., Pitrat, M., Schmit, J. and Boucher, C. 1978, A new class of fluidal avirulant mutants of *Pseudomonas solanacearum* unable to induce a hypersensitive reaction, in : Proceedings of the 4th International Conference on Plant Pathogenic Bacteria, Angers. INRA, pp. 822-833.

Moore, L.W. 1977, Prevention of crown gall on *Prunus* roots by bacterial antagonists, Phytopath. 67: 139-144.

Mukhopadhyay, P., Williams, J. and Mills, D. 1988, Molecular analysis of pathogenicity locus in *Pseudomonas syringae* pv. *syringae*, J. Bacteriol. 170: 5479.

Murata, H., Fons, M., Chatterjee, A., Collmer, A. and Chatterjee, A.K. 1990, Characterization of transposon insertion *Out*-mutants of *Erwinia carotovora* subsp. *carotovora* defective in enzyme export and of a DNA segment that complements *Out* mutations in *E. carotovora* subsp. *carotovora, E. carotovora* subsp. *atroseptica* and *Erwinia chrysanthemi*, J. Bacteriol. 172: 2970.

Negishi, H., Yamada, T., Shiraishi, T., Oku, M. and Tanaka, H. 1993, *Pseudomonas solanacearum*: plasmid PJTPSI mediates a shift from the pathogenic to nonpathogenic phenotype, Mol. Plant-Microbe. Interac. 6: 203.

Panagopoulos, C.G., Psallidas, P.G. and Alivizatos, P.S. 1979, Evidence of a breakdown in the effectiveness of biological control of crown gall, in : Soil Borne Plant Pathogens, B. Schippers and W. Gams, eds., Academic Press, London, pp. 569-578.

Payne, J.H., Schoedel, C., Keen, N.J. and Collmer, A. 1987, Multiplication and virulence in plant tissues of *Escherichia coli* clones producing pectate lyase isozymes PLb and PLe at high levels and of an *Erwinia chrysanthemi* mutant deficient in PLe, Appl. Environ. Microbiol. 53: 2315.

Pepper, E.H. 1967, Stewart's Bacterial Wilt of Corn Monograph No. 4, American Phytological Society., St. Paul.

Phelps, R.H. and Sequeira, L. 1967, Synthesis of indol-acetic acid by cell free systems from virulent and avirulent strains of *Pseudomonas solanacearum*, Phytopath. 57: 1182-1190.

Pilegeram, A.L. and Sands, D.C. 1996, Molecular biology of the biological control of plant bacterial diseases, in: Molecular Biology of the Biological Control of Pests and Diseases of Plants, M. Gurasekaran and D.J. Weber, eds., CRC Press, Boca Raton, pp. 39-55.

Powelson, M.L. and Apple, J.D. 1984, Soil and seed tubers as sources of inoculum of *Erwinia carotovora* pv *carotovora* for stem soft rot of potatoes, Phytopath. 74: 429-432.

Rahme, L.G., Mindrinos, M.N. and Panopoulos, N.J. 1991, The genetic and transcriptional organization of the *hrp* clusters of *Pseudomonas syringae* pathovar *phasiolicola*, J. Bacteriol. 170: 5479.

Resibois, A., Colet, M., Faelen, M., Schoonejans E. and Toussaint, A. 1984, PEC2, a new generalized transducing phage of *Erwinia chrysanthemi*, Virology, 137: 102.

Reverchon, S., Van Gijsegem, F., Roure, M., Kotoujansky, A. and Robert-Baudouy, J. 1986, Organization of a pectate lyase gene family in *Erwinia chrysanthemi*, Gene 49: 215.

Ried, J.L. and Collmer, A. 1988, Construction and characterization of an *Erwinia chrysanthemi* mutant with directed deletions in all of the pectate lyase structural genes, Mol. Plant-Microbe Interact. 1: 32.

Rosen, H.R. 1928, Variations within a bacterial species, I, Morphologic variations, Mycologia 20: 251-275.

Ryder, M.H., Slota, J., Scarim, A. and Farrand, S.K. 1987, Genetic analysis of agrocin 84 production and immunity in *Agrobacterium* spp., J. Bacteriol. 169: 4184-4189.

Scher, F.M. and Baker, R. 1982, Effect of *Pseudomonas putida* and a synthetic iron chelator on induction of soil suppresiveness to *Fusarium* wilt pathogens, Phytopath. 72: 1567-1573.

Schoonejans, E., Expert, D. and Toussaint, A. 1985, Bacteriocin-resistant mutants of *Erwinia chrysanthemi* possible involvement of iron acquisition in phytopathogenicity, J. Bacteriol. 163: 221.

Schroder, G., Waffenschmidt, S., Weiter, E.W. and Schroder, J. 1984, The T- region of Ti plasmids codes for an enzymes synthesising indole- 3- acetic acid, Eur. J. Biochem. 138: 387-391.

Schroth, M.N. and Moller, W.J. 1976, Crown gall controlled in the field with a nonpathogenic bacterium, Plant. Dis. Rep. 60: 275-278.

Schroth, M.N., Hildebrand, D.C. and Starr, M.P. 1981, Phytopathogenic members of the genus *Pseudomonas*, in: The Procaryotes, M.P. Starr, H. Stolp, H.G. Trupea, A. Bollows, and H.G. Schlegal, eds., Springer Verlag, Berlin, 701.

Schroth, M.N., Loper, J.E. and Hildebrand, D.C. 1984, Bacteria as biocontrol agents of plant diseases, in: Current Perspectives in Microbial Ecology, M.J. King and C.A. Reddy, eds., American Society for Microbiology, Washington D.C., pp. 362-369.

Schroth, M.N., Thomson, S.V. and Hildebrand, D.C. 1974, Epidemiology and control of fireblight, Ann. Rev. Phytopath. 12: 389-412.

Shim, J.S., Farrand, S.K. and Kerr, A. 1987, Biological control of crown gall: Construction and testing of new biocontrol agents, Phytopath. 77: 463.

Smidt, M. and Kosuge, T. 1978, The role of indole-3-acetic acid accumulation by alpha methyl tryptophan-resistant mutants of *Pseudomonas savastanoi* in gall formation on oleanders, Physiol. Plant Pathol. 13: 203-214.

Smith, J.H. and Powell, D. 1966, Characteristics of a yellow bacterium isolated from the tissue of Jonathan apple trees, Trans. Illinois State Acad. Sci. 59: 358-363.

Sneh, B., Dupler, M., Elad, Y. and Baker, R. 1984, Chlamydospore germination of *Fusarium oxysporum* f. sp. *cucumerinum* as affected by fluorescent and lytic bacteria from *Fusarium*-suppressive soil, Phytopath. 74: 1115-1124.

Staskawicz, B., Dahlbeck, D. Miller, J. and Damm, D. 1983, Molecular analysis of virulence genes in *Pseudomonas solanacearum*, in : Molecular Genetics of Bacteria - Plant Interactions, A. Puhler, and Heidelberg, eds., Springer verlag, Berlin, pp. 345-352.

Stromber, V.K., Orvos, D.R. and Scanferlato, V.S. 1990, *In planta* competition among cell degrading enzyme mutants and wild-type strains of *Erwinia carotovora*, Proc. 7th Int. Conf. Plant Pathogenic Bacteria. Z. Klements, ed., Akedimia Kiado, Budapest, 721.

Surico, G, Sparapano, L., Lerario, P., Durbin, R.D. and Iacobellis, N. 1975, Cytokinin-like activity in extracts from culture filtrates of *Pseudomonas savastanoi*, Experientia 31 : 929-930.

Surico, G., Comai, L. and Kosuge, T. 1984, Pathogenicity of strains of *Pseudomonas syringae* pv. *savastanoi* and their indole acetic acid-deficient mutants on olive and oleander, Phytopath. 74: 490.

Suslow, T.V. 1982, Role of root-colonizing bacteria in plant growth, in: Phytopathogenic Prokaryotes Vol. 1, M.S. Mount and G.H. Lacy, eds., Academic Press, New York, pp. 187-223.

Suslow, T.V. and Schroth, M.N. 1982, Rhizobacteria of sugar beets: Effects of seed application and root colonization on yield, Phytopath. 72: 199-206.

Thomson, S.V., Hilderbrand, D.C. and Schroth, M.N. 1977, Identification and nutritional differentiation of the *Erwinia* sugar beet pathogen from members of *Erwinia carotovora* and *Erwinia chrysanthemi*, Phytopath. 71: 1037.

Thurn, K.K. and Chatterjee, A.K. 1985, Single-site chromosomal Tn5 insertions affect the export of pectolytic and cellulolytic enzymes in *Erwinia chrysanthemi* EC16, Appl. Environ. Microbiol. 50: 894.

Trigalet, A. and Trigalet-Demery, D. 1986, Invasiveness in tomato plants of Tn5- induced avirulent mutants of *Pseudomonas solanacearum*, Physiol. Mol. Plant Pathol. 28: 423.

Trigalet, A. and Trigalet-Demery, D. 1990, Use of avirulent mutants of *Pseudomonas solanacearum* for the biological control of bacterial wilt of tomato plants, Physiol. Mol. Plant Pathol. 36: 27.

Trigalet, A., Frey, P. and Trigalet-Demery, D. 1994, Biological control of bacterial wilt caused by *Pseudomonas solanacearum*: State of the art and understanding, Bacterial Wilt: The Disease and its Causative Agent, *Pseudomonas solanacearum*, A.C. Hayward and G.L. Hartman, eds., CAB International Wallingford, CT. 225.

Van Larebeke, N., Genetello, C. , Schell, J., Schiperoort, R.A., Hermans, A.K., Hernalsteens J.P., and Van Mortagu M. 1975, Acquisition of tumor-inducing ability by non-oncogenic bacteria as a result of plasmid transfer, Nature 255: 742-743.

Vandenbergh, P.A., Gonzalez, C.F., Wright, A.M. and Kunka, B.S. 1983, Iron-chelating compounds produced by soil pseudomonads : Correlation with fungal growth inhibition, Appl. Environ. Microbiol. 46 : 128-132.

Vanneste, J.L., Paulin, J.P. and Expert, P. 1990, Bacteriophage Mu as a genetic tool to study *Erwinia amylovora* pathogenicity and hypersensitive reaction on tobacco, J. Bacteriol. 172: 932.

Vanneste, J.L., Yu, J. and Beer, S.V. 1992, Role of antibiotic production by *Erwinia herbicola* Eh 252 in biological control of *Erwinia amylovora*, J. Bacteriol. 174: 2785.

Vrany, J., Vancura, V. and Stanek, M. 1981, Control of microorganisms in the rhizosphere of wheat by inoculation of seeds with *Pseudomonas putida* and by foliar application of urea, Folia Microbiol. 26: 45-51.

Watson, B., Currier, T.C., Gordon, M.P., Chilton M.D. and Nester, E.W. 1975, Plasmid required for virulence of *Agrobacterium tumefaciens*, J. Bacteriol. 123: 255-264.

Weller, D.M. and Cook, R.J. 1983, Suppression of take-all of wheat by seed treatments with fluorescent pseudomonads, Phytopath. 73: 463-469.

Whatley, M., Hunter, N., Cantrell, M.A., Hendrick, C., Kergstra, K. and Sequeira, L. 1980, Lipopolysaccharide composition of the wilt pathogen, *Pseudomonas solanacearum* correlation with the hypersensitive response in tobacco, Plant Physiol. 65: 557-559.

Willmitzer, L., De Beuckeleer, M., Lemment, M.V., Montagu, M. and Schell, J. 1980, DNA from Ti plasmid present in nucleus and absent from plasmids of crown gall plant cells, Nature 287 : 359-361.

Wilson, E.E. and Magie, A.R. 1963, Physiological serological and pathological evidence that *Pseudomonas tonelliana* is identical with *Pseudomonas savastanoi*, Phytopath. 53 : 653-659.

Wilson, M. and Lindow, S. 1993, Release of recombinant microorganisms, Ann. Rev. Microbiol. 47: 913.

Winans, S. 1992, Two-way chemical signalling in *Agrobacterium* plant interactions, Microbiol. Rev. 56: 12.

Wordzinski, R.S., Umhostz, T.E., Rundle, J.R. and Beer, S.V. 1994, Mechanisms of inhibition of *Erwinia amylovora* by *Erwinia herbicola in vitro* and *in vivo*, J. Appl. Bact. 76: 22.

Xu, G.W. and Gross, D.C. 1986a, Selection of fluorescent pseudomonads antagonistic to *Erwinia carotovora* and suppressive of potato seed piece decay, Phytopath. 76: 414-422.

Xu, G.W. and Gross, D.C. 1986b, Field evaluations of the interactions among fluorescent pseudomonas *Erwinia carotovora* and potato yield, Phytopath. 76: 423-430.

Yunis, H., Bashan, Y. and Henis, Y. 1980, Weather dependence, yield losses, and control of bacterial speck of tomato caused by *Pseudomonas tomato*, Plant Dis. 64: 937-939.

Zambryski, P., Tempe, J. and Schell, J. 1989, Transfer and function of T-DNA genes from *Agrobacterium* Ti and Ri plasmids in plants, Cell 56: 193.

PROTOPLAST FUSION IN DISEASE CONTROL

Sumeet and K.G. Mukerji

Applied Mycology Laboratory
Department of Botany
University of Delhi
Delhi 110007
INDIA

1. INTRODUCTION

The use of ecologically safe and environmentally friendly methods of protecting crops from pests and pathogens is gaining importance. Control methods using chemical pesticides may lead to development of resistant pathogenic strains, moreover, their accumulation can cause serious ecological and health problems. Therefore, the control of plant diseases by the use of existing organism in their natural habitat is an urgently needed component of plant disease management. Biocontrol of plant diseases offers answers to many persistant problems in agriculture including problems of resource limitation, non-sustainable agriculture system and over-reliance on pesticides (Cook and Baker, 1983). Biocontrol agents have potential to replace chemical pesticides (Mukerji and Garg, 1988 a,b). Moreover, they offer advantages in plant health management not possible with chemical pesticides. Despite these advantages, beneficial microbes are rarely used to control plant diseases primarly because biocontrol agents have been both less effective and more variable than competitive pesticides. A principal reason for this is the poor growth of bioprotectant. The ability of bioprotectant to proliferate in a competitive microenvironment is crucial.

The interaction between a biocontrol fungus and its target organism is complex, involving mycoparasitism, antibiosis and competition. These characters are likely to be under the control of large number of genes, therefore, the recombination of whole genomes is required for the generation of new strains. Genetic recombination is a powerful technique for developing effective biocontrol strains than selection or mutation and protoplast fusion provides a means for promoting recombination of whole genomes, even between incompatible strains. This chapter emphasizes the basic elements of protoplast fusion technology and its application in improvement of biocontrol agents.

Biotechnological Approaches in Biocontrol of Plant Pathogens
Edited by Mukerji *et al.*, Kluwer Academic / Plenum Publishers, New York, 1999

2. PROTOPLAST ISOLATION

The rigid cell wall can be partially or totally removed by enzymatic digestion leaving a protoplast or protoplasts that will remain intact provided conditions are suitably stabilized with an osmoticum. Protoplasts are recognised as investigatory structures and providing a means of studying a wide range of cellular, biochemical and genetic processes. One of the important consideration in these investigations is to maximize yield, which in totality is capable of retaining the ability to regenerate. The number of factors *viz.*, lytic enzymes, osmotic stabilizer, physiological status of an organism, cultural conditions, pH, buffer, temperature and time of protoplasting are of prime importance in isolation of protoplasts. There are enormous examples of protoplast isolation from different fungi (Table 1).

Table1. Isolation of protoplasts from fungi since 1988.

Organism	Reference
Alternaria alternate	Cary and Stovall (1992)
A. eichhorniae	Shabana and Charudattan (1997)
Aspergillus awamori	Ogawa *et al.* (1988)
A. nidulans	Lynch *et al.* (1989), Vagvolgyi and Ferenczy (1991)
	Vainstein and Peberdy (1991)
A. oryzae	Kitamato *et al.* (1988), Ushijima *et al.* (1991)
A. parasiticus	Zvereva *et al.* (1996)
A. sojae	Ushijima *et al.* (1991)
A. usamii mut *shirousamii*	Ogawa *et al.* (1988)
Candida albicans	Kavanagh *et al.* (1991), Law *et al.* (1994)
C. apicola	Rlike *et al.* (1992)
C. boidinii	Kobori *et al.* (1991)
C. fennica	Andrade *et al.* (1992)
C. tropicalis	Kobori *et al.* (1991), Chang *et al.* (1995)
Claviceps purpurea	Didek- Brumec *et al.* (1993)
Cochliobolus lunatus	Znidarsic *et al.* (1992)
Coriolus versicolor	Bok *et al.* (1990)
Flammulina velutipes	Kitamato *et al.* (1988)
Gibberella fujikuroi	Kumar and Lansane (1988), Vagvolgyi *et al.* (1990)
Gliocladium virens	Shin and Cho (1994)
Leptosphaeria maculans	Farman and Oliver (1988)
Lyophyllum decastes	Bok *et al.* (1994)
Malbranchea sulfurea	Gautam *et al.* (1996)
Microsporum gypseum	Chadegani *et al.* (1989)
Mucor circinelloides	Nagy *et al.* (1994)
Neurospora crassa	Kitamato *et al.* (1988)
Ophiostoma ulmi	Richards (1994)
Oudemansiella mucida	Homolka (1988)
Pachysolen tannophilus	Kavanagh and Whittaker (1991), Heluane *et al.* (1993)
Penicillium citrinum	Maliszewsk and Zboinska (1996)
P. expansum	Dias *et al.* (1997)
P. grisoroseum	Dias *et al.* (1997)
Phytophthora megasperma f.sp. *glycinea*	Layton and Kuhn (1988)
P. sojae	Wagner and Wilkinson (1993)
Pleurotus ostreatus	Peng *et al.* (1993)
Pseudocercosporella anguioides	Hocart and McNaughton (1994)
P. herpotrichoides	Hocart and McNaughton (1994)
Puccinia graminis tritici	Huang *et al.* (1990)
Rhizoctonia solani	Chung *et al.* 1992, Yang *et al.* 1992, Yang *et al.* 1994
Saccharomyces cerevisiae	Skata *et al.* (1988)
	Kavanagh and Whittaker (1991), Kavanagh *et al.* (1991)
	Gabriel *et al.* (1992), Heluane *et al.* (1993), Chang *et al.* (1995),
S. fermentati	Skata *et al.* (1988)
S. fibuligera	Nga *et al.* (1992)
Schizophyllum commune	Mizuno *et al.* (1997)
Schizosaccharomyces pombe	Kobori *et al.* (1989)
Septoria nodorum	Cooley *et al.* (1988)
Suillus granulatus	Dias *et al.* (1996)
Trichoderma harzianum	Stasz *et al.* (1988), Sandhu *et al.* (1989)
	Goldman *et al.* (1990), Pe'er and Chet (1990), Shin and Cho (1994),
	Manczinger *et al.* (1995), Thrane *et al.* (1995), Migheli *et al.* (1995)
T. pseudokoningii	Furlaneto and Pizzirani-Kleiner (1992)
T. reesei	Kumari and Panda (1992)
T. viride	Kirimura *et al.* (1989)
Trichosporon cutaneum	Oberma *et al.* (1990), Sienko *et al.* (1992)
Ustilago maydis	Eidtmann and Schauz (1992)
Yarrowia lipolytica	Nga *et al.* (1992)

2.1. Lytic Enzymes

Owing to the diverse chemical nature of fungal cell wall, the choice of lytic enzymes for protoplast isolation is largely empirical. Enzymes suitable for isolation of protoplasts from fungi are commercially available or obtained from the microorganisms.

Commercial enzymes are crude or partially purified preparations with multiple activities, the main component being chitinase, α- & β-glucanase and protease. They are obtained from microbes, fungi or animals (Table 2).

Table 2. Commercially available enzymes used in isolation of protoplasts.

Enzyme Preparation	Source Organism
Cellulase CP	*Penicillium funiculosum*
Cellulase "Onozuka" R-10	*Trichoderma viride*
Cellulase RS	"
Cellulase (Mayvil)	*Aspergillus niger*
Driselase	*Irpex lacteus*
β-Glucanase	*P. emersoni*
β-Glucuronidase	*Helix pomatia*
Glusulase	"
Helicase	"
Novozym 234	*T. harzianum*
Rhozym HP150	Unknown
Zymolyase 5000	*Arthrobacter leuteus*
Zymolyase 60,000	"
Sulphatase	*H. pomatia*

Commercial enzymes are generally used in combinations of two or more enzymes. In mixture, they are effective in release of protoplasts than used either alone. The mixture of rhozyme HP150, driselase and cellulase CP was effective in release of protoplasts from *Pseudocercosporella herpotrichoides*. When used singly, both driselase and cellulase CP produced significantly less number of protoplasts. Rhozyme HP150 while inactive against *P. herpotrichoides* mycelium when used alone, but greatly increased the protoplast yield with either of the other two enzymes (Hocart *et al.*, 1987). Novozym 234, which is widely used to produce high yield of protoplasts from large number of fungi (Bok *et al.*, 1994; Dias *et al.*, 1997; Heluane *et al.*, 1993; Migheli *et al.*, 1995; Nga *et al.*, 1992; Peng *et al.*, 1993; Sandhu *et al.*, 1989; Shabana and Charudattan, 1997; Sienko *et al.*, 1992; Stasz *et al.*, 1988; Yang *et al.*, 1994) having high activities of protease, chitinase and β-1,3-glucanase which make it suitable for releasing protoplasts without supplementation of other enzymes (Hamlyn *et al.*, 1981). However, in few instances the supplementation with other enzymes was found to be more effective than used either of them alone.

Although commercial enzymes are readily available and convenient to use. Many workers found that the lytic enzymes produced from microorganisms are more effective in obtaining protoplasts. These mycolytic enzymes are produced by a suitable microorganisms grown on purified fungal cell wall as the sole carbon source derived from autolytic cultures of specific fungus (lynch *et al.*, 1985a; Reyes *et al.*, 1984). Number of inducible substrates such as cellulose, chitin and laminarin have also been used to promote the synthesis of lytic enzymes (Hocart and Peberdy, 1989; Witkowska *et al.*, 1997).

A wide range of microorganisms have been used to produce enzymes capable of totally or partially digesting the cell walls of fungi. *Micromonospora* and *Streptomyces* belongs to actinomycetes are most commonly used for production of lytic enzymes (Villanueava and Garcia-Acha, 1971). Enzymes isolated from them are referred as strepzymes and are used to isolate protoplasts from different fungi. Number of fungi have also been used successfully for production of lytic enzymes including *Penicillium purpurogenum* (Musilkova *et al.*, 1969), *P. italicum* (Benitez *et al.*, 1975b) and *Trichoderma* spp. (deVries and Wessels, 1972; Ogawa *et al.*, 1987). The lytic enzyme preparation from different species of *Trichoderma* is effective in isolation of protoplasts from wide range of fungi.

Enzymes present in the incubation mixture are known to exert a profound effect on protoplast release and viability so it is necessary to adjust the concentration of enzymes to reduce the incubation time, as longer incubation may results in the low yield due to lysis of protoplasts (Quigley *et al.*, 1987).

2.2. Osmotic Stabilizers

Isolated protoplasts are osmotically sensitive, therefore a suitable osmotic stabilizer is required to prevent lysis. There is no universal single osmotic stabilizer for all groups of fungi. The choice of stablizer will greatly influence the yield and stability of protoplasts. For stabilization of protoplasts sugars, sugar alcohols and inorganic salts such as KCl, NaCl, $MgCl_2$, $(NH4)_2SO_4$ are most frequently used as osmotic stabilizers. The problem may be encountered using $MgSO_4$ as osmotic stabilizer, in that during harvesting process, the protoplasts become highly vacuolated and fail to pellet during centrifugation and remain in suspension. The increase in buoyancy of protoplasts has long been recognised (deVries and Wessels, 1972; Peberdy, 1979). Thus after centrifugation at a relatively high speed with concentrations of $MgSO_4$ between 0.8M to 1.2M, protoplast buoyancy may be sufficiently enhanced to permit the concentrations of protoplasts at the surface of suspension (Hocart and Peberdy, 1989).

Different inorganic salts, surgars and sugar alcohols have been used to study their effect on protoplast yield from different fungi. In *Alternaria alternata* 1.2M KCl was most effective than NaCl, $MgSO_4$ and sorbitol whereas no protoplasting occurred with sucrose and mannitol (Cary and Stovall, 1992). High yield of protoplasts of *Candida apicola* were obtained with 1.2M mannitol, 0.8M KCl and 1.2M sorbitol (Rlike *et al.*, 1992). In *Trichoderma reesei* 0.9M KCl promoted the protoplast release as compared to $MgCl_2$, $CaCl_2$, NaCl, sucrose and mannitol (Kumari and Panda, 1992), whereas 0.6M KCl was found to be effective stabilizer for high yield of protoplasts from *T. harzianum* (Sandhu *et al.*, 1989).

The osmotic stabilizers are generally effective only at optimum concentration and any change in the concentration may decrease the protoplast yield. It is evident from the release of protoplasts from giant cells of *Aspergillus niger*, at low concentration of NaOH (0.1- 0.2M) protoplast began to swell and rupture after a short time. The stable protoplasts were released at 0.5-0.7M NaOH and above this protoplasts shrank after release (Davis *et al.*, 1977).Similarly, in *A. sojae* 0.8M sorbitol showed a good positive effect on protoplasting as compared to 0.7M and 1.0M sorbitol (Ushijima and Nakadai, 1987). Likewise in *Colletotrichum gloeosporioides* maximum number of protoplasts were obtained with 1.2M mannitol as compared to 0.8M and 1.0M mannitol (TeBeest and Weidemann, 1990).

Complete separation of protoplasts from hyphal debris may require more complicated process. Hahiba and Yamada (1982) described the liquid-liquid two phase system of 0.6M mannitol and 0.6M sucrose for obtaining homogenous suspension of protoplast from crude preparation of protoplasts. After centrifugation, the protoplasts free of hyphal fragments were located at the interface of two solutions.

The optimum stablizer for protoplast release may not be the most suitable for the subsequent stabilization of the isolated protoplasts. It may be necessary to change the stabilizer after isolation of protoplasts to prevent the gradual loss of protoplasts thorugh lysis. Therefore, a number of alternatives of organic and inorganic salts should be tested.

2.3. The Organism

The next most important aspect affecting the protoplast isolation is the nature of the fungal mycelium. Both filamentous fungi and single celled yeasts may respond to different environmental conditions in a variety of ways, therefore the nature of growth medium and cultural conditions are of prime importance in considering protoplast isolation. Experiments with *Aspergillus nidulans* indicated that the protoplast yield was higher from the mycelium grown on glucose-salt rich medium than from the mycelium grown on the same medium supplemented with yeast extract (Peberdy, 1976). The presence of sulphur containing amino acids or 2-deoxyglucose (2DG) in the growth medium enhanced the protoplast yield from log phase cells of *Schizosaccharomyces pombe* (Housset *et al.*, 1975) and *Rhodotorula rubra* (Evans and Conrad, 1987). Stasz *et al.* (1989) also reported the high yield of protoplasts from *Trichoderma* spp. when the hyphae for protoplasting were grown in the presence of 2DG. Although hyphae grew slowly in the presence of 2DG but the older cultures were also protoplasted adequately.

The physiological age of cultures also exerts a profound influence on protoplast yield. In general, protoplast yield is much higher from mycelium in the exponential phase of growth (Peberdy, 1979). In most species, the region of cell wall most susceptible to lysis is near the hyphal tip where the wall is thinnest and wall synthesis actively taking place. Consequently, the maximum yield of protoplasts is obtained during the period of growth coinciding with the occurrence of maximum branching and cell wall synthesis (Hocart and Peberdy, 1989). Depending upon the strain the period of maximum growth varies (Table 3).

Table 3. Optimum growth of different organisms for maximum yield of protoplasts.

Organism	Effective period of growth	Reference
Alternaria alternata	20h	Cary and Stovall (1992)
Candida apicola	20h	Rlike *et al.* (1992)
Gibberella fujikuroi	25h	Kumar and Lansane(1988)
Microsporum gypseum	2-4 days	Chadegani *et al.* (1989)
Pleurotus ostreatus	5-6 days	Peng *et al.* (1993)
Pseudocercosporella herpotrichoides	44h	Hocart *et al.* (1987)
Puccinia graminis tritici	3 days	Huang *et al.* (1990)
Trichoderma harzianum	16h	Sandhu *et al.* (1989)
T. reesei	20h	Kumari and Panda (1992)

The period of maximum susceptibility to lysis may be relatively narrow with the result that protoplast yield from mycelium cultured for just a few hours longer than optimum can be drammatically reduced. Deutch and Parry (1974) demonstrated that the cell wall build up resistance to lysis in the transitional growth period between exponential and stationary phase of growth and longer incubation in the lytic enzyme mixture also failed to increase the protoplast yield. It has been suggested that the ability of enzyme to hydrolyze fungal cell wall is inversely related to melanin content (Bloomfield and Alexander, 1967). The build up of melanin in the later stages of growth of *Aspergillus nidulans* prevents the action of lytic enzyme on the myclial cell wall (Bull, 1970) and in *Alternaria alternata* the protoplasts are released mainly from hyphal tips because the apical portion develops very little melanin (Cary and Stovall, 1992).

2.4. Pretreatment of Fungi

Pretreatment of yeast cells or fungal mycelium with appropriate thiol compound or detergent accelerates the rate of protoplast release. The effectiveness of pretreatment on protoplasts isolation is influenced by number of factors including the age and growth of culture, strain of an organism, the pretreatment medium and subsequent treatment conditions, and the length of time for pretreatment and treatment. As a means of enhancing the release of protoplasts, preincubation of mycelium of *Microsporum gypseum* in a mixture of dithiothreitol (DTT) and β-mercaptoethanol was found to be more effective but when used along with lytic enzymes, protoplast formation was inhibited and no protoplasting occurred when mycelium was pretreated with SDS, UMS and Na-thioglycolate (Chadegani *et al.*, 1989). Protoplast isolation from stationary phase cells is not widely reported. The reducing agents like cysteine, mercaptoethanol and DTT are essential for forming protoplasts from stationary phase cells in *Candida apicola* (Rlike *et al.*, 1992).

2.5. Buffer

The choice of buffer in the incubation mixture can affect the rate of protoplast release. There is no universal buffer used for protoplast release from different groups of fungi. Different types of buffers including glycine-HCl, tris-HCl, acetate, succinate, citrate, maleate, phosphate and imidazole buffers are used in the incubation mixture for protoplast release. The molarity of buffer also affects the protoplast yield. In case of *T.reesei* there was a gradual increase in the yield between 10mM and 25mM. Above 25 mM of phosphate buffer, protoplast yield gradually decreased and was minimum at 150mM phosphate buffer (Kumari and Panda, 1992).

Kavanagh and Whittaker (1991) observed that increasing buffer viscosity retarded the cell wall degradation in *Pachysolen tannophilus* and *Saccharomyces cerevisiae* as measured by the appearance of osmotically fragile cells in the population. It indicated that to maintain protoplast integrity, a relationship exist between the rate of protoplast liberation and inherent viscosity of the osmoticum employed.

2.6. pH of Incubation Mixture

pH of incubation mixture also affects the protoplast release from organisms. Depending upon the nature of cell wall, the optimum pH is different for all groups of fungi. The protoplast formation from *Aspergillus sojae* was maximum at pH 6.5 of 0.01M sodium - phosphate buffer and above that pH the protoplast yield was reduced (Ushijima and Nakadai, 1987), whereas in *Microsporum gypseum* pH 4.5 was required for maximum protoplast yield (Chadegani *et al.*, 1989). In *Trichoderma harzianum*, the protoplast count showed a sudden rise when pH was increased from 5 to 6 and after that yield was decreased (Sandhu *et al.*, 1989) and Kumari and Panda (1992) found pH 6.5 was most suitable for protoplasting in *T. reesei*. In *Hyphoderma roseum* the largest number of protoplasts were observed at relatively low pH ranging from 3.8 to 4.6 (Dlugonski *et al.*, 1984).

2.7. Temperature

Effect of temperature is related to the activities of the lytic enzyme system and higher temperature caused a significant decrease in protoplast yield. In general, temperature ranging from 28°C - 31°C is suitable for protoplasting. Kovac and Subik (1970) observed that unfavourably high temperature may cause agglutination of certain organelles in isolated protoplasts and low temperature may affect stability of protoplast membrane.

2.8. Time of Contact with Lytic Enzyme

Another important factor in protoplast isolation is time of contact with lytic enzyme. Maximum yield of protoplasts from *Trichoderma reesei* was obtained at 5h contact with lytic enzyme using a mixture of 1.66g dry mycelial wieght and 190ml lytic enzyme. There was a gradual decrease in yield with increasing the contact time with lytic enzyme. Increasing time of contact with lytic enzyme may increase the chance of lysis of already formed protoplasts, thereby decreasing the total protoplast yield (Kumari and Panda., 1992).

2.9. Release of Protoplasts

Usually protoplasts are released through small pores in the cell wall from the cell nearest to the hyphal tip and rarely the whole mycelium is converted into protoplasts. In both instances, the resulting population will show considerable homogenity for size, organelle content and physiological activity. Protoplasts derived from spores are expected to be more homogenous and also uninucleate than those from hyphae. The uninucleate protoplasts are considered to facilitate genetic analysis after protoplast fusion. The susceptibility of conidial wall can be increased by altering the conditions under which they are produced or by preincubation prior to lysis.

3. REGENERATION OF PROTOPLASTS

The sequence of events leading from protoplasts to normal cell capable of reproduction is considered as protoplast regeneration and for genetic manipulation of fungi, protoplasts must be regenerated and reverted to normal organims. The regeneration of protoplasts is affected by the culture medium used, osmotic stablizer and pH of the regenerating medium. The frequency of protoplast regeneration is highly variable in different fungi (Table 4).

In the presence of suitable osmotic stabilizer, isolated protoplasts rapidly regenerated into cell wall material, vegetative mycelium and subsequently developing normal colonies on regenerating medium. The most appropriate osmotic stabilizer for the regeneration need not necessarily be the same that which is used in the isolation of protoplasts. Therefore, a number of inorganic salts, sugars and sugar alcohols should be tested.

Regeneration of protoplasts on solid medium depends on several factors like effect on plating technique and concentration of agar. Overlayering with 0.5% agar on regenerating medium gave the highest regeneration frequency in *Microsporum gypseum* as compared to dilution plate and spread plate techniques (Chadegani *et al.*, 1989). Similarly, in *Mucor pucillus* and *M. miehei*, the best protoplast regeneration was obtained by overlayering with 0.5% soft agar (Ohnuki *et al.*, 1982). Concentration of agar in regeneration medium has tremendous effect on regeneration frequency of protoplasts. The reducing agar concentration from 2.0% to 0.5% doubled the regeneration frequency in *Microsporum gypseum* (Chadegani *et al.*, 1989). *Fusarium culmorum* protoplasts showed better regeneration frequency in 0.2% agar instead of 2.0% agar (Garcia Acha *et al.*, 1966).

Prolonged incubation of cells during protoplast isolation also reduced their regeneration frequency. In *Candida fennica*, the highest regeneration frequency was obtained when cells were treated for 60 min. with Novozym 234 in the presence of 0.8M KCl (Andrade *et al.*, 1992). Similarly in *Trichoderma reesei* increase in duration of treatment of mycelium resulted in decrease in the number of regenerated protoplasts (Picatagio *et al.*, 1983). In *Rhodotorula rubra* regeneration frequency was 80% after 30 min. and decreased to 15% after 5 hrs of incubation (Evans and Conrad, 1987).

Table 4. Regeneration frequency of fungal protoplasts

Organism	Regeneration frequency	Reference
Absidia glauca	35%	Wostemeyer and Brockhausen-Rohdemann (1987).
Alternaria alternata	11.2±2.9%	Cary and Stovall (1992)
A. cassiae	27.8±2.2%	"
A. macrospora	28.7±2.2%	"
Aspergillus awamori var. *kawachi*	10 - 20%	Ogawa *et al.* (1988)
A. niger	12.0%	Kirimura *et al.* (1989)
A. oryzae	18.6%	Kitamato *et al.* (1988)
Candida apicola	16 - 30%	Rlike *et al.* (1992)
C. fennica	8%	Andrade *et al,* (1992)
Colletotrichum gloeosporioides f.sp. *aeschynomene*	80 - 84%	TeBeest and Weidemann (1990)
Coprinus cinereus	10 - 20%	Morinaga *et al.* (1985)
C. macrorhizus	37%	Yanagi *et al.* (1985)
C. pellucidus	35 - 40%	Morinaga *et al.* (1985)
Coriolus versicolor	3.48%	Bok *et al.* (1990)
Flammulina velutipes	16.7%	Kitamato *et al.* (1988)
Fusarium oxysporum	9.7%	Lynch *et al.* (1985b)
F. tricinctum	16%	"
Gibberella zeae	10.2%	Adams *et al.* (1987)
Lyophyllum decastes	8.32%	Bok *et al.* (1994)
Malbranchea sulfurea	34.6%	Gautam *et al.* (1996)
Neurospora crassa	19%	Kitamato *et al.* (1988)
Penicillium citrinum	84.6%	Maliszewska and Zboinska (1996)
P. expansum	46.5%	Dias *et al.* (1997)
P. grisoroseum	56.9%	"
Phanerochaete chrysosporium	0.8 - 5.0%	Gold *et al.* (1983)
Phycomyces blakesleeanus	20%	Suarez *et al.* (1985)
Pleurotus ostreatus	12.89%	Peng *et al.*(1993)
Pseudocercosporella herpotrichoides	57.5%	Hocart *et al.* (1987)
Puccinia graminis tritici	19 - 39%	Huang *et al.* (1990)
Pyricularia oryzae	20 - 24%	Kobayashi *et al.* (1985)
Rhizoctonia solani	20%	Hashiba and Yamada (1982)
Saccharomyces cerevisiae	20 - 30%	Svoboda and Piedra (1983)
S. diastaticus	20 - 30%	Sakai *et al.* (1986)
Schizosaccharomyces versatilis	50%	Gabriel and Kopecka (1983)
Talaromyces flavus	70.75%	Santos and DeMelo (1991)
Trichoderma harzianum	10%	Stasz et al. (1988)
	24%	Manczinger *et al.* (1995)
	10 - 30%	Thrane *et al.* (1995)
T. reesei	95%	Picataggio *et al.* (1983)
	9 - 15%	Ogawa *et al.* (1987)
T. viride	43%	Kirimura *et al.* (1989)

The regeneration frequency was also influenced by the enzymes used for isolation of protoplasts. Presence of cellulase, chitinase and zymolyase 60,000 in incubation mixture resulted in higher regeneration frequency of protoplasts of *Coprinus macrorhizus* as compared to incubation mixture containing Novozym 234 (Yanagi *et al.*, 1985). The reduced proteinase activity in cell wall lytic enzyme from *Trichoderma harzianum* resulted in improved regeneration of protoplasts by 1.8 times (Kitamato *et al.*, 1988).

3.1. Monitoring of Protoplast Regeneration

The process of regeneration was monitored by using certain dyes e.g. calcoflour. It was used to study the regeneration of protoplasts in *Geotrichum candidum* (Dooijewaard - Kloosterziel *et al.*, 1973), *Neurospora crassa* (Quigley *et al.*, 1987) and *Pythium* PRL 2142 (Sietsma and DeBoer, 1973) by determining the cell synthesising calcoflour binding material i.e. glucan and chitin.

3.2. Patterns of Development

In protoplast regeneration different patterns of developments are found. In the first type, protoplast give rise to a chain of yeast - like cells. The terminal cell produces normal hyphae. In the second type, protoplasts first develop a new wall into which the cytoplasm migrates to form a primary cell that later produce a normal germ tube (Peberdy and Gibson, 1971; Anne *et al.*, 1974) whereas, in the third pattern, the regenerating protoplasts remained spherical for a long time. A germ tube like hyphae develops from the spherical cell and forms normal structure (Garcia - Acha *et al.*, 1966). The variation from this pattern was observed in *Trichoderma reesei* (Benitez *et al.*, 1975a,b) where the aberrant tube emerged from the protoplasts, it bursted thorugh the opposite end of original protoplasts. At the same time, a normal germ tube was produced on the other side of original protoplasts. However Picataggio *et al.* (1983) observed in *T. reesei* that after prolonged incubation the protoplasts developed into giant cells containing numerous large vacuoles and hyphae were emerging from these giant cells. Similar giant cells were also observed in *Pyricularia oryzae* (Teraoka *et al.*, 1992).

3.3. Composition of Cell Wall

Isolated protoplasts have been used to study structure and biochemical composition of cell walls. The giant protoplasts of *Pyricularia oryzae* had higher metabolic activity in biosynthesis of protein, DNA, RNA, lipid and cell wall than the original small sized protoplasts (Teraoka *et al.*, 1992). In *Schizophyllum commune* it was reported that there are three main cell wall polymers formed during regeneration and reversion. The amount of S-glucan and chitin increased from the beginning of regeneration. The accumulation of R-glucan started after 3h of starting of regeneration. Hyphal tubes emerged only after accumulation of R-glucan. The active protein synthesis occurred during the early stages of regeneration. The amount of nucleic acid was also increased (deVries and Wessels, 1975).

The chemical analysis of normal and aberrant tubes of *Trichoderma viride* showed that the normal cell walls were rich in protein and chitin as compared to aberrant tubes (Benitez *et al.*, 1975).

In *Saccharomyces cerevisiae* (Gabriel *et al.*, 1992; Kopecka and Gabriel, 1995) and in *Schizosaccharomyces pombe* (Kobori *et al.*, 1989) actin was found to be associated with formation of cell wall during protoplast regeneration.

4. PROTOPLAST FUSION

Protoplast fusion allows genetic recombination between species where no sexual cycle exists or difficult to demonstrate and even between incompatible strains. It can be achieved chemically by polyethylene glycol (PEG) or physically by electrofusion.

4.1. PEG Induced Fusion

The fusogenic activity of PEG was first demonstrated by Kao and Michayluk (1974) using plant protoplasts and was soon applied to induce fusion in fungal protoplasts.The first report using PEG/Ca2+ and fungal protoplasts to induce fusion showed fusion frequencies 1000 times higher than previously achieved (Ferenczy *et al.*, 1975; Anne and Peberdy, 1976). This combination has been widely used to produce intraspecific, interspecific and integeneric hybrids in both yeasts and filamentous fungi (Table 5).

Table 5. Recent reports of protoplast fusion in fungi.

Organism	Reference
I. Intraspecific Protoplast Fusion	
Absidia *glauca*	Wostemeyer and Brockhausen - Rohdemann (1987)
Aspergillus awamori var. *kawachi*	Ogawa *et al.* (1988)
A. oryzae	Benkhemmar *et al.* (1985)
A. nidulans	Kevei and Peberdy (1984)
A. sojae	Ushijima and Nakadai (1987)
Candida albicans	Sarachek *et al.* (1981), Pesti and Ferenczy (1982) Kavanagh *et al.*(1991), Law *et al.* (1994)
C. fennica	Andrade *et al.* (1992)
Cephalosporium acremonium	Hamlyn *et al.* (1985)
Calviceps purpurea	Didek-Brumec *et al.* (1985)
Fusarium graminearum	Bu'lock *et al.* (1982)
Gibberella fujikuroi	Yanagida *et al.* (1996)
Microsporum gypseum	Chadegani and Ahmadjian (1991)
Mucor pusillus	Goto - Hamamoto *et al.* (1986)
Penicillium chrysogenum	Chang *et al.* (1981)
Phanerochaete chrysosporium	Gold *et al.* (1983)
Phycomyces blakesleeanus	Suarez *et al.* (1985)
Phytophthora megasperma f.sp. *glycinea*	Layton and Kuhn (1988)
Pseudocercosporella herpotrichoides	Hocart *et al.* (1993)
Saccharomyces cerevisiae	Maraz and Subik (1981), Kavanagh *et al.* (1991) Vidoli *et al.* (1982)
S. diastaticus	Sakai *et al.* (1986)
Schwanniomyces alluvius	Wilson *et al.* (1982)
Trichoderma harzianum	Stasz *et al.* (1988), Pe'er and Chet (1990) Manczinger *et al.* (1995), Migheli *et al.* (1995)
T. pseudokoningii	Furlaneto and Pizzirani-Kleiner (1992)
T. reesei	Toyama *et al.* (1984a) Manczinger and Ferenczy (1985), Ogawa *et al.* (1987)
T. viride	Bojnanska *et al.* (1980)
Verticillium lecanii	Jackson and Heale (1987)
II. Interspecific Protoplast Fusion	
Aspergillus nidulans + *A. rugulosus*	Bradshaw *et al.* (1983)
A. usamii mut *shirousamii* + *A. niger*	Ogawa *et al.* (1989)
A. terreus + *A. usamii*	Kirimura *et al.* (1997)
A. sp. + *A. flavus*	Solis *et al.* (1997)
Beauveria bassiana + *B. sulfurescens*	Couteaudier *et al.* (1996)
Candida albicans + *C. tropicalis*	Kucsera and Ferenczy (1986)
C. tropicalis +*C. boidinii*	Kobori *et al.* (1991)
Mucor pusillus + *M. miehei*	Ohnuki *et al.* (1982)
Pseudocercosporella hepotrichoides + *P. anguioides*	Hocart and McNaughton (1994)
Saccharomyces cerevisiae + *S. diastaticus*	deFigueroa *et al.* (1984)
Schizosaccharomyces pombe + *Schizo.malidevorans*	Sipiczki *et al.* (1982)
Schwanniomyces castellii + *Sch. alluvius*	Dhawale and Ingledew (1983)
Trichoderma harzianum + *T. longibrachiatum*	Lalithakumari *et al.* (1996)
T. reesei + *T. koningii*	Toyama *et al.* (1984a)
T. reesei + *T. viride*	Toyama *et al.* (1984b)
Verticillium albo-atrum + *V. dahliae*	Typas (1983)
III. Intergeneric Protoplast Fusion	
Aspergillus niger + *Trichoderma viride*	Kirimura *et al.* (1989)
Coprinus cinereus + *Schizophyllum commune*	Zhao and Chang (1995)
Gliocladium virens + *T. harzianum*	Shin and Chao (1994)
Kluyveromyces fragillis + *Saccharomyces cerevisiae*	Farahnak *et al.* (1986)
Pacchysolen tannophilus + *S. cerevisiae*	Heluane *et al.* (1993)
Pleurotus ostreatus + *Schizo. commune*	Zhao and Chang (1996)
Rhodotorula rubra + *S. cerevisiae*	Evans and Conrad (1987)
S. cerevisiae + *Candida tropicalis*	Chang *et al.* (1995)
S. cerevisiae + *Hansenula wingei*	Yamashita *et al.* (1981)
S. cerevisiae + *Sch. castellii*	Tamaki (1986)
S. cerevisiae + *Zygosaccharomyces fermentati*	Pina *et al.* (1986)
Saccharomycopsis fibuligera + *S. diastaticus*	Choi *et al.* (1997)
Sac. fibuligera + *Yarrowia lipolytica*	Nga *et al.* (1992).

The effect of PEG on protoplast aggregation and fusion was observed immediately after PEG was added to the protoplast suspension. The memberane of two or more protoplasts were observed to merge into one large structure. The time required for complete fusion of protoplast depends upon the reaction of the fusogenic factor and osmotic stablizers present in the incubation mixture with the protoplast membrane. Generally, 25-40% of PEG (mw 4000 and 6000) are used in incubation mixture (Anne and Peberdy, 1975; 1976; Ferenczy, 1976). At lower concentrations, the protoplast donot aggregate well so that fusion frequency decreased and at higher concentrations the protoplasts aggregate shrink thereby lowering the fusion frequency (Chadegani and Ahmadjian, 1991).

The effect of Ca^{2+} ions on fusion frequency is significant. Without Ca^{2+} ions the fusion frquency is negligible despite the fact that good aggregation of protoplasts is observed. On the other hand, as little as 1mM $CaCl_2$ is effective in stimulating the fusion process (Ferenczy *et al.*, 1975a). $CaCl_2$ promotes the fusion of fungal protoplasts at concentrations ranging from 0.1M to 0.01M. Presence of K^+, Na^+ and Mg^{2+} in addition to $CaCl_2$ reduced the fusion frequency in *Penicillium chrysogenum* and *Aspergillus nidulans*. These cations could be preferentially bound to the membranes thus preventing the stimulation by Ca^{2+}. The pH of the medium is also critical. Protoplast fusion could occur in PEG solution at pH level lower than pH 7.0 in the presence of 0.01M $CaCl_2$ but at high pH i.e. pH 9.0, 0.6M $CaCl_2$ was required to obtain the highest protoplast fusion frequency (Anne and Peberdy, 1975; 1976; Ferenczy *et al.*, 1975b). However in *Microsporum gypseum* 1M KCl as osmotic stabilizer along with pH 6.1 of PEG incubation mixture was required for the survival of protoplasts (Chadegani and Ahmadjian, 1991).

Although $CaCl_2$ has been most frequently employed source of the requisite exogenous Ca^{2+} for the PEG induced fusion but the attempts have been made to use other Ca^{2+} salts in fusion frequency. Calcium acetate, calcium nitrate and calcium propionate in the incubation also enhanced the hybrid yield in *Candida apicola*. The pH of the incubaion mixture also influenced the fusion frequency, at pH 4.0 approximately equal number of hybrids formed in both $CaCl_2$ and calcium acetate but at pH 7.0 calcium acetate facilitated the formation of hybrids. The size of anion also affects the formation of hybrids, the use of calcicum propionate in conjuction with PEG enhanced the fusion frequency as compared to $CaCl_2$ and calcium acetate (Kavanagh *et al.*, 1991).

The mechanism whereby PEG promotes membrane fusion is still unresolved. Constable and Kao (1974) assumed that PEG acts as a molecular bridge between adjacent membranes either directly by hydrogen bonds or indirectly by Ca^{2+}. The Ca^{2+} is considered to potentiate fusion by creating areas of membrane instability at the points of closest contact thus facilitating the coalesence of adjacent protoplasm. The fusogenic potential of the PEG is attributed to its ability to dehydrate protoplasts and to induce the formation of aggregates in which intimate membrane contact is achieved between adjacent protoplasts. Subsequent resuspension of the protoplasts in a hypertonic buffer assures rehydration which results in expanding of the inter-protoplast cytoplasmic continuities. Following protoplast fusion, karyogamy may proceed or nuclei may co-exist in the cytoplasm as an heterokaryon (Kavanagh *et al.*, 1991; Sato *et al.*, 1993).

4.2. Electrofusion

In this process, protoplasts of parental strains are mixed and subjected to a high frequency alternating electric field which causes the adhesion between protoplasts and formation of a pearl like chain of protoplasts. This ensures the close membrane contact essential for fusion. Then the fusion is achieved by the application of one to several short duration direct current pulses which momentarily destabilize the protoplast membrane at sites of membrane contact (Lynch *et al.*, 1989). This two step process can be followed microscopically and the strength and duration of dielectrophoretic field and pulses can be controlled. It is possible to investigate the precise effect of these factors on fusion frequency. High frequency fusion depends on (i) the establishment of strong inter-membrane contact and (ii) optimum membrane destabilization. For high frequency electrofusion the protoplasts of both strains should be uniform in size and should not contain high proportions of small protoplasts or cell debris, as a large disparity in protoplasts size between the two parental strains results in their dielectrophoretic separation with a corresponding reduction in fusion frequency (Hocart and Peberdy, 1989). Lynch *et al.* (1989) studied the fusion events induced by electrofusion of protoplasts from *Aspergillus nidulans*. Similarly Tsoneva *et al.* (1989) achieved the high fusion frequencies between the protoplasts of non-fragile and fragile strains of yeast. Electrofusion has also been used in interspecific hybridization of protoplasts between *A. oryzae* and *A. sojae* (Ushijima *et al.*, 1991).

4.3. Selection of Fusion Products

After protoplast fusion, some method of selection of fusion products must be employed to ensure that only the fusion products are able to regenerate into colonies. For such studies, mutants with auxotrophic, antibiotic, temperature sensitive, respiration deficient, morphological or colour markers are used. Among these complementary auxotrophic mutants have been used frequently as selection tool in protoplast fusion. The fused protoplast suspension is plated on an osmotically stabilized minimal medium lacking the nutritional supplement required by the parental strains so that only fusion products can grow. Strains able to grow on this medium, are prototrophic and selected as putative fusant progeny.

Direct visual identification of fusion products may be possible if the parental strains carry suitable colour markes i.e. either by spore colour (Ogawa *et al.*, 1987) or by release of pigment in the medium (Didek - Brumec *et al.*, 1993). In most of the cases pigmentation markers are used along with auxotrophs (Pe'er and Chet, 1990; Ushijima *et al.*, 1990). Resistance to antimicrobial compounds can be used as a method of selection of fusion products. Protoplasts from strains carrying different fungicide resistant markers were fused with unmarked fungicide - sensitive strains and the selection of recombinant types is done by growing on minimal medium containing the inhibitor (Hocart *et al.*, 1993; Layton and Kuhn, 1988). Antibiotic markers can also be used along with auxotrophic or respiratory deficient markers. In several yeasts complementation of respiratory deficient petite mutants have been proved effective. These mutants lack a functional system for mitochondrial protein synthesis and fail to grow on non-fermentable carbon source such as starch and lactate. Thus in a cross involving petite mutant and respiratory competent strain, selection against parental types was achieved by recovery of fusion products on minimal medium containing starch and glycerol (de Figueroa *et al.*, 1984).

5. PROTOPLAST FUSION IN DISEASE CONTROL

Although protoplast fusion is achieved in large number of fungi but there are few studies related to fungal biocontrol agent. Much emphasis is given on one of the important fungal biocontrol agent i.e. *Trichoderma* spp. Using protoplast fusion technology, it would be possible to combine different traits into single superior strain. The protoplast fusion has been used to improve biocontrol ability of *T. harzianum* (Harman, 1991; Harman *et al.*, 1989; Lalithakumari *et al.*, 1996; Migheli *et al.*, 1995; Pe'er and Chet, 1990; Sivan and Harman, 1991; Stasz *et al.*, 1988). The diversity created by this process can provide a source of improved biocontrol strain. Recently, few strains with improved biocontrol ability have been generated by using protoplast fusion. Only 2-4% of the fusant progency with improved biocontrol ability were obtained from a fusion between *T. harzianum* strains T12 and T95 (Harman and Stasz, 1991). Similarly Pe'er and Chet (1990) found only one progeny strain among 14 tested from fusion between various mutant strains of *T. harzianum* ATCC32173 gave significant biocontrol of *Rhizoctonia solani*.

The progency derived from protoplast fusion between T12 and T95 strains of *T. harzianum* (Stasz *et al.*, 1988) exhibit better biocontrol activity and more effectively colonize the rhizosphere of several crops *viz.*, cotton, maize, pea and cucumber. (Harman *et al.*, 1989; Sivan and Harman, 1991). Both parental strains have ability to control several plant pathogenic fungi. Strain T12 is resistant to competitive seed colonizing *Pseudomonas* sp. and strain T95 is rhizosphere competent and benomyl resistant (Ahmad and Baker, 1987; Hader *et al.*, 1984; Harman and Taylor, 1988). The thalli originating from the fusion were unstable and grew slowly. As cultures sectored, a wide diversity of subprogeny developed and some of these were stable. Most of them were no better than the parental strain, but one strain 1295-22 was isolated with improved biocontrol efficacy. This strain was prototrophic and grew more rapidly than either of the parental strains. The strain 1295-22 was effective against an array of plant pathogenic fungi *viz., Pythium ultimum, Fusarium graminearum, Rhizoctonia solani, Sclerotium rolfsii, Sclerotinia homoeocarpa* and *Botrytis cinerea* (Harman *et al.*, 1989). This strain is substantially more rhizosphere competent than the parental strains. After application as a seed treatment strains T12, T95, 1295-22 increased the root elongation rate in cotton, maize, pea and cucumber (Harman *et al.*, 1989). The protoplast fusion progeny 1295-22 resulted in longest roots in maize and cotton as compared to parental strains T12 and T95. All the strains were present at higher levels in the upper portion of the root and near the location of root, decrease in central root portion and then increase near the root tip region. However, the strain 1295-22 derived from protoplast fusion between strains T12 and T95 maintained a constant high level along the root. The main improvement in the rhizosphere colonisation obtained with 1295-22 on both crops was on the middle portion of the root, the contribution of root exudates to the improved rhizosphere competence of 1295-22 remains unclear.

In soils infested with *Pythium ultimum* seed treatment with either parent or fusant strains of *T. harzianum* provided increased stands compared with untreated control and were effective as treatment with thiram. Further, the post-emergence damping off was less with 1295-22. The fusant progeny 1295-22 induced robust growth of cucumber seedlings than the parental strains (Harman *et al.*, 1989). Thus, this strain has effective biocontrol potential with increased rhizosphere competence than either of the parental strain.

In another study, the parental strains, mutant strains and six fusant strains of *T. harzianum* significantly reduced grey mould incidence on grape caused by *Botrytis cinerea*. These strains were significant or highly significant in complete control of disease. A highly significant strain effect of parental strain, mutant and two of the fusants were studied on percentage plant stand and plant fresh weight in comparison with the control of *P. ultimum* on lettuce (Migheli *et al.*, 1995).

Intraspecies fusion between *T. harzianum* (an efficient antagonist) and *T. longibracheatum* (tolerant to copper sulphate and carbendazim) resulted in the recovery of numerous fusants. Two fuscants F1 and F2 showed improved antagonistic potential against *Rhizoctonia solani, Fusarium oxysporum* f.sp. *lycoperici, Venturia inaequalis, Curvularia lunata* and *Bipolaris oryzae* and also tolerant to copper sulphate and carbendazim (Lalithakumari *et al.*, 1996).

Protoplast fusion may provide an attractive method for the genetic improvement of biocontrol efficiency in entomopathogenic fungi. The genetic improvement of two strains of the entomopathogenic fungus *Beauveria bassiana* for more effective control of *Ostrinia nubilalis* and *Leptinotarsa decemlineate* was obtained by crosses with the insecticidal toxin producing strain *B. sulfurescens.* Protoplast fusion between di-auxotrophic mutants resulted in the recovery of some stable prototrophic fusion products with enhanced biocontrol efficiency (Couteaudier *et al.*, 1996).

6. FUTURE STRATEGIES

Protoplast fusion is an invaluable tool for strain improvement. Strain with suitable mycoparasitic or antagonistic attributes for potential as suitable biocontrol organisms, often lack features which are necessary, if they are to be fully exploited, viz., (i) mass production of suitable inoculum (ii) host specificity (iii) rhizosphere colonising ability (iv) greater tolerance to environment stress.

Increase the effectiveness of an organism may extend its useful range to new crop situations. As the basis of biocontrol interactions becomes better understood, breeding programmes to improve strains will become more common. Crosses between isolates with particular desirable characters can be made to construct superior strains in which the desirable features are combined. The development of resistant pathogenic strains against pesticides has clearly shown the adaptable nature of pest and pathogens. With the increasing use of biocontrol organisms, there will be acontinuedneed to update the strains in use, as the pathogen populations change to meet the challenge.

7. CONCLUSIONS

Biological control of plant pathogens has been considered as potential control strategy. Practically biological control will not immediately nor completely replace chemical pesticides but the judicious use of biocontrol agents with chemical pesticides in integrated disease management can significantly enhance the quality of environment. Biocontrol agents should proliferate in a competitive microenvironment. Genetic manipulation for improved biocontrol ability may reduce the use of pesticides. Biocontrol acitivity is under the control of large number of genes, therefore, genetic recombination is required for generation of superior strains.

Although protoplast fusion has been performed in large number of fungi but there are few reports of improved biocontrol strains with enhanced performance. It undoubtedly plays an important role in developing superior strains. The effective protoplast fusion inturn depends on high yield of protoplasts with ability to regenerate. Owing to the diverse chemical nature of cell well, there is a need to standardized the effective conditions for protoplasting for each and every species.

REFERENCES

Adams, G., Johnson, N., Leslie, J.F. and Hart, L.P. 1987, Heterokaryons of *Gibberella zeae* formed following hyphal anastomosis or protoplast fusion, Exp. Mycol. 11 : 339-353.

Ahmad, J.S. and Baker, R. 1987, Rhizosphere competence of *Trichoderma harzianum*, Phytopathology 77 : 182-189.

Andrade, C.M.M., Oliveira, F.M. and Valter, R.L. 1992, *Candida fennica* : enhancement of protoplast formation and fusion, Can. J. Microbiol. 38 : 807 - 810.

Anne, J., and Peberdy, J.F. 1975, Conditions for induced fusion of fungal protoplasts in polyethylene glycol solutions, Arch. Microbiol. 105 : 201-205.

Anne, J., and Peberdy, J.F. 1976, Induced fusion of fungal protoplasts following treatment with polyethylene glycol, J. Gen. Microbiol. 92 : 413-417.

Anne, J., Eyssen, H. and de Somer, P. 1974, Formation and regeneration of *Penicillium chrysogenum* protoplasts, Arch. Microbiol. 98 : 159-166.

Benitez, T., Ramos, S. and Garcia Acha, I. 1975a, Protoplasts of *Trichoderma viride* : formation and regeneration, Arch. Microbiol. 103 : 199-203.

Benitez, T., Villa, T.G. and Garcia Acha, I. 1975b, Chemical and structural differences in mycelial and regeneration walls of *Trichoderma viride,* Arch. Microbiol, 105 : 277-282.

Benkhemmar, O., Gaudemer, F. and Bouvier-Fourcade, I. 1985, Heterokaryosis between *Aspergillus oryzae* cyclopiazonic acid-defective strains : method for estimating the risk of inducing toxin production

among cyclopiazonic acid-defective industrial strains, App. Environ. Microbiol. 50 : 1087-1093.
Bloomfield, B.J. and Alexander, M. 1967, Melanins and resistance of fungi to lysis, J. Bacteriol. 93 : 1267-1280.
Bojnanska, A., Sipiczki, M. and Ferenczy, L. 1980, Characterization of conidiation mutants in *Trichoderma viride* by hyphal anastomosis and protoplast fusion, Acta. Microbiol. Acad. Sci. Hung. 27: 305-307.
Bok, J., Kim, J., Jin, M., Choi, E. and Kim, B. 1994, Studies on protoplast formation and regeneration of *Lyophyllum decastes*, Korean J. Mycol. 22 : 130-137.
Bok, J., Park, S., Chol, E., Kim, B. and Yoo, Y. 1990, Studies on protoplast formation and regeneration of *Coriolus versicolor*, Korean J. Mycol. 18 : 115-126.
Bradshaw, R.E., Lee, K. and Peberdy, J.F. 1983, Aspects of genetic interaction in hybrids of *Aspergillus nidulans* and *Aspergillus rugulosus* obtained by protoplast fusion, J. Gen. Microbiol. 129 : 3525-3533.
Bu'lock, J.D., Wright, C.E. and Mooney, J.E. 1986, Use of a protoplast fusion test to establish the status of mycotohin genes in an edible *Fusarium*, Biotechnol. Lett. 8 : 621-624.
Bull, A.T., 1970, Chemical composition of wild type and mutant *Aspergillus nidulans* cell walls : the nature of the polysaccharide and melanin constituents, J. Gen. Microbiol. 63 : 75-94.
Cary, J.W. and Stovall, M.E. 1992, Optimum conditions for release and regeneration of protoplasts of *Alternaria alternata*, Lett. Appl. Microbiol. 14 : 100-103.
Chadegani, M., and Ahmadjian, V., 1991, Protoplast fusion between mutant strains of *Microsporum gypseum*, Mycologia 83 : 779-786.
Chadegani, M., Brink, J.J., Shehata, A. and Ahmadjian, V. 1989, Optimization of protopast formation, regeneration and viability in *Microsporum gypseum*, Mycopathologia 107 : 33-50.
Chang, L.T., Terasaka, D.T. and Elander, R.P. 1981, Protoplast fusion in industrial fungi, Dev. Ind. Microbiol. 23 : 21-29.
Chang, S.Y., Li, C.T., Hiang, S.Y. and Chang, M.C. 1995, Intraspecific protoplast fusion of *Candida tropicalis* for enhancing phenol degradation, Appl. Microbiol. Biotechnol. 43 : 534-538.
Choi, S.H., Sung, C., Oh, M.J. and Kim, C.J. 1997, Intergeneric protoplast fusion in *Saccharomycopsis fibuligera* and *Saccharomyces cerevisiae*, J. Ferement. Bioengg. 84 : 156-161.
Chung, H., Him, S. and Ahm, H. 1992, Protoplast formation and fusion between anastomosis groups of *Rhizoctonia solani*, Korean J. Mycol. 20 : 44-50.
Cook, R.J. and Baker, K.F. 1983, The Nature and Practice of Biological Control of Plant Pathogens, 2nd ed., American Phytopathological Society, St. Paul, Minnesota, USA.
Cooley, R.N., Shaw, R.K., Frankli, F.C.H. and Caten, C.E. 1988, Transformation of the phytopathogenic fungus *Septoria nodorum* to hygromycin resistance, Curr. Genet. 13 : 383-389.
Constable, F. and Kao, K.N. 1974, Agglutination and fusion of plant protoplasts by polyethylene glycol, Can. J. Bot. 52 : 1603-1606.
Couteaudier, Y., Viaud, M. and Riba, G. 1996, Genetic nature, stability, and improved virulence of hybrids from protoplast fusion in *Beauveria*, Microbial Eco. 32 : 1-10.
Davis, B., d'Avillez Paixao, M.T., Deans, S.G. and Smith, J.E. 1977, Protoplast formation from giant cells of *Aspergillus niger*, Trans. Br. Mycol. Soc. 69 : 207-212.
de Figueroa, L.I., de Richard, M.F. and de van Broock, M.R. 1984, Interspecific protoplast fusion of the baker's yeast *Saccharomyces cerevisiae* and *Saccharomyces diastaticus*, Biotechnol. Lett. 6 : 269-274.
de Vires, O.M.H. and Wessels, J.G.H. 1972, Release of protoplasts from *Schizophyllum commune* by a lytic enzyme preparation from *Trichoderma viride*, J. Gen. Microbiol. 73 : 13-22.
de Vries, O.M.H. and Wessels, J.G.H. 1975, Chemical analysis of cell wall regeneration and reversion of protoplasts from *Schizophyllum commune*, Arch. Microbiol. 102 : 209-218.
Deutch, C.E. and Parry, J.M. 1974, Sphaeroplast formation in yeast during the transition from exponential phase to stationary phase, J. Gen. Microbiol. 80 : 259-268.
Dhawale, M.R. and Ingledew, W.M. 1983, Interspecific protoplast fusion of *Schwanniomyces* yeasts, Biotechnol. Lett. 5 : 825-830.
Dias, E.S., Araujo, E.F., Guimaraes, W.V., Coelho, J.L.C. and Silva, D.O. 1997, Production and generation of *Penicillium expansum* and *Penicillium grisoroseum* protoplasts, Revista de Microbiol. 28 : 116-120.
Didek-Brumec, M., Gaberc-Porekar, V., Alacevic, M. and Socic, H. 1993, Strain improvement of *Claviceps purpurea* by protoplast fusion without introducing auxotrophic markers, Appl. Microbiol. Biotechnol. 38 : 746-749.
Dlugonski, J., Sedlaczek, L. and Jaworski, A. 1984, Protoplast release from fungi capable of steroid transformation, Can. J. Microbiol. 30 : 57-62.
Dooijewaard-Kloosterziel, A.M.P., Sietsma, J.H. and Wouters, J.T.M. 1973, Formation and regeneration of *Geotrichum candidum protoplasts*, J. Gen. Microbiol. 74 : 205-209.

Eidtmann, A. and Schauz, K. 1992, Cryopreservation of protoplasts from sporidia of *Ustilago maydis*, Mycol. Res. 96 : 318-320.
Evans, C.T. and Conrad, D. 1987, An improved method for protoplast formation and its application in the fusion of *Rhodotorula rubra* with *Saccharomyces cerevisiae*, Arch. Microbiol. 148: 77-82.
Farahnak, F., Seki, T., Ryu, D.D.Y. and Ogrydziak, D. 1986, Construction of lactose-assimilating and high-ethanol-producing yeast by protoplast fusion, Appl. Environ. Microbiol. 51 : 362-367.
Farman, M.L. and Oliver, R.P. 1988, The transformation of protoplasts of *Leptosphaeria maculans* to hygromycin B resistance, Curr. Genet. 13 : 327-330.
Ferenczy, L. 1976, Some characteristics of intra-and intrspecific protoplast fusion products of *Aspergillus nidulans* and *Aspergillus fumigatus*, in: Cell Genetics in Higher Plants, D. Dudits, G.L. Farkas and P. Maliga, eds., Akademiai Kiado, Budapest.
Ferenczy, L. Kevei, F. and Szegedi, M. 1975a, Increased fusion frequency of *Aspergillus nidulans* protoplasts, Experimentia 31 : 50-52.
Ferenczy, L., Kevei, F. and Szegedi, M. 1975b, High frequency fusion of fungal protoplasts, Experimentia 31 : 1028-1030.
Furlaneto, M.C. and Pizzirani-Kleiner, A.A. 1992, Intraspecific hybridization of *Trichoderma pseudokoningii* by anastomosis and by protoplast fusion, FEMS Microbiol. Lett. 90 : 191-196.
Gabriel, M. and Kopecka, M. 1983, Regeneration of the cell walls of protoplast of the fission yeast *Schizosaccharomyces versatilis* in liquid media and their reversion to cells, Folia Microbiol. 28 : 452-457.
Gabriel, M., Kopecka, M. and Svoboda, A. 1992, Structural rearrangement of the actin cytoskeleton in regenerating protoplasts of budding yeasts, J. Gen. Microbiol. 138 : 2229-2234.
Garcia Acha, I., Lopez-Belmonte, F. and Villanueva, J.R. 1966, Regeneration of mycelial protoplasts of *Fusarium culmorum*, J. Gen. Microbiol. 45 : 515-523.
Gautam, S.P., Gupta, A.K., Shrivastava, R. and Awasthi, M. 1996, Protoplast formation from the thermophilic fungus *Malbranchea sulfurea*, using the thermostable chitinase and laminarinase of *Paecilomyces varioti*, World J. Microbiol. Biotechnol. 12 : 99-100.
Gold, M.H., Cheng, T.M. and Alic, M. 1983, Formation, fusion and regeneration of protoplasts from wild-type and auxotrophic strains of the white rot basidiomycete *Phanerochaeta chrysosporium*, Appl. Environ. Microbiol. 46 : 260-263.
Goldman, G.H., Montagu, M.V. and Herrera-Estrella, A. 1990, Transformation of *Trichoderma harzianum* by high voltage electric pulse, Curr. Genet. 17 : 169-174.
Goto-Hamamoto, M., Ohnuki, T., Uozumi, T. and Beppu, T. 1986, Intraspecific hybridization by protoplast fusion in Mucorales producing milk-clotting proteases, Agric. Biol. Chem. 50 : 1467-1473.
Hader, Y., Harman, G.E. and Taylor, A.G. 1984, Evaluation of *Trichoderma koningii* and *T. harzianum* from New York soils for biological control of seed rot caused by *Pythium* spp., Phytopathology 74 : 106-110.
Hamlyn, P.F., Birkett, J.A., Perez, G. and Peberdy, J.F. 1985, Protoplast fusion as a tool for genetic analysis in *Cephalosporium aceremonium*, J. Gen. Microbiol. 131 : 2813-2823.
Hamlyn, P.F., Bradshaw, R.E., Mellon, F.M., Santiago, C.M., Wilson, J.M. and Peberdy, J.F. 1981, Efficient protoplast isolation from fungi using commercial enzymes, Enz. Microb. Technol. 3 : 321-325.
Harman, G.E. 1991, Seed treatment for biological control of plant disease, Crop Protecion 10 : 166-171.
Harman, G.E. and Stasz, T.E. 1991, Protoplast fusion for the production of superior biocontrol fungi, in : Microbial Control of Weeds, D.O. TeBeest, ed., Chapman and Hall, New York.
Harman, G.E. and Taylor, A.G. 1990, Development of an effective biocontrol seed treatment system, in : Biological Control of Soil-Borne Plant Pathogens, D. Hornby, ed., C.A.B. International, Wallingford.
Harman, G.E., Taylor, A.G. and Stasz, T.E. 1989, Combining effective strains of *Trichoderma harzianum* and solid matrix priming to improve biological seed treatment, Plant Dis. 73 : 631-637.
Hashiba, T. and Yamada, M. 1982, Formation and purification of protoplasts from *Rhizoctonia solani*, Phytopathology 72 : 849-853.
Heluane, H., Spencer, J.F.T., Spencer, D., de Figueroa, L. and Challieri, D.A.S. 1993, Characterization of hybrids obtained by protoplast fusion, between *Pachysolen tannophilus* and *Saccharomyces cerevisiae*, Appl. Microbiol. Biotechnol. 40 : 98-100.
Hocart, M.J. and McNaughton, J.E. 1994, Interspecific hybridization between *Pseudocercosporella herpotrichoides* and *P. anguioides* achieved through protoplast fusion, Mycol. Res. 98 : 47-56.
Hocart, M.J. and Peberdy, J.F. 1989, Protoplast technology and strain selection, in: Biotechnology of Fungi for Improving Plant Growth, J.M. Whipps and R.D. Lumsden, eds., Cambridge University Press, Cambridge.
Hocart, M.J., Lucas, J.A. and Peberdy, J.F. 1987, Production and regeneration of protoplasts from *Pseudocercosporella herpotrichoides* (Fron) Deighton, J. Phytopathology 119 : 193-205.
Hocart M.J., Lucas, J.A. and Peberdy, J.F. 1993, Parasexual recombination between W and R pathotypes of

Pseudocercosporella herpotrichoides through protoplast fusion, Mycol. Res. 97 : 977-983.
Homolka, L. 1988, Preparation reversion, mutagenesis and intra-species fusion of protoplasts of *Oudemansiella mucida,* Folia Microbiol. 33 : 298-308. .
Housset, P., Nagy, M. and Schwencke, J. 1975, Protoplasts of *Schizosaccharomyces pombe* : an improved method for their preparation and the study of their guanine uptake, J. Gen. Microbiol. 90 : 260-264.
Huang, D., Staples, R.C., Bushnell, W.R. and Maclean, D.J. 1990, Preparation and regeneration of protoplasts from axenic mycelia derived from the wheat stem rust fungus, Phytopathology 80 : 81-84.
Jackson, C.W. and Heale, J.B. 1987, Parasexual crosses by hyphal anastomosis and protoplast fusion in the entomopathogen *Vertilcillium lecani.* J. Gen. Microbiol. 133 : 3537-3547.
Kavanagh, K. and Whittaker, P.A. 1991, Effect of buffer viscosity on the rate of protoplast formation in yeast, Biomedical Lett. 46 : 57-62.
Kavanagh, K., Walsh, M. and Whittaker, P.A. 1991, Enhanced intraspecific protoplast fusion in yeast, FEMS Microbiol. Lett. 81 : 283-286.
Kevei, F. and Peberdy, J.F. 1984, Further studies on protoplast fusion and interspecific hybridization within the *Aspergillus nidulans* group, J. Gen. Microbiol. 130 : 2229-2236.
Kirimura, K., Imura, M., Lee, S.P., Kato, Y. and Usami, S. 1989, Intergeneric hybridization between *Aspergillus niger* and *Trichoderma viride* by protoplast fusion, Agric. Biol. Chem. 53 : 1589-1596.
Kirimura, K., Sato, T., Nakanishi, N., Terada, M. and Usami, S. 1997, Breeding of starch utilizing and itaconic-acid-producing kojimolds by interspecific protoplast fusion between *Aspergillus terreus* and *Aspergillus usamii,* Appl. Microbiol. Biotechnol. 47 : 127-131.
Kitamoto, Y., Mori, N., Yamamoto, M., Ohiwa, T. and Ichikawa, Y. 1988, A simple method for protoplast formation and improvement of protoplast regeneration from various fungi using an enzyme from *Trichoderma harzianum,* Appl. Microbiol. Biotechnol. 28 : 445-450.
Kobayashi, I., Ohkuma, T., Ishizaki, H. and Kunoh, H. 1985, Hardness of culture media affecting regeneration and reversion of *Pyricularia oryzae* protoplasts and a thin- layer agar-plate method devised for numerical evaluation of regeneration and reversion, Exp. Mycol. 9 : 161-169.
Kobori, H., Takata, Y. and Osumi, M. 1991, Interspecific protoplast fusion between *Candida tropicalis* and *Candida boidinii* : characterization of the fusants, J. Ferment. Bioengg. 72 : 439-444.
Kobori, H., Yamada, N., Taki, A. and Osumi, M. 1989, Actin is associated with the formation of cell wall in reverting protoplasts of the fission yeast *Schizosaccharomyces pombe,* J. Cell Sci. 94 : 635-646.
Kopecka, M. and Gabriel, M. 1995, Actin cortical cytoskeleton and cell wall synthesis in regenerating protoplasts of the *Saccharomyces cerevisiae* actin mutant DBY 1693, Microbiology 141 : 1289-1299.
Kovac, L. and Subik, J. 1970, Some aspects of the preparation of yeast protoplasts and isolation of mitochondria, Acta. Fac. Med. Univ. Brun. 37 : 91-94.
Kucsera, J. and Ferenczy, L. 1986, Interspecific hybridization between *Candida albicans* and *Candida tropicalis,* FEMS Microbiol. Lett. 36 : 315-318.
Kumar, P.K.R. and Lansane, B.K. 1988, Kinetics of mycelial protoplast formation in *Gibberella fujikuroi* and *Trichoderma reesei,* Folia Microbiol. 33 : 183-187.
Kumari, J.A. and Panda, T. 1992, Studies on critical analysis of factors influencing improved production of protoplasts from *Trichoderma reesei* mycelium, Enz. Microb. Technol. 14 : 241-248.
Lalithakumari, D., Mrinalini, C., Chandra, A.B. and Annamalai, P. 1996, Strain improvement by protoplast fusion for enhancement of biocontrol potential integrated with fungicide tolerance in *Trichoderma* spp., Zeitschrift fur Pflanzenkrankeiten und Pflanzenschutz 103 : 206-212.
Law, C., Kavanagh, K. and Whittaker, P. 1994, Analysis of hybrids of *Candida albicans* formed by protoplast fusion, FEMS Microbiol. Lett. 115 : 77-82.
Layton, A.C. and Kuhn, D.N. 1988, Heterokaryon formation by protoplast fusion of drug- resistant mutants in *Phytophthora megasperma* f.sp. *glycinea,* Exp. Mycol. 12 : 180-194.
Lynch, P.T., Collin, H.A. and Issac, S. 1985a, Use of autolytic enzyme for isolation of protoplasts from *Fusarium tricinctum* hyphae, Trans. Br. Mycol. Soc. 84 : 473-478.
Lynch, P.T., Collin, H.A. and Issac, S. 1985b, Isolation and regeneration of protoplasts from *Fusarium tricinctum* and *F.oxysporum,* Trans.Br. Mycol. Soc. 85 : 135-140.
Lynch, P.T., Issac, S. and Collin, H.A. 1989, Electrofusion of protoplasts from *Aspergillus nidulans,* FEMS Microbiol. Lett. 59 : 225-228.
Maliszewsk, I.H. and Zboinska, E.M. 1996, *Penicillium citrinum* protoplasts : preparaton, regeneration and lipolytic activity of regenerants, Acta Biotech. 16 : 193-198.
Manczinger, L. and Ferenczy, L. 1985, Somatic cell fusion of *Trichoderma reesei* resulting in new genetic combinations, Appl. Microbiol. Biotechnol. 22 : 72-76.
Manczinger, L., Antal, Z. and Ferenczy, L. 1995, Isolation of uracil auxotrophic mutants of *Trichoderma harzianum* and their transformation with heterologus vectors, FEMS Microbiol. Lett. 130 : 59-62.
Maraz, A. and Subik, J. 1981, Transmission and recombination of mitochondrial genes in *Saccharomyces cerevisiae* after protoplast fusion, Mole.Gen. Genet. 181 : 131-133.

Migheli, Q., Whipps, J.M., Budge, S.P. and Lynch, J.M. 1995, Production of inter-and intra-strain hybrids of *Trichoderma* spp. by protoplast fusion and evaluation of their biocontrol activity against soil-borne and foliar pathogens, J. Phytopathology 143 : 91-97.

Mizuno, K., Kimura, O. and Tachiki, T. 1997, Protoplast formation from *Schizophyllum commune* by a culture filterate of *Bacillus circulans* KA-304 grown on cell-wall preparations of *S. commune* as a carbon source , Biosci. Biotech. Biochem. 61 : 852-857.

Morinaga, T., Kikuchi, M. and Nomi, R. 1985, Formation and regeneration of protoplasts in *Coprinus pellucidus* and *Coprinus cinereus,* Agric. Biol. Chem. 49 : 523-524.

Mukerji, K.G. and Garg, K.L. 1988a, Biocontrol of Plant Diseases, Vol. I, CRC Press, BocaRaton, Florida.

Mukerji, K.G. and Garg, K.L. 1988b, Biocontrol of Plant Diseases, Vol. II, CRC Press, BocaRaton, Florida.

Musilkova, M., Fencil, Z. and Seichertova, O. 1969, Release of *Aspergillus niger* protoplasts by *Penicillium purpurogenum enzymes,* Folia Microbiol. 14 : 47-50

Nagy, A., Vagvolgyi, Cs., Balla, E. and Ferenczy, L. 1994, Electrophoretic karyotype of *Mucor circinelloides,* Curr. Genet. 26 : 45-48.

Nga, B.H., Abu Baker, F.D., Loh, G.H., Chiu, L.L., Harashima, S., Oshima, Y. and Heslot H. 1992, Integeneric hybrids between *Saccharomycopsis fibuligera* and *Yarrowia lipolytica,* J. Gen. Microbiol. 138 : 223-227.

Oberma, H., Stobinska, H. and Kregiel, D. 1990, Obtaining and regeneration of *Trichosporon cutaneum* TrNu18 protoplasts, Acta Aliment Pol. 16 : 83-96.

Ogawa, K., Brown, J.A. and Wood, T.M. 1987, Intraspecific hybridization of *Trichoderma reesei* QM 9414 by protoplast fusion using colour mutants, Enz. Microb. Technol. 9 : 229-232.

Ogawa, K., Ohara, H. and Toyama, N. 1988, Intraspecific hybridization of *Aspergillus awamori* var. *kawachi* by protoplast fusion, Agric. Biol. Chem. 52 : 337-342.

Ogawa, K., Tsuchimochi, M., Taniguchi, K. and Nakatsu, S. 1989, Interspecific hybridization of *Aspergillus usamii* mut *shirousamii* and *Aspergillus niger* by protoplast fusion, Agric. Biol. Chem. 53 : 2873-2880.

Ohnuki, T., Etoh, Y. and Beppu, T. 1982, Intraspecific and interspecific hybridization of *Mucor pusillus* and *M. miehei* by protoplast fusion, Agric. Biol. Chem. 46 : 451-458.

Peberdy, J.F. 1976, Isolation and properties of protoplasts from filamentous fungi, in: Microbial and Plant Protoplasts, J.F. Peberdy, A.H. Rose, H.J. Rogers and E.C. Cocking, eds., Academic Press, London.

Peberdy, J.F. 1979, Fungal protoplasts : isolation, reversion and fusion, Ann. Rev. Microbiol. 33 : 21-39.

Peberdy, J.F. and Gibson, R.K. 1971, Regeneration of *Aspergillus nidulans* protoplasts, J. Gen. Microbiol. 69 : 325-330.

Pe'er, S. and Chet, I. 1990, *Trichoderma* protoplast fusion : a tool for improving biocontrol agents, Can. J. Microbiol. 36: 6-9.

Peng, M., Lemke, P.A. and Shaw, J.J. 1993, Improved conditions for protoplast formation and transformation of *Pleurotus ostreatus,* Appl. Microbiol. Biotechnol. 40 : 101-106.

Pesti, M. and Ferenczy, L. 1982, Protoplast fusion hybrids of *Candida albicans* sterol mutant differing in nystatin resistance, J. Gen. Microbiol. 128 : 123-128.

Picataggio, S.K., Schamhart, D.M.J., Montenecourt, B.S. and Eveleigh, D.E. 1983, Sphaeroplast formation and regeneration in *Trichoderma reesei,* Eur. J. Appl. Microbiol. Biotechol. 17 : 121-128.

Pina, A. Calderon, I.L., and Benitez, T. 1986, Integeneric hybrids of *Saccharomyces cerevisiae* and *Zygosaccharomyces fermentati* obtained by protoplast fusion, Appl. Environ. Microbiol. 51 : 995-1003.

Quigley, D.R., Taft, C.S., Stark, T. and Selitrennikoff, C.P. 1987, Optimal conditions for the release of protoplasts of *Neurospora* using Novozym 234, Exp. Mycol. 11 : 236-240.

Reyes, F., Perez-Leblic, M.I., Martinez, M.J. and Lahoz, R. 1984, Protoplast production from filamentous fungi with their own autolytic enzymes, FEMS Microbiol. Lett. 24 : 281-283.

Richards, W.C. 1994, An enzyme system to liberate spore and mycelial protoplasts from a dimorphic fungal plant pathogen *Ophiostoma ulmi* (Buism.) Nannf, Physiol. Mole. Plant Pathol. 44 : 311-319.

Rlike, O., Baum, A., Weiss, J., Hommel, R. and Kleber, H.P. 1992, Kinetics of enzymatic lysis, formation and regeneration of protoplasts of *Candida* (Torulopsis) *apicola,* World J. Microbiol. Biotechnol. 8 : 14-20.

Sakai, T., Koo, K., Saitoh, K. and Katsuragi, T. 1986, Use of protoplast fusion for the development of rapid starch fermenting strains of *Saccharomyces diastaticus,* Agric. Biol. Chem. 50 : 297-306.

Sandhu, D.K., Bawa, S. and Bagga, P.S. 1989, Optimization of protoplast production from *Trichoderma harzianum,* Ind. J. Exp. Biol. 27 : 180-181.

Santos, T.M.C. and DeMelo, I.S. 1991, Preparation and regeneration of protoplasts of *Talaromyces flavus,* Rev. Bras. Genet. 14 : 835-840.

Sarachek, A., Rhoads, D.D. and Schwarzhoff, R.H. 1981, Hybridization of *Candida albicans* through fusion of protoplasts, Arch. Mcrobiol. 129 : 1-8.

Sato, M., Toya, M., Shimura, Y., Kobori, H. and Osumi, M. 1993, Ultrastructural changes during the interspecific protoplast fusion between *Candida tropicalis* and *Candida boidinii* observed by TEM, Plant Morphology 5 : 117-126.

Shabana, Y.M. and Charudattan, R. 1997, Preparation and regeneration of mycelial protoplasts of *Alternaria eichhorniae*, J. Phytopathology 145 : 335-338.

Shin, P. and Cho, M. 1994, Intergeneric protoplast fusion between *Gliocladium virens* and *Trichoderma harzianum*, Korean. J. Mycol. 21 : 323-331.

Sienko, M., Stepien, P.P. and Paszewski, A. 1992, Generation of genetic recombinants in *Trichosporon cutaneum* by spontaneous segregation of protoplast fusants, J. Gen. Microbiol. 138 : 1409-1412.

Sietsma, J.H. and deBoer, W.R. 1973, Formation and regeneration of protoplasts from *Pythium* PRL 2142, J. Gen. Microbiol. 74 : 211-217.

Sipiczki, M., Kucsera, J., Ulaszewski, S. and Zsolt, J. 1982, Hybridization studies by crossing and protoplast fusion within the genus *Schizosaccharomyces* Linder, J.Gen. Microbiol. 128 : 1989-2000.

Sivan, A. and Harman, G.E. 1991, Improved rhizosphere competence in a protoplast fusion progeny of *Trichoderma hazianum*, J. Gen. Microbiol. 137 : 23-29.

Skata, J., Luty, J. and Kotylak, Z. 1988, Interspecific protoplast fusion between the yeasts *Saccharomyces cerevisiae* and *Saccharomyces fermentati*, Curr. Genet. 13 : 101-104.

Solis, S., Flores, M.E. and Huitron, C. 1997, Improvement of pectinase production by interspecific hybrids of *Aspergillus* strains, Lett. Appl. Microbiol. 24 : 77-81.

Stasz, T.E., Harman, G.E. and Gullino, M.L. 1989, Limited vegetative compatibility following intra- and interspecific protoplast fusion in *Trichoderma*, Exp. Mycol. 13 : 364-371.

Stasz, T.E, Harman, G.E. and Weeden, N.F. 1988, Protoplast preparation and fusion in two biocontrol strains of *Trichoderma harzianum*, Mycologia 80 : 141-150.

Suarez, T., Orejas, M. and Eslava, A.P. 1985, Isolation, regeneration and fusion of *Phycomyces blakesleeanus* spheroplasts, Exp. Mycol. 9 : 203-211.

Svoboda, A. and Piedra, D. 1983, Reversion of yeast protoplasts in media containing polyethylene glycol, J. Gen. Microbiol. 129 : 3371-3377.

Tamaki, H. 1986, Genetic analysis of intergeneric hybrids obtained by protoplast fusion in yeasts, Curr. Genet. 10 : 491-494.

TeBeest, D.O. and Weidemann, G.J. 1990, Preparation and regeneration of protoplasts of *Colletotrichum gloeosporioides* f.sp. *aeschynomene*, Mycologia 82 : 249-255.

Teraoka, T., Shimura, Y., Hosokawa, D. and Watanabe, M. 1992, Giant protoplasts of *Pyricularia oryzae* Cavara, Ann. Phytopathol. Soc. Jpn. 58 : 726-733.

Thrane, C., Lubeck, M., Green, H., Degefu, Y., Allerup, S., Thrane, U. and Jensen, D.F. 1995, A tool for monitoring *Trichoderma harzianum* : I. Transformation with the GUS gene by protoplast technology, Phytopathology 85 : 1428-1435.

Toyama, H., Yamaguchi, K., Shinmyo, A. and Okada, H. 1984a, Protoplast fusion of *Trichoderma reesei*, using immature conidia, Appl. Environ. Microbiol. 47 : 363-368.

Toyama, H. Yokoyama, T., Shinmyo, A. and Okada, H. 1984b, Interspecific protoplast fusion of *Trichoderma*, J. Biotech. 1 : 25-35.

Tsoneva, I., Doinov, P. and Dimitrov, D.S. 1989, Electrofusion of fragile mutants of *Saccharomyces cerevisiae*, FEMS Microbiol. Lett. 60 : 61-66.

Typas, M.A. 1983, Heterokaryon incompatibility and interspecific hybridization between *Verticillium albo-atrum* and *Verticillium dahliae* following protoplast fusion and microinjection, J. Gen. Microbiol. 129 : 3043-3056.

Ushijima, S. and Nakadai, T. 1987, Breeding by protoplast fusion of koji mold, *Aspergillus sojae*, Agric. Biol. Chem. 51 : 1051-1057.

Ushijima, S., Nakadai, T. and Uchida, K. 1990, Further evidense on the interspecific protoplast fusion between *Aspergillus oryzae* and *Aspergillus sojae* and subsequent haploidization, with special reference to their production of some hydrolyzing enzymes. Agric. Biol. Chem. 54 : 2393-2399.

Ushijma, S. Nakadai, T. and Uchida, K., 1991, Interspecific electrofusion of protoplasts between *Aspergillus oryzae* and *Aspergillus sojae*, Agric. Biol. Chem. 55 : 129-136.

Vagvolgyi, Cs. and Ferenczy, L. 1991, Isolation of nuclei from *Aspergillus nidulans* protoplasts, FEMS Microbiol. Lett. 82 : 247-252.

Vagvolgyi, Cs., Bruckner, B. and Franko, A. 1990, Size dependent regeneration of *Gibberella* protoplasts, Acta Microbiol. Hungarica 37 : 375-378.

Vainstein, M.H. and Peberdy, J.F. 1991, Location of invertase in *Aspergillus nidulans* : release during hyphal wall digestion and secretion by protoplasts, Mycol. Res. 95 : 1270-1274.

Vidoli, R., Yamazak, H., Nasim, A. and Veliky, I.A. 1982, A novel procedure for the recovery of hybrid products from protoplast fusion, Biotechnol. Lett. 12 : 781-784.

Villanueva, J.R. and Garcia Acha, I. 1971, Production and use of fungal protoplasts, in: Methods in

Microbiology, Vol. IV, C. Booth, ed., Academic Press, New York.

Wagner, R.E. and Wilkinson, H.T. 1993, Spheroplast and protoplast formation from cysts of *Phytophthora sojae* treated with digestive enzymes, Mycologia. 85 : 131-136.

Wilson, J.J., Khachatourians, G.G. and Ingledew, W.M. 1982, Protoplast fusion in the yeast *Schwanniomyces alluvius*, Mole.Gen. Genet. 186 : 95-100.

Witkowska, D., Stempniewicz, R. and Wojatowicz, M. 1997, Production of lytic enzymes using *Trichoderma reesei* M7-1 in yeast protoplast formation, Pol. J. Food and Nutrition Sci. 6 : 51-60.

Wostemeyer, J. and Brockhausen-Rohdemann, E. 1987, Inter-mating type protoplast fusion in the zygomycete *Absidia glauca*, Curr. Genet. 12 : 435-441.

Yamashita, K., Fukuda, H., Murata, K. and Kimura, A. 1981, Transfer of mitochondria of *Hansenula wingei* into protoplasts of *Saccharomyces cerevisiae* by mini-protoplast fusion, FEBS Lett. 132 : 305-307.

Yanagi, S.O., Monma, M., Kawasumi, T., Hino, A., Kito, M. and Takebe, I. 1985, Conditions for isolation and colony formation by mycelial protoplasts of *Coprinus macrorhizus*, Agric. Biol. Chem. 49 : 171-179.

Yanagida, T., Okamura, C., Itoh, K., Oki, T. and Tachibana, S. 1996, A trial for formation of a fusant by protoplast fusion of *Gibberella fujikuroi* strains, Bull. Ehime Univ. Forest C (34) : 93-106.

Yang, H.A., Tommerup, I.C., Sivasithamparam, K. and O'Brien, P.A. 1992, Heterokaryon formation with homokaryons derived from protoplasts of *Rhizoctonia solani* anastomosis group eight, Exp. Mycol. 16 : 268-278.

Yang, H.A., Zhou, J., Sivasithamparam, K. and O'Brien, P.A. 1994, Variation in culture morphology and pathogenicity among protoplast-regenerated strains of *Rhizoctonia solani*, FEMS Microbiol. Lett. 115 : 83-86.

Zhao, J. and Chang, S.T. 1995, Intraspecific hybridization between *Coprinus cinereus* and *Schizophyllum commune* by PEG-induced protoplast fusion and electrofusion, World J. Microbiol. Biotech. 11 : 585-590.

Zhao, J. and Chang, S.T. 1996, Intergeneric hybridization between *Pleurotus ostreatus* and *Schizophyllum commune* by PEG induced protoplast fusion, World J. Microbiol. Biotech. 12 : 573-578.

Znidarsic, P., Pavko, A. and Komel, R. 1992, Laboratory - scale biosynthesis of *Trichoderma* mycolytic enzymes for protoplast release from *Cochliobolus lunatus*, J. Ind. Microbiol. 9 : 115-119.

Zvereva, E.A., Obraztsova, I.N., DeMina, V.A. and Nutsubidge, N.N. 1996, Effect of protoplast formation and cell regeneration on the biosynthesis capability of the mycelial fungus *Aspergillus parasiticus*, Biochem. (Moscow) 60 : 1383-1386.

ROLE OF TISSUE CULTURE IN PLANT DISEASE CONTROL

P. S. Srivastava[1], Muhammad Iqbal[2] and M. Haseeb Mughal[2]

[1]Centre for Biotechnology, [2]Department of Botany, Faculty of Science, Hamdard University, Hamdard Nagar, New Delhi 110 062 (INDIA)

1. INTRODUCTION

Agriculture, the backbone of all developments is one of the oldest vocation pursued from time immemorial. Unfortunately, the damage caused by pathogenic fungi, bacteria and viruses is in terms of millions of dollars annually. The indiscriminate use of pesticides to control such diseases has proven hazardous to other forms of life including animals. The growing number of health problems related to these pesticides has made it imperative to look for eco-friendly and sustainable programmes to combat this global menace. Disease control through the heavy application of synthetic/chemical bactericides, fungicides, insecticides and pesticides has caused tremendous damage to the terrestrial and aquatic ecosystems. It has wiped out hundreds of ecologically important species of plants and animals, causing a tremendous loss of biodiversity (Aziz *et al.*, 1992; Langeweg, 1989). Another problem associated with the excess use of chemotherapy in plant disease is its regular / seasonal applications which renders many pest/pathogens resistant.

The thrust therefore has been to meet the most important challenges before the mankind: the production of more food, fibres and fuels from inelastic land area to feed the ever-increasing demand of growing population. It is in these endeavours that Plant Tissue Culture, an important component of Biotechnology, has come to the rescue and helped the scientists break new grounds.

2. PLANT DISEASES

Plant diseases are the major threat to agricultural productivity. Fungi, bacteria, viruses and insect pests are the main causes of economic losses across the globe. Despite the frequent use of pesticides, 20-30% of crops are lost due to pathogens and pests annually (Estruch *et al.*, 1997; Peferoen, 1997). Inevitably, this has necessitated the programme for resistant varieties at global level but success has been met only in limited cases. The cost incurred for practices and management of chemical control runs nearly into 10 billion US $ per annum (Goswami *et al.*, 1998; Oerke, 1994).

In modern agriculture, protection technology for crops is of paramount importance. On one hand, there is enormous increase in the population and on the other agricultural land is shrinking at a much faster pace than hitherto considered. If this trend continues, time is not far when, inspite of the best technologies and infrastructure, it would be extremely difficult to feed the growing population (Kher and Larka, 1995).

Search for alternative pesticides or resistant lines of crops has become a serious concern and needs a highly directional and purposeful approach. The problem is grave and a reliable solution could be possible only through the appropriate use of biotechnological approaches. With the advancement in the biotechnological techniques, it has become feasible to develop resistant and disease-free plants for future use.

Development of transgenics particularly for better resistance against various diseases and pests is a fast growing and a successful field that holds great promise. Identification and subsequent introduction of certain tolerant genes from microorganisms or naturally resistant plant species into the susceptible host confer resistance against specific diseases.

Tissue culture has become the crucial platform for all the genetic manipulations being carried out with plant systems. From mere callus induction and regeneration of roots and shoots, the technique has helped transformation experiment's leading to transgenics. Plant tissue culture techniques are based on the property of totipotency (the capacity to grow and regenerate) and the exploitation of the nutritional status of various explants. This aseptic technique offers certain advantages over the conventional methods. Production of haploids, creation of usable genetic variability (somaclonal variation), somatic hybrids, genetically manipulated (genetically engineering) plants and micropropagation are some novel applications of tissue culture.

In vitro methods thus offer a tool for cellular and DNA-mediated intervention, variant selection and multiplication of desired clones. The effective way to eliminate the virus has been the culture of apical meristems.

Infact, repeated subcultures of tomato roots *in vitro* by White (1934) and subsequent work by Limasset and Cornuet (1949) led Morel and Martini (1952) to postulate that it is possible to isolate the apical meristem of systemically infected plants and grow it *in vitro* to raise virus free plants. Ever since, the technique of apical meristem culture has been successfully employed to a large number of valuable cultivars of different plants (Quack, 1977). In other cases, the terminal and axillary/lateral buds can also be available for meristem culture. However, there are reports that suggest the superiority of apical meristem culture over lateral-bud culture, for example *Chrysanthemum* (Hollings and Stone 1968). With bulbs and corms also it is possible to raise virus-free plants. However, this may not be universally applicable.

3. ERADICATION TECHNIQUES

Three methods have been recognized as efficient ways to eliminate virus and raise virus-free plants. First, by isolating the apical meristem or the root tips from infected plants. Second, use of antimetabolites either applied to the infected plant part before bud excision, or incorporation into the nutrient medium. Third, infected plants are subjected to temperature treatment (37°C) which retards or inhibits the multiplication of viruses. It has often been observed that heat-treated apical meristems are better and are even more efficient in eliminating the viruses which can't be achieved through meristem-culture only (Quack, 1977).

3.1. Chemotherapy

An array of chemical groups have been used to eliminate the virus from the plants. These include amino acids, antibiotics, growth regulators purine and pyrimidine analogues etc. But only limited success has been achieved in eliminating/suppressing or inactivating the viruses. A whole plant treatment by chemicals have failed to eradicate the virus from a number of plants (Tomlinson, 1982). 'Virazol' (9-ß-D-ribofuranosyl-1,2,4-triazole-3-corboxamide) has been the most effective agent to eliminate a wide spectrum of viruses (Shepard, 1977). Chemotherapy, however, has been successful only in a limited number of viruses (Table 1).

Table 1 : Some successful examples of chemotherapy against virus diseases

Chemical compound	Active against		Reference (s)
	Virus	Host	
Actinomycin-D	TMV in	Chinese cabbage	Renaudin and Bove, 1977
Malachite Green	PVX	Potato	Norris, 1954
Ribavirin	ACLSV	Apple	Hansen and Lane, 1985
	CMV	Tobacco	Cassells and Long, 1980
	EMCV	Egg plant	Raj *et al.*, 1991
	LSV	Lily	Blom-Barnhoorn and Aartrijik, 1984
	ORS	Cymbidium	Debergh *et al.*, 1990
	PVS, PVX	Potato	Griffiths *et al.*, 1990
	PVY, PVM	Potato	Wang, 1985
2-Thiouracil	PVY	Tobacco	Kassanis and Tinsley, 1958
Vidarabine	OMV	Ornithogalum	Veelar *et al.*, 1992

3.2. Thermotherapy

Tobacco mosaic virus, carnation mottle virus, potato virus-X, cucumber mosaic virus etc. have been reported even from the apical meristem (Hollings and Stone, 1964) and a simple meristem-culture of such infected plants does not help to get disease-free plants. Thermotherapy-the heat treatment of the infected parts/plant, seem to have proved helpful in eradicating the virus. In such cases, a combination of thermotherapy followed by apical-meristem culture has produced disease-free plants. Prior to the excision of explants, mother plant is subjected to a high temperature treatment. This technique has proved successful in several potato cultivars (MacDonald, 1973; Slack *et al.*, 1987). The frequency as well as the intensity of the heat treatment determines the degree of sterilization (being free from viruses). Virus-free plants in white rose variety of potato increased from 50% after a heat treatment for 8 weeks to 100% after 18 wks. Not in all cases shoot-tip culture have proved essential but even heat-treated axillary buds have generated disease-free plants (Stace-Smith and Mellor, 1989). It has also been shown that daily treatments of donor shoot at 39°C, 22°C for 6 hrs, for 28-40 days proved effective in elimination of grapevine fan leaf virus (GFLV) but not the arabis mosaic virus (Monette, 1986).

Paludan (1985) has reported a cold treatment (-50°C) of excised shoot tips for one month to raise virus-free plants of *Chrysanthemum morifolium in vitro*. The effectiveness of the cold treatments at 5°C to 15°C were earlier used to eradicate potato spindle tuber virus (Lizarraga *et al.*, 1980).

3.3. Breeding for Resistance

Man has always looked for better yield and resistant varieties of crops. The initial success in crop improvement depended on selection of high yielding and resistant varieties. Unfortunately, the selection process did not provide enough commercial advantage for sustained exploitations. Breeding programmes, however, produced the desired results in the early decades of this century. Rust resistant varieties of wheat is one such example. Repeated backcross was desired to achieve the goal of transferring the gene from resistant to susceptible species. The genes introduced into wheat from various species to confer resistance to usual rusts, and in potato to *Phytophthora infestans,* for example 'broke down' very easily. Similarly, resistance achieved through mutations (e.g., in Mint against wilt caused by *Verticillium alboratum*) by the treatment of mutagens, requires exhaustive selection that has proved to be laborious and time consuming and no better than the conventional breeding. (Parlevliet 1995; Singh and Singh, 1995). Moreover breeding programmes are also restricted by species barrier and incompatibility factors. The future thus lies with the biotechnological techniques only.

3.4. Ecological Impact of These Techniques

Be it thermo-or chemotherapy, there is never 100% application of these techniques that too with equally good/encouraging results in all the plants. Thermotherapy of delicate and young explants often results in irreversible physiological damages that may render them non-responsive to *in vitro* differentiation.

Chemotherapy has also not proved its effectiveness at universal level. Probably, the hazardous effect on the physiological status of the excised explants are the main bottle necks of this technique. Besides, the effectiveness of different chemicals varies greatly with different conditions, explants and plant species. Chemical treatment to get virus-free plant material also has a serious adverse effect on environment and thus can not be used indiscriminately on a large scale.

India alone loses 6000 crores of agricultural produce annually, due to insect pests, pathogens and parasites (Raj *et al.*, 1994). The use of pesticides has become absolutely essential to save the food grains. But, the heavy and indiscriminate use of chemical pesticides has resulted in resistance, immunity and resurgence of insects and diseases, causing problems of loss of produce. This has also led to soil, air and water contamination beyond corrigible limits. Conceivably pesticides and their harmful residues accumulate in consumable and non-consumable objects which prove detrimental to all life forms.

4. BIOTECHNOLOGICAL APPROACHES

4.1. Meristem Culture

Meristem-the group of actively dividing cells, located at the growing points of the plant offers contamination-free material for *in vitro* multiplication. Meristem culture has been successful more with the vegetatively propagated crops like potato and sugar cane etc. Much of the work in this regard has been carried out with *Solanum tuberosum* (Potato) cultivars (Bhojwani and Razdan, 1983; Hartmann *et al.*, 1990; Slack, 1980). Potato being the chief vegetable crop that grows across the globe, is unfortunately infected by over 20 viral diseases that adversely affect the yield. Tissue culture technique has been utilized for both basic and applied purposes in potato programs (Dodds, 1988). Meristem culture has been successfully employed to maintain disease-free and parent strain of potatoes (Bryan, 1988; Jones, 1988). Plantlets thus obtained provide a base for mass multiplication (Bryan, 1988) and for chemical and heat therapy procedures that improve the virus elimination efficiency (Sanchez *et al.*, 1991). Nearly 160 wild-species of potato have been recognized but the global gene pool is relatively untapped (Hakes 1978). Commercial cultivars are tetraploids (2n=4x=48) and extremely heterozygous, with simple inheritance (Aaaa) for many characters (Howard, 1978). Tissue culture provides the experimental system to which technique of genetic manipulations can be applied. Potato is one of the most amenable crop for tissue culture. Complete plantlets can be regenerated from a range of tissues (roots, leaves, tubers, stems) and even from single cells (Karp *et al.*, 1988). Meristematic bud culture under aseptic culture has been described by Mellor and Stace-Smith (1977) who listed 136 virus-free potato cultivars produced by meristem tip culture.

4.2. Shoot-tip Culture

Shoot tip culture has also been considered as an alternative approach to multiply trees which is conventionally done through seeds. Infection of seeds due to one or the other reason which may hamper the propagation of healthy plants can be overcome by using shoot tip culture (Roy *et al.*, 1996). This method would also eliminate the possibility of genetic variability and heterozygosity that spreads through the seeds. Shoot tip culture have more plus points compared to seeds, it can give rise to a number of shoots which is not possible otherwise with a normal seed.

In vitro techniques provide important tools to understand several intricate details of host cell-pathogen interrelationships at cellular levels (Takebe and Otsuki, 1974). Single cell clones of pathogenic fungi, bacteria and other plants have been well established. Culture of such clones have enabled us to study the intricate relationship of host and pathogen. Tissue culture studies have helped reveal the details of life-cycle and pathogenesis of various pathogens (Hilderbrandt, 1977).

4.3. Callus Line Selection

It has been observed that the callus from various explants of the susceptible hosts treated with different mutagens (physical/chemical) induce a degree of resistance in the plants regenerated from such selected calli. Gamma irradiated resistant rice calli have regenerated plants showing tremendous resistance against *Pyricularia oryzae* (Palit and Reddy, 1987). Further studies have established the role of tissue culture in raising the resistant plants from selected lines of callus *in vitro*. Table 2 represents some successful reports of resistance achieved through the selection of *in vitro* raised callus/cultures.

Table 2: Plants with inherent resistance selected through tissue cultures

Explant/tissue source	Host plant	Resistant against	Reference(s)
Suspension cultures	*Apium gravoelens*	*Fusarium oxysporium*	Health-Pagliuso and Rappaport, 1990
Stem callus	*Brassica compestris*	*Phoma lingam*	Sacristan, 1982, 1985
Leaf callus	*Fragaria chiloensis*	*Fusarium oxysporium*	Toyoda *et al.*, 1991
Leaf callus	*Lycopersicum esculentum*	*Fusarium oxysporium*	Miller *et al.*, 1985; Fulton, 1986
Cotyledon protoplast	*L. esculentum*	*Fusarium oxysporium*	Shahin and Spivey, 1986
Callus	*Medicago sativa*	*Fusarium oxysporium*	Hartmann *et al.*, 1984
Callus	*M. sativa*	*Fusarium* spp.	Hrubocova *et al.*, 1992
Shoot tip	*Musa paradisica*	*Fusarium oxysporium*	Hwang, 1991
Callus	*Nicotiana tabacum*	*Peronospora tabacina* Tobacco Mosaic Virus	Atanassove, 1985
Callus	*Saccharum officinarum*	*Sclerospora sacchari*	Liu and Chen, 1978; Larkin and Scoworoft, 1981
Mesophyll protoplasts	*Solanum tuberosum*	*Alternaria solani*	Matern *et al.*, 1978
-do-	*S. tuberosum*	*Phytophthora infestans*	Shepard, 1981
Embryo callus	*Zea mays*	*Helminthsporium maydis*	Brettel *et al.*, 1978

4.4. DNA Introduction and Transgenics

Introduction of the desired genes (DNA) into plant cell has become a routine process with its significant applicability in agricultural and medicinal traits. The untiring capacity to grow and regenerate *in vitro*, and express the foreign genes have made plants a suitable system for genetic manipulations. Several methods have been developed to transfer the desired gene/DNA into the host cells. Most common of them are:

4.4.1. *Agrobacterium*-mediated

Most exploited *Agrobacterium tumefaciens* and lesser known *A. rhizogenes* have the capacity to transform multicellular explants as well as single cell protoplasts. *A. tumefaciens*, a Gram-negative soil bacterium that causes crown gall disease in plants (gall at the site of infection). Since the possibility of *Agrobacterium* medicated transformation is restricted to susceptible plants, this has limited the application to dicots only, although transformation in monocots have been achieved. This bacterium is able to transfer a portion of its megaplasmid (T_1 plasmid) to the plant cell where it gets stably integrated into the nuclear DNA of host-cell. Ti-plasmid is 20-25 Kb and consists of genes with promoters that can be read by the plant transcription/translation system. This DNA carries two significant advantages- (a) high endogenous synthesis of auxins and cytokinins, thus causing non-directional proliferation of infected cells ('crown gall') and (b) synthesis of tumour specific opines that serve as C-and N-source for this agrobacterium. An appropriate term, 'Genetic colonization' has been used for this type of parasitism where the parasite genetically transforms its host in order to draw an exclusive access to host photosynthetic machinery. The procedure underlying such transformations is:

- Introduction of desired gene into *Agrobacterium* T-DNA
- Transformation of the plant/plant cells/protoplasts with this modified T-DNA.
- *In vitro* culture of the transformants on the appropriate medium to regenerate the whole plant from these transformed cells. Success so far achieved is represented in Table 3.

Agrobacterium rhizogenes, a Gram-negative bacterium causing hairy root disease has also been exploited as a gene transfer system in higher plants. The strategy used is similar to *A. tumefaciens* (Mendel and Hansch, 1995; Visser *et al.*, 1989).

4.4.2. Direct DNA transfer

The introduction of DNA directly into protoplasts (cells and spheroplasts from pro- and eukaryotic organisms) has been made possible by polyethylene glycol (PEG) or calcium method. Direct DNA transfer method has unique advantage of being applicable to protoplasts of any plant source (monocot or dicot) and is not restricted to the choice of hosts. This technique has been successfully applied to both monocots and dicot species (Potrykus, 1991).

4.4.2.1. PEG method

Addition of plasmid DNA to protoplasts does not help the cell to take foreign DNA. It's possible only by the addition of PEG- a compound that facilitates the process of DNA uptake.

Table 3 : Some successful reports of *Agrobacterium*-mediated transformation in plants

Plant Name Botanical	English	Reference (s)
Dicots	Pepino	Atkinson and Gardner, 1991
Allocasurina	—	Phelep *et al.*, 1991
Apium gravoelens	Celery	Catlin *et al.*, 1988
Arabidopsis thaliana	Arabidopsis	Lloyd *et al.*, 1986
Armortia lapthifolia	Horseradish	Noda *et al.*, 1987
Beta vulgaris	Sugarbeet	Botterman and Leemans, 1989
Brassica olearacea	Cauliflower	David and Tampe, 1980
Chrysanthemum spp.	Chrysanthemum	Ledger *et al.*, 1991
Cucumis melo	Muskmelon	Fang and Grumet, 1990
Cucumis sativus	Cucumber	Trulson *et al.*, 1986
Fragaria chiloensis	Strawberry	Nehra *et al.*, 1990
Gossypium hirsutum	Cotton	Umbeck *et al.*, 1987
Helianthus annuus	Sunflower	Everett *et al.*, 1987
Ipomea violacea	Morning glory	Tepfer, 1984
Juglans regia	Walnut	McGranahan *et al.*, 1988
Kalanchoe laciniata	Kalanchoe	Jai *et al.*, 1989
Lactuca sativa	Lettuce	Michelmore *et al.*, 1987
Linum usitatissimum	Flax	Basiran *et al.*, 1987
Lycopersicum esculentum	Tomato	McCormick *et al.*, 1986
Medicago sativa	Alfalfa	Kuchuk *et al.*, 1990
Nicotiana tabacum	Tobacco	Horsch *et al.*, 1985
Populus spp.	Poplar	Phythoud *et al.*, 1987
Petunia axillaris	Petunia	Horsch *et al.*, 1985
Pisum sativum	Pea	Puonti-Kaerlas *et al.*, 1990
Pyrus malus	Apple	James *et al.*, 1989
Solanum melongena	Eggplant	Rotino and Gleddie, 1990
Solanum tuberosum	Potato	Stiekema *et al.*, 1988
Stylosanthes humilis	Lucerne	Manners and Way, 1989
Vicia narbonnensis	Norbonne vetch	Pickardt *et al.*, 1991
Monocots		
Asparagus	Asparagus	Bytebier *et al.*, 1987
Chlorophytum capense	—	Mendel and Hansch, 1995
Dioscorea bulbifera	Potato yam	Mendel and Hansch, 1995
Gladiolus	Gladiolus	Bytebier *et al.*, 1987
Narcissus	Jonquil	Mendel and Hansch, 1995
Oryza sativa (japonica)	Rice	Zhang and Wu, 1988
Oryza sativa (indica)	Rice	Datta *et al.*, 1990
Zea mays	Maize	Mullner, 1991

4.4.2.2. Electroporation

In electroporation protoplasts are subjected to electric impulse of 500-3000 V/cm for milli or microseconds. This creates transient pores in membrane through which DNA enters the cells (Mendel and Hansch, 1995). Electroporation is being used instead of PEG or in association with it. However, it has been observed that PEG is more successful in most cases and causes less damage to fragile protoplasts than caused by electrophoresis. Rice, maize and some other cereals have been successfully transformed through these methods (Mendel and Hansch, 1995).

4.4.3. DNA transfer by microprojectiles

Many plants have proved recalcitrant to gene transfer through protoplast methods. In all such cases direct DNA delivery by high-velocity microprojectiles is being attempted (Mendel and Hansch, 1995). This method uses the minute metal balls coated with DNA that are shot directly into the live target cell. Microprojectiles typically made up of high density metals, like tungstun and gold (diameter 1-2 µm), are accelerated to high velocities so that they are able to penetrate the cell walls. After entering the cell, microparticles may hit plastids, mitochondria or nucleus or may be found in the cytoplasm. Thereafter DNA is released and in subsequent steps it randomly integrates to the cell genome. Maize, rice, wheat, barley, sugarcane, papaya, spruce, poplar, cotton, soybean, *Phaseolus* etc. have been successfully transformed by using microprojectile method (Gordon-Kamm *et al.,* 1990; McCabe *et al.,* 1988).

Selection/recognition of transformed cells has been a difficult job but the identification of the antibiotic resistance genes and their introduction into the host cells along with transfer of desired gene/s enables the identification of transformants. This is achieved by supplementing the culture media that helps in selecting resistant strains/cells-i.e. transformants. The most commonly used selectable markers have been listed in Table 4.

Table 4 : Gene markers that have been used for selecting transformed cells

Enzyme(s)	Resistant against
Bleomycin resistance	Bleomycin
Gentamycin acetyltransferase	Gentamycin
Hygromycin phosphotransferase	Hygromycin
Neomycin phosphotransferase	Kanamycin
Dihydrofolate reductase phosphinotricine	Methotrexate
Phosphinotricine acetyltransferase	Phosphinotricine

5. RESISTANCE ACHIEVED AGAINST

5.1. Fungi and Bacteria

There have been attempts to develop plants resistant to fungi and bacteria. The strategy mainly has been by the transfer of genes coding enzymes involved in organic compounds-phytoalexins that are more aggressive against fungi, by introduction of genes that are responsible for the synthesis of enzymes degrading cell wall of the fungal hyphae, and by the introduction of genes encoding low-molecular weight proteins that have a direct antifungal and antibacterial effect (Willington, 1993). It has been observed that fungal and bacterial infection cause a re-adjustment in the biosynthetic machinery (Lamb *et al.,* 1989). *In vitro* methods have proved beyond doubt that there indeed is the formation of low-molecular weight organic compound which possess strong phytopathogenic action against fungi, -Phytoalexins (Bailey, 1987).

One such approach to develop transgenic plants resistant to fungi and bacteria has been the modification of phytoalexin by transferring a gene coding stilbene synthases from groundnut to tobacco (Hain *et al.,* 1990).

Stilbenes, synthesized in a restricted group of plants, are one of the major groups of secondary metabolites that possess tremendous phytoalexinic and antifungal properties. Introduction of stilbene genes to susceptible plants will prove effective against large number of fungal and bacterial diseases. However, it has been seen that in many plants, precursors necessary for stilbenes are already there, addition of stilbene synthase to the medium during the *in vitro* culture may prove an alternative to the gene transfer in such plants, e.g. in tobacco plant.

Chitinases and glucanases are another group of compounds that display strong antimicrobial activity (Schlumbaum *et al.*, 1986). There has been several attempts to overexpress the chitinase genes in transgenic plants to provide them resistance against fungi (Lund *et al.*, 1989; Neuhaus *et al.*, 1991). The chitinase genes obtained from *Serratia marcescens*, a soil bacteria, and beans have been introduced into tobacco, making it resistant to *Alternaria longipes* (brown spot disease) and *Rhizoctonia solani*, respectively. Acetyl transferase gene has also been introduced into tobacco making it resistant to *Pseudomonas syringe* (wild fire disease). The expression of chitinase gene of bean in tobacco caused resistance against *Rhizoctonia* (Broglie *et al.*, 1991).

Osmotins, thionins and zeamatins are group of proteins that display a clear antifungal activity in vitro (Bohlmann and Apel, 1991; Roberts and Selitrennikoff 1990; Vigers *et al.*, 1991; Woloshuk *et al.*, 1991). However, whether their expression or overexpression in the transgenic plants can lead to the enhanced resistance, is to be ascertained.

It has usually been proposed that similar lines of action may be used to raise transgenic plants resistant to bacterial diseases. Nevertheless, slightly modified methods have also proved highly specific for developing bacterial resistance. The general strategy has been to discover more and more of enzymes/inhibitors that cause degradation of cell wall or interfere with metabolism and thus prevent bacterial infection. Lysozyme, a T_4 bacteriophage encoded enzyme has already been successfully transferred to transgenic plants (During, 1989). Anzai and co-workers (1989) achieved resistance against *Pseudomonas syringae* pv. *tabaci*; the tabtoxin, the toxin produced by this bacterium acts by inhibiting glutamine synthetase. Transfer of this gene modified by suitable plant regulatory sequences into tobacco plants resulted in resistance in the transgenic plants.

This success has paved the way for several other manipulations with assured success. Antibiotic principle seems to work out the problem. Identification of antibiotic producing genes and their insertion in the plant system would prove highly profitable in offering resistance to wide variety of bacteria, like Gram negative and Gram positive.

This pioneering work with transgenics aimed at developing resistant lines against fungi and bacteria has opened new vistas in tissue culture research. The exploitation of already available resistant lines, their multiplication through *in vitro* methods or modification in the genetic make-up of the plants through mutagenesis, genetic engineering, enzyme technologies in the transgenic plants has met with commendable success. Further, research in this direction will add significantly to the development of resistant transgenic plants in large numbers.

5.2. Insects

Since the control of insect pests was largely being done by synthetic organic chemicals which has led to the elevated environmental degradation and forced scientists to develop genetically resistant strains of plants that would enable the elimination of use of such hazardous reagents. The first to be actively considered was the transfer of *Bt* gene from *Bacillus thuringienes* to the desired crops. This resulted in the expression of insecticidal proteins that ultimately protected these plants from insects (Brunke and Meeusen, 1991; Van Rie, 199). This method of inducing plants to produce their own protein that provides complete escape/ resistance from insect pest, is highly recommendable. This approach has of course advantages over the conventional chemotherapy, like the absence of non-proteinaceous residues in soil or ground water, high specificity with respect to the target organism, and protection of different parts of plant which are difficult to reach by conventional methods.

δ-endotoxin: These proteins are present as crystals in spores of *B. thuringienes* (*Bt*), on digestion these proteins get solubilized in the mid-gut of the insect due to the alkaline pH and some proteases in gut. Due to this the protoxin gets converted into toxin. And eventually disrupts the cells of mid-gut (Hofte and Whiteley, 1989). The commercial *Bt* insecticides are effective against more than 50 lepidopteran pest insects. Cloned *Bt* gene has been successfully introduced into tomato (Fischaff *et al.*, 1987), tobacco (Barton *et al.*, 1987; Vaeck *et al.*, 1987) and other crops.

Nearly 15% of the total world's crop are lost due to insects. Introduction of chimeric gene encoding for an orzyacystatin-gus fusion protein into rice protoplasts has been successfully carried out by Hosoyama *et al.*, (1995). The resultant transgenic plants carrying such fusion-protein have shown resistance against insects. Transfer of corncystatin into protoplasts of Nipponbare by Irie *et al.*, (1996) has shown that resultant transgenic plants have inhibitory action against commercial proteinases and proteinases of different insect pests.

Xu *et al.*, (1996) have been able to get expression of cowpea trypsin inhibitor gene in rice. In field trials the transgenic plants expressing this gene showed significant levels of resistance against two major insect pests, stripped stem borer and pink stem borer (*Sesamia viferens*).

Recently, role of lectins in providing insect resistance is gaining ground (Rao *et al.*, 1997). Transfer of snowdrop lectin gene has proved this fact. PEP carboxylase promoter gene from maize to drive expression of B^+ gene, cry IA (b) has recently been achieved by Ghareyazie *et al.*, (1997). Results of this experiment show high expression of B^+ protein in leaves and not in grains. Transformation of indica rice by synthetic B^+ gene, cry IA (c) with multiple toxin domain has been reported by Nayak *et al.*, (1997). Stable integration and transmission of the transgene has been confirmed by molecular analysis. High degree of resistance has been recorded against yellow skin borer.

5.3. Viruses

The practical control of virus infection depends on the control methods that prevent or restrict infection. Introduction of coat protein (Cp) gene delays the disease and the subsequent spread of systemic infection. Coat protein gene expression in tobacco mosaic virus (Abel *et al.*, 1986; Nelson *et al.*, 1987), alfalfa mosaic virus (Tumer *et al.*, 1987), potato virus X (Hemenway *et al.*, 1988) and cucumber mosaic virus (Cuozzo *et al.*, 1988) have resulted in tremendous protection of the transgenics against viral infections. On a similar front introduction of alfalfa mosaic virus-coat protein into tobacco has also reflected a high degree of resistance to infection (Van Dun *et al.*, 1987). The extent of resistance in all such cases has been shown to be directly related to the expression of Cp-gene.

In case of tomato, tobacco and potato, numerous field experiments which achieved control over the target insects, have been performed in USA and several countries in Europe (Delanney *et al.*, 1989). Significant control of fruitworm (*Heliothis zea*) and pinworm (*Keiferia lycopericella*), the two common pests of tomato, have already been achieved (Brunke and Meeusen 1991; Perlak *et al.*, 1991).

Plant diseases caused by viruses affect a substantial reduction in food, forage and fibre crops throughout the world. Irish famine can not be ignored while discussing plant losses caused by pathogens. There indeed have been cultural practices to remove the infected debris, host weeds, checking of vectors, sterilization of propagules (seed/tubers etc.) and breeding for resistance. But a limited success has been reported so far (Culver, 1995; Fraser, 1990).

Three general approaches have been identified so far to raise the transgenics resistant against viral infections:

(i) coat protein-mediated infection,

(ii) via expression of satellite RNA and

(iii) via anti-sense RNA

Of these approaches, the first has proved to be the best, justifying the use of this method for large scale DNA introduction (for details see Resistance Mechanisms)

5.4. Herbicides

Development of herbicide resistance in plants has been one of the prime aims of genetic manipulations. Development of herbicide resistance in plants has definite economic and ecological significance:

i) Herbicide resistant plants have proved a significant tool against the complete elimination of the weed and thus prevent a huge loss to the crops caused by the potential threat from weeds.

ii) Resistance in plants offer a permanent protection against the physiological or biochemical stress offered by various synthetic herbicides.

Herbicide resistant plants produced through genetic engineering basically use two distinct strategies (Mazur and Falco, 1989; Oxtoby and Hughes, 1990):

i) Target modification: since in many cases the target of herbicide is a protein, modification in the level and/or sensitivity to the herbicide, are the best possible approaches available.

ii) Herbicide transformation: Introduction of a gene in transgenic plants that encodes the enzyme which causes degradation of herbicide into non-toxic components.

Target-proteins for herbicides have been identified in higher plants and biotechnological applications are underway to find a permanent solution to the herbicide stress. Triazines inhibit photosynthesis whereas glyophosate, inidiazolinones, phosphinoctricine and sulfontricine groups of herbicides cause inhibition in amino acid biosynthesis.

Expression of herbicide detoxifying enzymes in the transgenic plants has been very successful in nullifying or chemically transforming the herbicides to non-toxic and harmless chemicals (Willmitzer, 1993).

6. RESISTANCE MECHANISM: THROUGH

6.1. Coat Protein-Mediation

Since it has been observed that inception by a mild virus, prior to infection by a potentially infective and an aggressive virus results in the decrease of the symptoms normally evoked by the infective virus. It is believed that the two viruses are somehow related, which has given way to the speculation that coat proteins are responsible for this phenomenon (Sequeua, 1984; Sherwood and Fulton, 1982). The reason behind this is based on the assumption that mild virus which infected the cells first, produced excess of coat protein in the cellular pool. The free floating coat protein inhibited the uncoating of RNA of the aggressive virus. There are indeed growing number of reports regarding the transfer of coat protein gene to various crops that in turn offers a strong resistance against infective viruses (Willmitzer, 1993). The basic mechanism at molecular levels is still not clearly understood. Nelson *et al.* (1988) conducted a field experiment on transgenic plants (Tomato) expressing TMV coat protein gene and Beachy *et al.* (1990) on transgenic plants (Potato) expressing coat protein of potato virus X and potato virus Y. These experiments resulted in the dramatically reduced infection by the viruses under question. Thus an overall loss of 30% in these crops was saved (Kamiwski *et al.*, 1990).

6.2. Anti-sense RNA

Introduction of the gene that encodes anti-sense RNA to the viral RNA into transgenic plants has been done to obtain resistance against viral infections. The method did not prove as effective as coat-protein (Cuozzo *et al.*, 1988; Powell *et al.*, 1989). However, after a few more years of investigations it became evident that anti-sense RNA corresponding to the viral genome can led to a significant level of resistance (Day *et al.*, 1991).

Introduction of some non-structural protein genes into transgenic tobacco plants have resulted in complete resistance against viruses (Golemboski *et al.*, 1990). The success is related to the advancement in genetic engineering. Recombinant DNA technology and enzyme engineering are destined to play an important role in future investigations. Identification of resistant lines after genetic manipulation and their mass multiplication through *in vitro* methods is of immense significance in agriculture and medicine.

Rice blast caused by *Pyriculria oryzae* is wide spread in the major rice growing regions of the world. Gamma irradiation of the pollen callus has resulted in the isolation of a resistant callus that ultimately gave rise to resistant rice plantlets (Palit and Reddy, 1987).

Hayakawa and others (1992) transformed two japonica cultivars with coat protein (Cp) gene of rice stripe virus. Transgenic plants expressing Cp gene exhibited a significant degree of resistance against the virus. Recently, Huntley and Hall (1996) showed that expression of various regions of the brume mosaic virus (BMV) genome in transgenic plants resulted in substantial level of resistance against virus infection. Zheng and his group (1997) successfully transformed rice by rice dwarf virus coat protein (S8), using biolistic-mediated gene delivery. Molecular analysis has confirmed the integration of S8 into the host genome. The first success in obtaining resistance against RTBV and RTSV has been reported more recently by Fauquet *et al.*, (1997).

6.3. Sense RNA

Resistance offered by the transgene expression of the TMV and AMV coat protein has clearly been shown to be associated with production of coat protein. It has also been observed that coat protein open reading frames (ORF) unable to translate protein did not confer resistance when expressed in transgenic plants, although a high levels of RNA transcripts accumulated (Powell *et al.*, 1989; 1990). However, a TEV coat protein ORF sense construct, unable to produce coat protein, provided a high degree of resistance which was greater than resistance offered by coat protein accumulation in similar plants (Lindbo *et al.*, 1992; Singh and Singh, 1995). Protoplasts from these TEV sense plants did not support viral replication, helping us to believe that resistance is due to the inhibition of some early steps during replication. Transgenics resistant against R/Y and PLRV express little or no detectable amount of coat protein but have still proved resistant (Kawchuk *et al.*, 1990). Therefore, it is assumed that resistance in these coat protein systems may be due to an RNA sense-mediated mechanism (Culver, 1995).

Expression of other sense regions of viral genome has also resulted in different degrees of resistance. For example transgene constructs containing complete or partial ORF of the RX 165-kDa replicase protein did confer resistance. No such protein could be detected in these constructs, suggesting that resistance is due to interfering properties of sense RNA (Day *et al.*, 1991)

6.4. Satellite RNA

Small, extragenomic RNA molecules which are not autonomous and require the presence of intact viral genome for its growth, replication and propagation within plants and other host plants, have turned to be useful for generating resistant plants. These RNA molecules have a tendency to modulate the symptoms produced by an infective virus. For example, cucumber mosaic virus associated RNA, alters the symptoms induced by cucumber mosaic virus on tobacco and similarly satellite RNA of tobacco ringspot virus results in alteration of the symptoms produced by tobacco ringspot virus on tobacco (Harrison *et al.*, 1987).

It has now been revealed that there exists very little difference between benign and virulent satellite RNA species. It has, therefore, become hazardous to use such satellite RNA at field levels.

6.5. Proteinase Inhibitors

Proteinase inhibitors are widely distributed in higher plants. Mature propagation tissues (seeds, tubers) accumulate highest concentration of proteinase inhibitors, whereas other organs can be made to synthesize such proteins under stress conditions that mimic the attack of a herbivorous insect, e.g. by wounding. In vitro stress through different means can be a good inducer for high proteinase inhibitor activity. Now it has been found that most proteinase inhibitors display an activity not against the endogenous proteinases but against the proteinase of microbial and mammalian origin. This has proved beyond doubt that proteinase inhibitors represent a strong defense mechanism (Richardson, 1977).

It has also been found that proteinase inhibitors have anti-metabolic properties in a wide range of insects. This offers a clue, that if proteinase inhibitor genes are successfully introduced into the plant system, such plants would be resistant to herbivorous insect pests. However, there are only few successful reports to this effect. Expression of plant-derived protease inhibitor (the cow pea trypsin inhibitor) in transgenic tobacco has led to tremendous resistance against tobacco budworm (*Heliothis virescens*) (Hilder *et al.*, 1987). Supplementation of proteinase inhibitor along the diet has shown remarkable impact on enzymatic activity in a varied groups of insects like *Spodoptera litoralis, Heliothis* spp. and *Diabrotica undecimpunctata* (Hilder *et al.*, 1990) etc.

6.6. Ribozyme Mediated

Ribozymes, the RNA based RNA restriction enzymes can catalytically break down RNA molecules at specific sites. Their role in ribosomal RNA production, intron splicing and satellite RNA replication is well established (Culver, 1995). These authors have also reported the genetically engineered ribozymes directed against a large number of RNA molecules. Besides, they have demonstrated the activity of these ribozymes to cleave. The ability of ribozymes to cleave at a specific site has given a way to develop potentially useful for control of viral diseases. Since the majority of plant viruses have RNA genomes, this technique offers the best rescue against such diseases.

Transgene expression of ribozymes designed to cleave viral RNAs could thus be used specifically to disrupt viral replication machinery and ultimately the disease development.

Ribozymes with different sequences and structures have been identified (Edington and Nelson, 1992). Structurally, ribozymes are composed of a highly conserved region which functions as catalytic domain and there are flanking regions which pair with substrate (RNA). Cleavage occurs at 3‘ GUC sequence and results in a free 5‘ hydroxyl and 3‘terminal 2‘,3‘-cyclic phosphate. Presence of a GUC sequence seems to be the only requirement for the action of ribozymes. A high degree of stability in a temperature range of 37-40°C suggests its potential applications in *in vivo* systems. As much as 90% resistance (accumulation of pathogenic virus) was achieved within 24 hrs in ribozymes directed against the replicase ORF of TMV (Edington and Nelson, 1992). Thus, further investigations to design the ribozymes aiming or targetting a broad group of viruses will make an efficient tool to control viral diseases in a permanent manner.

6.7. Trypsin Inhibition

Many plant species have shown the presence of certain proteins which cause inhibition of the mammalian digestive protease (Gatehouse and Boulter, 1983). Synthesis of such protein inhibitors in response to insect/pest attack has been reported in tomato, cowpea etc. An 80-amino acid polypeptide has been identified as an insecticidal component of *Vigna unguiculata*, that has shown tremendous inhibitory action against *Callosobruchus maculatus* (Bruchid beetle) (Gatehouse *et al.*, 1979; 1991) Hilder and co-workers (1987) have successfully characterized the cowpea trypsin inhibitor (C_pTI) gene, cloned it and finally introduced this gene to tobacco plants. This transgenic showed enhanced resistance against *Heliothis virescens*. Subsequent experimentations with C_pTI have shown its tremendous insecticidal potentials along with the high degree of heritability (Hilder and Gatehouse, 1990; Hilder *et al.*, 1993).

6.8. Secondary Metabolites

Many plants especially legumes produce secondary metabolites which are used to make transgenics insecticide resistant. 2,5 dihydroxymethyl-3,4dihydroxypyrolidine (DMDP) derived from *Lonchocarpous* spp., Castanospermine obtained from *Castanospermum australe*, P-aminophenylalanine derived from *Vigna* sp., Pectosans from *Phaseolus vulgaris* are resistant against *Callosobruchus maculatus* and *C. chinesis*. Heteroploysaccharides derived from *Phaseolus vulgaris* makes plant resistant against *Acanthoscelidus obtectus*. Arcelin/LLP obtained from *Phaseolus vulgaris* is resistant against *Zabrotes subfasciatus* insect.

7. CONCLUSIONS

The unprecedented growth in population and consequent urbanization coupled with environmental degradation have caused tremendous pressure on productivity of food crops to meet the ever-increasing demand. Agriculture is also affected because of shrinking in agricultural land and depletion in yield. Since it is not possible to change the environment, it would be worthwhile to consider manipulating the orgranism through genetic tailoring. The techniques of biotechnology have proved successful in introducing the genes of the interest into large number of plants. Enhanced yield and disease resistance has been the prime aim of such genetic manipulations. Such plants can be raised and propagated through plant tissue culture.

The impetus to these techniques was generated because of inherently built quality of totipotency in plant cells. A large number of high yielding and disease resistance plants could therefore be produced. Chemo- and thermotherapy that were used for eradication of diseases proved effective only in limited cases. Such eradication methods also proved hazardous to the eco-balance causing widespread air, water and soil pollution. Besides, chemotherapy has resulted in resistance and resurgence of large number of diseases. Breeding for resistance has also proved laborious, time consuming and less effective in getting the desired results. Through the applications of tissue culture technique it has become possible to regenerate and micropropagate a large number of plants and establish the expression of inserted genes in the transgenics. Thus, it is now well established that tissue culture techniques have helped in biotechnological approaches to obtain disease-free plants.

Progress in technology during the couple of years have been quite impressive enabling further exploitation of the technique of tissue culture in raising disease resistant plants. Identification, cloning and insertion of *Bt* gene has not only served to raise resistant plants but has equally helped to understand the gene expression in plants. Further transformation by the tailored DNA and its subsequent expression in regenerated tissues, plants will significantly add to the field efficacy of transgenics. Development in the metabolic engineering will help in deciphering the genes coding insecticidal/pesticidal chemical/enzymes like trypsin, proteinase, vegetable insecticidal proteins, chloresterol oxidase and some lectins. Characterization, cloning and transfer of such genes carry a great promise for future . Commendable success has already been achieved in trypsin and proteinase inhibitor genes. Co-cultures of fungi will also lead to the development of resistant varieties. Some of the successful examples have already been cited in Table 5. Metabolic modulations and *in vitro* selection of callus lines are other areas destined to an important role in achieving resistant lines. Secondary metabolites have recently been exploited for pesticidal properties, and *in vitro* system carry tremendous scope for the induction of desired levels of such compounds in plants. It is therefore logical to suggest that the success already met with tissue culture techniques is a significant achievement and ultimately it will provide a permanent and cost effect means to fight the plant disease along with sustaining the environmental quality. This will also improve production and save million of US $ annually on account of toxic pesticides/ fungicides.

Table 5 : Resistance expression in co-cultures

Fungus	Host
Alternaria triticina	*Triticum aestivum*
Botrytis cinerea	*Cicer arietinum*
Claviceps fusiformis	*Pennisetum typhoides*
Cronartium ribicola	*Pinus monticola*
Phytophthora megasperma f. sp. *medicaginis*	*Medicago sativa*
P. megasperma f.sp. *sojae*	*Glycine max*
P. parasitica f.sp. *nicotianae*	*Nicotiana tabacum*
Peronospora tabacina	*Nicotiana debneyi*
Phaseolus schweinitzii	Conifers
Phoma lingam	*Brassica napus*
Phytophythora infestans	*Solanum tuberosum*
Ustilago scitamineae	*Saccharum officinarum*

ACKNOWLEDGMENTS

Authors are thankful to Dr. Deepshikha Pande, Dr. Nisha Bharti, Dr. Gayoor Ali, Ms Saba, Ms. Harleen Bajwa, Ms. Alka Narula and Sanjeev Kumar for their selfless help during the preparation of this manuscript. MHM gratefully acknowledges the financial assistance from UGC in the form of SRF.

REFERENCES

Abel, P.P., Nelson, R.S., De, B., Hoffmann, N., Rogers, S.G., Fralely, R.T. and Beachy, R.N. 1986, Delay of disease development in transgenic plants that express tobacco mosaic virus coat protein gene, Science 232:738-43.

Anzai, H., Yoneyama, K. and Yamaguchi, I. 1989, Transgenic tobacco resistant to a bacterial disease by the detoxification of a pathogenic toxin, Mol. Gen. Genet. 219: 492-494.

Atkinson, R. and Gardner, R. 1991, *Agrobacterium* mediated transformation of pepino and regeneration of transgenic plants, Plant Cell Rep. 10: 208-212.

Aziz, A., Kadir, S.A. and Barlow, H.S. (eds) 1992, Pest Management and the Environment in 2000, CAB International, Oxon.

Bailey, J.A. 1987, Phytoalexins: a genetic view of their significance, in : Genetics and Plant Pathogenesis, P. Day and G. Ellis, eds., Oxford; Blackwell Scientific Publishers, pp. 233-244.

Barton, K.A., Whiteley, H. and Yang, N. 1987, *Bacillus thuringenes* delta-endotoxin expressed in transgenic *Nicotiana tabacum* provides resistance to Lepidopteran insects, Plant Physiol. 85: 1103-1109.

Basiran, N., Armitage, P., Scott, R. and Draper, J. 1987, Genetic transformation of flax (*Linum usitatissimum*) by *Agrobacterium tumefaciens*: Regeneration of transformed shoots via a callus phase, Plant Cell Rep. 6: 396-399:

Beachy, R., Loesch-Fries, S. and Tumer, N. 1990, Coat-protein mediated resistance against virus infection, Ann. Rev. PhytoPathol. 28: 447-451.

Bhojwani, S.S. and Razdan, M.K. (eds) 1983, Plant Tissue Culture: Theory and Practice, Elsevier, Ammsterdam.

Blom-Barnhoom, G.J. and Van Aartrijik, J. 1984, The regeneration of plants free of LSV and TBV from infected *Lilium* bulbs-scale explants in the presence of virazole, Acta Hortic. 164: 163-168.

Bohlmann, H. and Apel, K. 1991, Thionins. Ann. Rev. Plant Physiol. Plant Mol. Biol. 72: 227-240.

Botterman, J. and Leemans, J. 1989, Field testing of insect and herbicide resistant crops, Vortr. Kflanzenzuchtung 16: 455-461.

Brettel, R.I.S., Thomas, E. and Ingram, D.S. 1979, Tissue culture in the production of novel disease-resistant crop plants, Biol. Rev. 54: 329-345.

Broglie, R., Broglie, K., Chet, I., Roby, D. and Holliday, M. 1991, Chitinase expression in transgenic plants: a molecular approach to fungal disease resistance, J. Cell Biol. 15: 9.

Brunke, K. and Meeusen, R. 1991, Insect control with genetically engineered crops, Trends Biotechnol. 9: 197-200.

Bryan, J.E. 1988, Implementation of rapid multiplication and tissue culture methods in third world countries, Am. Potato J. 65: 199-207.

Bytebier, B., DeBoeck, F., DeGreve, H., Van Montagu, M. and Hernalsteens, J.P. 1987, T-DNA organization in tumor callus and transgenic plants of the monocotyledon *Asparagus officinalis*, Proc. Natl. Acad. Sci. USA 84:5345.

Cassells, A.C. and Long, R.D. 1980, The regeneration of virus-free plants from cucumber mosaic virus and potato virus Y-infected tobacco plants cultured in the presence of virazole, Z. Naturforsch. 35: 350-351.

Catlin, D., Ochoa, O., McCormick, S. and Quiros, C. 1988, Celery transformation by *Agrobacterium tumefaciens*: Cytological and genetic analysis of transgenic plants, Plant Cell Rep. 7: 100-103.

Culver, J. N. 1995, Molecular strategies to develop virus-free plants, in : Molecular Methods in Plant Pathology, R.P. Singh and U.S. Singh, eds., CRC, Lewis Publishers, Boca, Raton, pp. 235-248.

Cuozzo, M., O'Connell, K.M., Kaniewski, W., Fang, R.X., Chua, N.H. and Tumer, N.E. 1988, Viral protection in transgenic tobacco plants expressing the cucumber mosaic virus coat protein or its antisense RNA, Biotechnology 6: 549-557.

Datta, S., Peterhans, A., Datta, K. and Potrykus, I. 1990, Genetically engineered lettile indica-rice recovered from protoplasts, Biotechnology 8: 736-740.

David, C., Tempe, J. 1988, Genetic transformation of cauliflower (*Brassica oleracea* L. var. *botrytis*) by *Agrobacterium rhizogenes*, Plant Cell Rep. 7: 88-91.

Day, A., Bejarano, E., Buck, K., Burrell, M. and Lichtenstein, C. 1991, Expression of an antisense viral gene in transgenic tobacco confers resistance to the DNA virus golden mosaic virus, Proc. Natl. Acad. Sci. USA 88: 672-675.

Debergh, P., Roggemans, J. and Stadaert-de-Metesenaere. 1990, Tissue culture in relation to ornamental plants, in: Plant Tissue Culture: Applications and Limitations, S.S. Bhojwani, ed., Elseviers, Amterdam, pp. 161-189.

Delanney, X., LaVallee, B., Proksch, R., Fuchs, R., Sims, S., Greenplate, J., Marrone, P., Dodson, R., Augustine, J., Layton, J. and Fischhoff, D. 1989, Field performance of transgenic tomato plants expressing the *Bacillus thuringenes* var. *kurstoki* insect control protein, Biotechnology 7: 1265-1269.

Dodds, J.H. 1988, Tissue culture technology: practical application of sophisticated methods, Am. Potato J. 65: 167-180.

During, K. 1989, Wunden'zierbare expression und sekretion von T_4-lysoxym und monok lonalen anti korpern in *Nicatiana tabacum*, Dissertation, University Sitata, Koln, FRG.

Estruch, J.J., Carozzi, N.B., Desai, N., Duck, N.B., Warren, G.W. and Koziel, M.G. 1997, Transgenic plants: an emerging approach to pest control, Nature Biotechnol. 15: 137-141.

Everett, N., Robinson, K. and Mascarenhas, D. 1987, Genetic engineering of sunflower (*Helianthus annuus* L), Biotechnology 5: 1201-1204.
Fang, G. and Grumet, R. 1990, *Agrobacterium* mediated transformation and regeneration of muskmelon plants, Plant Cell Rep. 9: 160-164.
Fauquet, C.M., Huet, H., Ong, C.A., Siwamani, E., Chen, L., Viegas, P., Marmey, V.P., Wang, P. D., de Kochko, A. and Beachy, R.N. 1997, Control of the rice tungro disease by genetic engineering is now a reality, in : General Meeting of the Intl. Prog. Rice Biotechnol., Malacca, Abstr, pp. 59.
Fischaff, D.A., Bowdish, K.S., Perlak, F.J., Marrone, P.G., McCormick, S.M., Niedermeyer, J.G., Dean, D.A., Kusano-Kretzmer, K., Mayer, E.J., Rochester, D.E., Rogesters, G.S. and Fraley, R.T. 1987, Insect tolerant transgenic tomato plants, Biotechnology 5: 807-813.
Gatehouse, A.M.R and Boulter, D. 1983, Assessment of the antimetabolic effects of trypsin inhibitors from cowpea (*Vigna unguiculata*) and other legumes on development of the bruchid beetle *Callosobruchus maculatus*, J. Sci. Food Agric. 34: 345.
Gatehouse (nee'Baker), A.M.R., Gatehouse, J.A., Dobie, P., Kilminster, A.M. and Boulter, D. 1979, Biochemical basis of insect resistance in *Vigna unguiculata,* J. Sci. Food. Agric. 30: 948-958.
Gatehouse, J.A., Hilder, V.A. and Gatehouse, A.M.R. 1991, Genetic engineering of plants for insect resistance, in : Plant Genetic Engineering, D. Grierson, ed., Blackie, Glasgow, pp. 105-135.
Ghareyazie, B., Alinia, F., Mengiuto, C.A., Rubia, L.G., dePalma, J.M., Linwanag, E.A., Chen, M.B., Khush, G.S. and Bennett, J. 1997, Enhanced resistance to two stem borers in an aromatic rice containing a synthetic cryIA (b) gene, Mol. Breed. 3: 401-414.
Golemboski, D., Lomonosoff, G. and Zaitin, M. 1990, Plant transformed with a tobacco mosaic virus non structural gene sequence are resistant to the virus, Proc. Natl. Acad. Sci. USA, 87: 6311-6315.
Gordon-Kamm, W., Spencer, T., Mangano, M., Adams, T., Danies, R., Kausch, A., Lemaux, P. 1990, Transformation of maize and regeneration of fertile trangenic plants, Plant Cell 2: 603-618.
Goswami, J., Sanyal, R., Mohan, J., Dhawan, V. and Lakshmikumarn, M. 1998, Genetic manipulation of crops for insect resistance, in : Plant Tissue Culture and Molecular Biology, P.S. Srivastava, ed., Narosa Publishing House, New Delhi, pp. 670-693.
Griffiths, M.H., Slack, S.A. and Dodds, J.H. 1990, Effect of chemical and heat therapy on virus concentration in *in vivo* potato plantlets, Can. J. Bot. 68: 1515-1521.
Hain, R., Bielseler, B., Kinde, H., Schroder, G. and Stoker, R. 1990, Expression of a stilbene synthase gene in *Nicotiana tabacum* results in synthesis of a phytoalexin, resveratrol, Plant Mol. Biol. 15: 325-335.
Hakes, J.G. 1978, Biosystematics of the potato, in : The Potato Crop: The Scientific Basis for Improvement, P.M. Hamis, ed., Chapman and Hall, London, pp. 15-69.
Hansen, A.J. and Lane, W.D. 1985, Elimination of apple chlorotic leafspot virus from apple shoot cultures by ribavirin, Plant Dis. 69: 134-135.
Harrison, B., Mayo, M. and Bailcomle, D. 1987, Virus resistance in transgenic plants that express cucumber mosaic virus satellite RNA, Nature 328: 799-802.
Hartmann, C.L., McCoy, T.J. and Knhous, T.R. 1984, Selection of alfalfa (*Medicago sativa*) cell lines and regeneration of plants resistant to the toxin(s) produced by *Fusarium oxysporum* f. sp. *medicaginis*, Plant Sci. Lett. 34: 183-194.
Hartmann, H.T., Kester, T.T. and Davies, T.T. 1990, Plant Propagation, (5th edn), Prentice Hall, Englewood cliffs, NJ.
Hayakawa, T., Zhu, Y., Itoh, K., Kimura, Y., Izawa, T., Shumamoto, K. and Torryama, S. 1992, Genetically engineered rice resistant to rice stripe virus, an insect-transmitted virus, Proc. Natl. Acad. Sci. USA 89:9865-9869.
Health-Pagliuso, S. and Rappaport, L. 1990, Somaclonal variant UCT3: the expression of *Fusarium* wilt resistance in progeny arrays of celery, Apium graveolens L. Theor. Appl. Genet. 80: 390-394.
Hemenway, C., Fang, R.X., Kaniewski, W.K., Chu, N.H. and Tumer, N.E. 1988, Analysis of the mechanism of protection in transgenic plants expressing the potato virus X coat protein or its antisense RNA, EMBO J. 7: 1273-1280.
Hilder, V.A., Gatehouse, A.M.R., Sheerman, S.E., Barker, R.F. and Boulter, D. 1987, A novel mechanism of insect resistance engineered into tobacco, Nature 300: 160-163.
Hilder, V.A. and Gatehouse, A.M.R. 1990, Transforming plants as a means of crop protection against insects, Outlook Agric. 19: 179-183.
Hilder, V., Gatehouse, A. and Boulter, D. 1990, Genetic engineering of crops for insect resistance using genes of plant origin, in : Genetic Engineering of Crop Plants, G. Lycett and D. Grerson, eds., Butterrworth, London- Boston, pp. 51-66.
Hilder, V.A., Angharad, M.R., Gatehouse, A.M.R. and Boulter, D. 1993, Transgenic plants conferring insect tolerance : Protease inhibitor approach, in : Transgenic Plants, Vol. 1: Engineering and Utilization, S.D. Kung and R. Wu, eds., Academic Press, San Diego, pp. 317-338.
Hilderbrandt, A.C. 1977, Single cell culture, protoplasts and plant viruses, in : Plant Cell Tissue and Organ

Culture, J. Reinert and Y.P.S. Bajaj, eds., Springer-Verlag, Berlin, pp. 581-597.

Hofte, H. and Whiteley, H. 1989, Insecticidal crystal proteins of *Bacillus thuringienes*, Microb. Rev. 53:242-255.

Hollings, M. and Stone, O.M. 1968, Investigation of carnation viruses, I, carnation mottle, Ann. Appl. Biol. 53: 103-118.

Horsch, R., Fry, J., Hoffmann, N., Eicholtz, P., Rogers, S. and Fraley, R. 1985, A simple and general method for transferring genes into plants, Science 227:1229.

Hosoyama, H., Irie, K., Abe, K. and Arai, S. 1995, Introduction of a chimeric gene encoding an oryzacystatin-b-glucornidase fusion protein into rice protoplasts and regeneration A transformed plants, Plant Cell Rep. 165:174-177.

Howard, H.W. 1978, The production of new varieties, in : The Potato Crop: The Scientific Basis for Improvement, P.M. Hamis, ed., Chapman and Hall, London, pp. 607-636.

Hrubocova, M., Cvikrova, M., Meravy, L., Eder, J. and Binarova, P. 1992, Phenolic accumulation and peroxidase activity an *in vitro* selected alfafa callus cultures resistant to filtrate of *Fusarium* spp., Biol. Plant. 43: 203-211.

Huntley, C.C. and Hall, T.C. 1996, Interferencne with brome mosaic virus replication in transgenic rice, Mol. Plant-Microbe Interact. 9: 164-170.

Hwang, S.C. 1991, Somaclonal resistance in Cavendish banana to *Fusarium* wilt, Plant Protection Bull. (Taiwan), 33: 124-132.

Irie, K., Hosoyama, H., Takeuchi, T., Iwabuchi, K., Watanabe, H., Abe, M., Abe, K. and Arai, S. 1996, Transgenic rice established to express corn cystatin exhibits strong inhibitory activity against insect gut proteinases, Plant Mol. Biol. 30: 149-157.

James, D., Passey, A., Barbara, D. and Bevan, M. 1989, Genetic transformation of apple (*Malus pumila* Mill) using a disarmed Ti-binary vector, Plant Cell Rep. 7: 658-661.

Jai, S., Yang, M., Ott, R. and Chua, N. 1989, High frequency transformation of *Kalanchoe laciniata*, Plant Cell Rep. 8: 336-340.

Jones, E.D. 1988, A current assessment of *in vitro* culture and other rapid multiplication methods in North America and Europe, Am. Potato J. 65: 209-220.

Kamiwski, W., Lawson, C., Sammons, B., Haley, L., Hart, J., Delannay, X. and Tumer, N. 1990, Field resistance of transgenic Russet burbank potato to effects of infection by potato virus X and potato virus Y, Biotechnology 8: 750-754.

Kassanis, B. and Tinsely, T.W. 1958, The freeing of tobacco tissue culture from potato virus Y by thiouracil, Proc. Third Conf. Potato Virus Dis. Lisse Wageningen, pp. 153-155.

Kawchuk, L.M., Martin, R.R. and McPherson, J. 1990, Resistance in transgenic potato expressing the potato leafroll virus coat protein gene, Mol. Plant-Microbe Interact. 3: 301-307.

Kher, S. and Larka, R.K. 1995, Alternatives to chemical pesticides, Invent. Intell. June: 267-273.

Kuchuk, N., Komarnitsky, I., Shakovsky, A. and Gleba, Y. 1990, Genetic transformation of *Medicago* species by *Agrobacterium tumefaciens* and electroporation of protoplasts, Plant Cell Rep. 8:660-663.

Lamb, C., Lawton, M., Dron, M. and Dixon, R. 1989, Signal and transduction mechanism for activation of plant defences against microbial attack, Cell 56: 215-224.

Langeweg, Ir.F. (ed.) 1989, Concern for Tomorrow: A National Environmental Survey 1985-2010, National Institute for Public Health and Environmental Protection, Bilthoven, the Netherlands, 350 p.

Larkin, P.J. and Scowcroft, W.R. 1981, Somaclonal variation - A novel source of variability from cell cultures for plant improvement, Theor. Appl. Genet. 60: 197-214.

Ledger, S., Deroles, S. and Given, N. 1991, Regeneration and *Agrobacterium*-mediated transformation of chrysanthemum, Plant Cell Rep. 10:195-199.

Limasset, E.M. and Cornuet, P. 1949, Organic growth factor requirements of tobacco tissue cultures, Physiol. Plant. 18: 100-127.

Liu, M.C. and Chen, W.H. 1978, Improvement in sugarcane by using tissue culture methods, 4th Intl. Congr. Plant Tissue and Cell Culture, Calagary, Canada, 163 (Abstr).

Lizarraga, R.E., Salazar, L.F., Roca, W.M. and Schilde Rentschler, L. 1980, Elimination of potato spindle tuber viroid by low temperature and meristem culture, Phytopathology 70: 754-755.

Llooyd, A., Barnason, A., Rogers, S., Byrne, M., Fraley, R. and Horsch, R. 1986, Transformation of *Arabidopsis thaliana* with *Agrobacterium tumefaciens*, Science 234: 464-466.

Lund, P., Lee, R. and Dunsmuir, P. 1989, Bacterial chitinase is modified and secreted in transgenic tobacco, Plant Physiol. 91: 130-135.

MacDonald, D.M. 1973, Heat treatment and meristem culture as a means of freeing potato varieties from virus X and S, Potato Res. 16: 263-269.

Manners, J. and Way, H. 1989, Efficient transformation with regeneration of the tropical pasture legume *Stylosanthes humilis* using *Agrobacterium rhizogenes* and a Ti-Plasmid binary vector system, Plant Cell Rep. 8: 341-345.

Matern, V., Strobel, G. and Shepard, J. 1978, Reaction to phytoxins in a potato population derived from mesophyll protoplasts, Proc. Natl. Acad. Sci. USA 75: 4935-4939.

Mazur, B. and Falco, S. 1989, The development of herbicide resistant crops, Ann. Rev. Plant Physiol. Plant Mol. Biol. 40: 441-470.

McCabe, D., Swain, W., Marinell, B. And Christou, P. 1988, Stable transformation of soybean *Glycine max* by particle acceleration, Biotechnology 6: 923-926.

McCormick, S., Niedermeyer, J., Fry, J., Barnason, A., Horsch, R. and Fraley, R. 1986, Leaf disc transformation of cultivated tomato using *Agrobacterium tumefaciens*, Plant Cell Rep. 5: 81-84.

McGranahan, G., Leslie, C. And Uratsu, S. 1988, Marmediated transformation of walnut somatic embryos and regeneration of transgenic plants, Biotechnology 6: 800-804.

Mellor, F.C. and Stace-Smith, R. 1977, Virus-free potatoes by tissue culture, in : Applied and Fundamental Aspects of Plant Cell, Tissue and Organ Culture, J. Reinert and Y.P.S. Bajaj, eds., Springer-Verlag, Berlin, pp. 616-636.

Mendel, R.R. and Hansch, R. M. 1995, Gene transfer to higher plants, in : Molecular Methods in Plant Pathology, R.P. Singh and U.S. Singh, eds., CRC, Lewis Publishers, Boca, Raton, pp. 189-202.

Michelmore, R., Marsh, E., Seely, S. and Landry, B. 1987, Transformation of lettuce (*Lactuca sativa*) mediated by *Agrobacterium tumefaciens*, Plant Cell Rep. 6:439-442.

Miller, S.A., Williams, G.R., Medina-Filho, H. and Evans, D.A. 1985, A somaclonal variant of tomato resistant to race 2 of *Fusarium oxysporum* f. sp. *lycopersici* (Abstr.), Phytopathology 75: 1354.

Monette, P.L. 1986, Elimination *in vitro* of two grapevine nepoviruses by an alternating temperature regime, J. Phytopathol. 116: 88-89.

Mullner, H. 1991, Herbicide tolerance, a contribution to integrated crop management, in : Pesticide Chemistry: Advances in International Research, Development and Legislation, H. Frehse, ed., Weinheim-New York-Basel-Cambridge : VCH, pp. 131-138.

Nayak, P., Debabrate, B., Das, S., Basu, A., Ghosh, D., Ram Krishnan, N.A., Ghosh, M. and Sen, S.K. 1997, Transgenic elite indica rice plants expressing CryIAcd-endotoxin of *Bacillus thuringenesis* are resistant against yellow stem borer (*Scripophaga incetulas*), Proc. Natl. Acad. Sci., USA 94: 2111-2116.

Nehra, N., Chibbar, R., Kartha, K., Datla, R., Crosby, W. and Stushnoff, C. 1990, *Agrobacterium* mediated transformation of strawberry calli and recovery of transgenic plants, Plant Cell Rep. 9:10-13.

Nelson, R., McCormick, S., Delannay, X., Dube, P., Layton, J., Anderson, E., Proksch, R., Horsch, R., Rogers, S., Fraley, R. and Beochy, R. 1988, Virus tolerance, plant growth and field performance of transgenic tomato plants expressing the coat protein gene of tobacco mosaic virus, Biotechnology 6: 403-409.

Nelson, R.S., Abel, P.P. and Beachy, R.N. 1987, Lesions and virus accumulation in inoculated transgenic tobacco plants expressing the coat protein gene of tobacco mosaic virus, Virology 158:126-132.

Neuhaus, J., Ahl-Goy, P., Hinz, U., Flores, S. and Meins, F. 1991, High levels of expression of a tobacco chitinase gene in *Nicotiana sylvestris*, Susceptibility of transgenic plants to *Cercospora nicotianae* infection, Plant Mol. Biol. 16: 141-151.

Noda, T., Tanaka, N., Mano, Y., Nabeshima, S., Ohkawa, H. and Matsui, C. 1987, Regeneration of horseradish hairy roots incited by *Agrobacterium rhizogenes* infection, Plant Cell Rep. 6:283-286.

Norris, D.O. 1954, Development of virus-free stock of green mountain potato by treatment with malachite green, Aust. J. Agric. Res. 5: 658-663.

Oxtoby, E. and Hughes, M. 1990, Engineering herbicide tolerance into crops, Trends Biotechnol. 8: 61-65.

Palit, R. and Reddy, G.M. 1987, Regeneration of blast resistant plantlets from irradiated rice callus, in : Plant Cell and Tissue Culture of Economically Important Plants, G.M. Reddy, ed., Osmania Univ., Hyderabad, pp. 141-146.

Paludan, N. 1985, Inactivation of viroids in chrysanthemum by low-temperature treatment and meristem-tip culture, Acta Hortic. 164: 181-186.

Peferoen, M. 1997, Progress and prospects for field use of *Bt* genes in crops, Trends Biotechnol. 15: 173-177.

Perlak, F., Fuchs, R., Dean, D., McPherson, S. and Fischhoff, D. 1991, Modification of the coding sequence enhances protein expression of insect control protein genes, Proc. Natl. Acad. Sci. USA 88: 3324-3328.

Phelep, M., Petit, A., Martin, L., Duhoux, E. and Tempe, J. 1991, Transformation and regeneration of a nitrogen-fixing tree, *Allocasusarina verticillata* Lam., Biotechnology 9: 461-466.

Phythoud, F., Sinkar, V., Nester, E. and Gordon, M. 1987, Increased virulence of *Agrobacterium rhizogenes* conferred by the vir-region of pTiBo542: Application to genetic engineering, Biotechnology 5: 1323-1328.

Pickardt, T., Meixner, M., Schade, V. and Schieder, O. 1991, Transformation of *Vicia narbonnensis* via *Agrobacterium* mediated gene transfer, Plant Cell Rep. 9: 535-538.

Potrykus, I. 1991, Gene transfer to plants: as assessment of published approaches and results, Ann. Rev. Plant Physiol. Plant Mol. Biol. 42: 205-225.

Powell, P., Stark, D., Sanders, P. and Beacy, R. 1989, Protection against tobacco mosaic virus in transgenic

plants that express tobacco mosaic virus anti-sense RNA, Proc. Natl. Acad. Sci., USA 86: 6949-6952.
Puonti-Kaerlas, J., Eriksson, T. and Engstrom, P. 1990, Production of transgenic pea plants by *Agrobacterium tumefaciens* mediated gene transfer, Theor. Appl. Genet. 80: 246-252.
Quack, F. 1977, Merstem culture and virus -free plants, in : Plant Cell, Tissue and Organ Culture, J. Reinert and Y.P.S. Bajaj, eds., Springer-Verlag, Berlin, pp. 598-615.
Raj, S.K., Aslam, A.M. and Singh, B.P. 1991, Elimination of eggplant mottled crinkle virus using virazole in explant cultures of *Solanum melongena*, Indian J. Exp. Biol. 29: 594-595.
Raj, H., Sharma, M.L. and Sharma, N.K. 1994, Biopesticides-the safer alternative to chemical pesticides, Invent. Intell. June: 291-294.
Rao, K.V., Powell, K.S., Rathore, K.S., Bharathi, M., Edmonds, H.S., Gatehouse, J.A. and Hodges, T.K. 1997, Expression of snowdrop lectin (GNA) in the phloem of transgenic rice plants confers Malacca, resistance to rice brown plant hopper (Abstr), in : General Meeting of the Intl. Progr. Rice Biotechnol., p.36.
Renaudin, J. and Bove, J. 1977, Effect of actimycin D on turnip yellow mosaic production by chinese cabbage protoplasts, C.R. Acad. Sci., Paris 284: 783-786.
Richardson, M. 1977, The proteinase inhibitors of plants and micro-organisms, Phytochemistry 16:159-169.
Roberts, W. and Selitrennikoff, C. 1990, Zeamatin, an antifungal protien from maize with membrane-permeabilizing activity, J. Gen. Microbial. 136: 1771-1778.
Rotino, G. and Gleddie, S. 1990, Transformation of eggplant (*Solanum melongena* L.) using a binary *Agrobacterium tumefaciens* vector, Plant Cell Rep. 9:26-29.
Roy, S.K., Islam, M.S., Sen, J. and Hadiuzzaman, S. 1996, Shoot tip an alternative to seed as a source of genetic material for conservation and propagation of forest trees, in : Plant Tissue Culture, A.S. Islam, ed., Oxford & IBH Publishing Co. Pvt. Ltd., New Delhi pp. 8-15.
Sacristan, M.D. 1982, Resistance responses to *Phoma lingam* of plants regenerated from selected cell and embryogenic cultures of haploid *Brassica napus*, Theor. Appl. Genet. 61: 193-200.
Sacristan, M.D. 1985, Selection for disease resistance in *Brassica* cultures, Hereditas Suppl. 3: 57-63.
Schlumbaum, A., Munch, F., Vogeli, U. and Boler, T. 1986, Plant chitinase are potent inhibitors of fungal growth, Nature 327: 365-367.
Sanchez, G.E., Slack, S.A. and Dodds, J.H. 1991, Response of selected *Solanum* species to virus eradication therapy, Am. Potato J. 68: 299-315.
Sequeua, L. 1984, Cross protection and induced resistance: their potential in plant disease control, Trends Biotechnol. 2:25-29.
Shahin, E.A. and Spivey, R. 1986, A single dominant gene for *Fusarium* wilt resistance in protoplasm-derived tomato plants, Theor. Appl. Genet. 73: 164-166.
Shepard, J.F. 1981, Protoplasts as sources of disease resistance in plants, Ann. Rev. Phytopathol. 19: 145-166.
Shepard, J.F. 1977, Regeneration of plants from protoplasts of potato virus X-infected tobacco leaves, Virology 78: 261-266.
Sherwood, J. and Fulton, R. 1982, The specific involvement of coat protein in tobacco mosaic virus protection, Virology 119: 150-158.
Singh, R.P. and Singh, U.S. (eds) 1995, Molecular Methods in Plant Pathology, CRC, Lewis Publishers, Boca Raton, Florida.
Slack, S.A. 1980, Pathogen-free plants by meristem-tip culture, Plant Dis. 64: 15-17.
Slack, S.A., Petrson, M.D., Rott, M. and Quinones, S.H. 1987, Effect of heat and chemical therapy treatments on potato virus S elimination in Desiree, Am. Potato J. 64: 457.
Stace-Smith, R. and Mellor, F.C. 1989, Eradication of potato viruses X and S by thermotherapy and axillary bud culture, Phytopathology 58: 199-203.
Stiekema, W., Heidekamp, F., Louwers, J., Verhoeven, H. and Dijkhuis, P. 1988, Introduction of foreign genes into potato cultivars Bintje and Desiree using an *Agrobacterium tumefaciens* binary vector, Plant Cell Rep. 7: 47-50.
Takebe, I. and Otsuki, Y. 1974, Multiple infection of tobacco mesophyll protoplasts by plant viruses, 3rd Intl. Congr. Plant Tiss. Cell Cult., Leicester Univ., Leicester, Abstr no 36.
Tepfer, D. 1984, Transformation of several species of higher plants by *Agrobacterium rhizogenes*: Sexual transmission of the transformed genotype and phenotype, Cell 37: 959-967.
Tomlinson, J.A. 1982, Chemotherapy of plant viruses and virus disease, in : Pathogens, Vectors and Plant Diseases: Approaches to Control, K.F. Harris and K. Maramorsch, eds., Academic Press, N.Y., pp. 23-44.
Toyoda, H., Horiloshi, K., Yamano, Y. and Ouchi, S. 1991, Selection for *Fusarium* wilt disease resistance from regenerants derived from leaf callus of strawberry, Plant Cell Rep. 10: 167-170.
Trulson, A., Simpson, R. and Shahin, E. 1986, Transformation of cucumber (*Cucumis sativus* L.) plants with *Agrobacterium rhizogenes*, Theor. Appl. Genet. 73: 11-15.

Tumer, N.E., O'Connell, K.M., Nelson, R.S., Sander, P.R., Beachy, R.N., Fraley, R.T. and Shah, D.M. 1987, Transgenic plants protected from insect attack, Nature 328: 33-37

Umbeck, P., Johnson, G., Barton, K. and Swain, W. 1987, Genetically transformed cotton (*Gossypium hirsutum* L.), Biotechnology 5: 263-266.

Vaeck, M., Reynaerts, A., Hofte, H., Jansens, S., De Beukeleer, M., Dean, C., Zebeau, M., Van Montagu, M. and Leemans, J. 1987, Transgenic plant protected from insect attack, Nature 328: 33-37.

Van Dun, C.M.P., Bol, J.F. and Van Vloten-Dting, L. 1987, Expression of alfalfa mosaic virus and tobacco rattle virus coat, Virus protein genes in transgenic tobacco plants, Virology 159:299-305.

Van Rie, J. 1991, Insect control with transgenic plants: resistance proof, Trends Biotechnol. 9: 177-179.

Veelar, B.M., Ferreira, D.I. and Niederwiesser, J.G. 1992, Elimination of Ornithogalum mosaic virus in the *Ornithgalum* cv. *Rogel* through meristem tip culture and chemotherapy, Plant Cell Tiss. Org. Cult. 29: 51-55.

Vigers, A., Roberts, W. and Selitrennikoff, C. 1991, A new family of anifungal proteins, Mol. Plant-Microbe Intract. 4: 315-323.

Wang, P.J. 1985, Producing pathogen-free plants using tissue culture, IAPTC Newslett. 46: 2-8.

White, R.P. 1934, Multiplication of the viruses of tobacco and ancuba mosaic in growing excised tomato roots, Phytopathology 24: 1003-1011.

Willmitzer, L. 1993, Transgenic plants, in : Biotechnology Vol. 2, Genetic Fundamental and Genetic Engineering, H.J. Reham and G. Reed, eds., VCH, Weinheim.

Woloshuk, C.P., Meulenhoff, J., Sela-Buurlage, M., Van Den Elzen, P. and Cornelissen, B. 1991, Pathogen-induced proteins with inhibitory activity toward *Phytophthora infestans*, Plant Cell 3: 619-628.

Xu, D., Xue, Q., McElroy, D., Mawal, Y., Hilder, VA. and Wu, R. 1996, Constitutive expression of a cowpea trypsin inhibitor gene, CpTi, in transgenic rice plants confers resistance to two major rice insect pests, Mol. Breed. 2: 167-173.

Zhang, W. and Wu, R. 1988, Efficient regeneration of transgenic rice protoplasts and correctly regulated expression of foreign genes in the plants, Theor. Appl. Genet. 76: 835-840.

Zheng, H.H., Li, Y., Yu, Z.H., Li, W., Chen, M.Y., Meng, X.T., Casper, R. and Chen, Z.L. 1997, Recovery of transgenic rice plants expressing the rice dwarf virus outer coat protein gene (58), Theor. Appl. Genet. 94: 522-527.

ABBREVIATIONS

ACLSV	Apple Cholrotic Leaf Spot Virus
Bt	*Bacillus thuringienes*
CMV	Cucumber Mosaic virus
Cp	Coat Protein
CpTI	Cowpea trypsin Inhibitor
EMCV	Egg Plant Mottled Crinkle Virus
GFLV	Grapevine Fan Leaf Virus
LSV	Lily Symptomless Virus
OMV	Ornithogalum Mosaic Virus
ORF	Open Reading Frame
ORS	Ornithogalum Ringspot Virus
PEG	Poly Ethylene Glycol
PEP	Phospho Enol Pyruvate
PLRV	Potato Leaf Roll Virus
PVM	Potato Virus M
PVS	Potato Virus S
PVX	Potato Virus X
PVY	Potato Virus Y
RTBV	Rice Tungro Bacilliform Virus
TEV	Tobacco Etch Virus
TMV	Tobacco Mosaic Virus

GENETIC MANIPULATION OF ANTAGONISTIC *FUSARIUM* SPP.

Quirico Migheli, M. Lodovica Gullino and Angelo Garibaldi

Dipartimento di Valorizzazione e Protezione delle Risorse Agroforestali - Patologia vegetale, Università di Torino, Via Leonardo da Vinci 44, I - 10095 Grugliasco (Torino)

1. INTRODUCTION

Saprophytic *Fusarium* spp., isolated from *Fusarium*-suppressive soils, are highly effective in reducing vascular wilts caused by formae speciales of *F. oxysporum* (Alabouvette, 1986; Garibaldi *et al.*, 1986, 1990, 1994; Lemanceau and Alabouvette, 1991; Minuto *et al.*, 1995, 1997; Tramier *et al.*, 1983). Several genetic transformation protocols were developed for pathogenic *Fusarium* spp., some of which have been used in monitoring antagonistic strains of *F. oxysporum* with resistance markers (Gullino *et al.*, 1995). Integrative transformation with non-homologous plasmids may be also used to elucidate the molecular basis of biocontrol in antagonistic *Fusarium* spp. strains by generating tagged mutations. Aim of this work was to evaluate the potential of genetic transformation for strain improvement in antagonistic *F. oxysporum*. Eighteen transformants, obtained by the insertion of plasmids pbenA33 and pAN7-1, were applied in biocontrol experiments against *F. oxysporum* f. sp. *dianthi* on carnation in order to evaluate if random integration of foreign DNA sequences may have modified the biocontrol potential of the recipient strains. In addition, both hygromycin B resistance and mitotic stability of pAN7-1 and pLD transformants were evaluated after *in vitro* growth in the absence of hygromycin B selective pressure and after release of the transformants in soil microcosms.

1.1. Fungal Strains

The following antagonistic *Fusarium oxysporum* strains were used in transformation experiments: 141 WT, 251 WT and 233 WT, isolated from the rhizosphere of carnation grown in a *Fusarium*-suppressive soil in Italy and sensitive to benomyl and to hygromycin B; the dark red-pigmented mutant 233/1 C5 is able to grow in the presence of > 30 μg ml^{-1} benomyl, but is sensitive to hygromycin B; this mutant was selected for altered pigmentation from the benomyl resistant (MIC > 30 μg ml^{-1}) mutant 233/1, which was obtained by a 30-min UV treatment from isolate 233 WT. The 9 transformants coded T8.1, T8.2, T8.3, T8.4, T8.5, T8.6, T8.7, T8.8 and T8.9, deriving from isolate 141 WT, are able to grow in the presence of > 30 μg ml^{-1} benomyl.

The 9 transformants T26.1, T26.6, T26.19, T26.40, T27.7, T27.11, T27.28, T28.9, and T28.26, all deriving from mutant 233/1 C5, are able to grow in the presence of > 100 μg ml^{-1} hygromycin B and > 30 μg ml^{-1} benomyl. A virulent strain of *F. oxysporum* f.sp. *dianthi* (pathotype 2), originally isolated from a wilted carnation plant grown in Liguria (Northern Italy) and coded as F75, was used throughout the biocontrol experiments.

1.2. Protoplast Preparation and Transformation

Protoplasts isolation and transformation were carried out by the method of Langin *et al.* (1990). The following plasmids were used in the experiments: pbenA33, which carries a β-tubulin gene (*benA*) of *Aspergillus nidulans* (May *et al.*, 1987), pAN7-1 (Punt *et al.*, 1987), which carries the hygromycin B phosphotransferase gene (*hph*) of *Escherichia coli*, and pLD (Kistler and Benny, 1992), which contains the *hph* gene and a terminal, tandemly repeated hexanucleotide telomeric sequence (TTAGGG)18 of *F. oxysporum*.

1.3. Mitotic Stability and Molecular Analysis of Transformants

DNA isolated as described (Migheli *et al.*, 1996) from 9 pAN7-1 transformants grown *in vitro* under selective pressure or after 4 serial transfers on non-selective medium, and reisolated after 4 weeks from inoculation in carnation rhizosphere were digested with the restriction enzyme *Eco*RI, which cuts twice in pAN7-1. Digested DNAs were subjected to agarose gel electrophoresis and transferred onto Hybond N^+ nylon membranes by following standard methods (Sambrook *et al.*, 1989). Probe labelling, hybridization and chemiluminescent detection of the hybridising bands with Lumigen PPD (Boehringer) were all performed according to manufacturer recommendation.

2. BIOLOGICAL CONTROL EXPERIMENTS

Six biological control experiments against *F. oxysporum* f.sp. *dianthi* on carnation were carried out under contained conditions as described previously (Migheli *et al.*, 1996) by using 9 pbenA33 and 9 pAN7-1 transformants.

2.1. Transformation of Antagonistic *Fusarium oxysporum*

Over 20 independent experiments, PEG-mediated transformation of antagonistic *F. oxysporum* was accomplished with a maximum efficiency of 3.2 and 8.4 transformants per μg of plasmid DNA in the case of plasmids pbenA33 and pAN7-1, respectively. Transformation efficiency was considerably higher (up to 1,000 transformants per μg of DNA) for plasmid pLD (Fig. 1) (Migheli *et al.*, 1995, 1996). When undigested DNA from the transformants was hybridized with the vector DNA, hybridization occurred only with high molecular weight DNA, indicating that the transforming plasmids pbenA33 and pAN7-1 had integrated into the chromosomes. On the contrary, plasmid pLD did not integrate in the chromosomes of *F. oxysporum*.

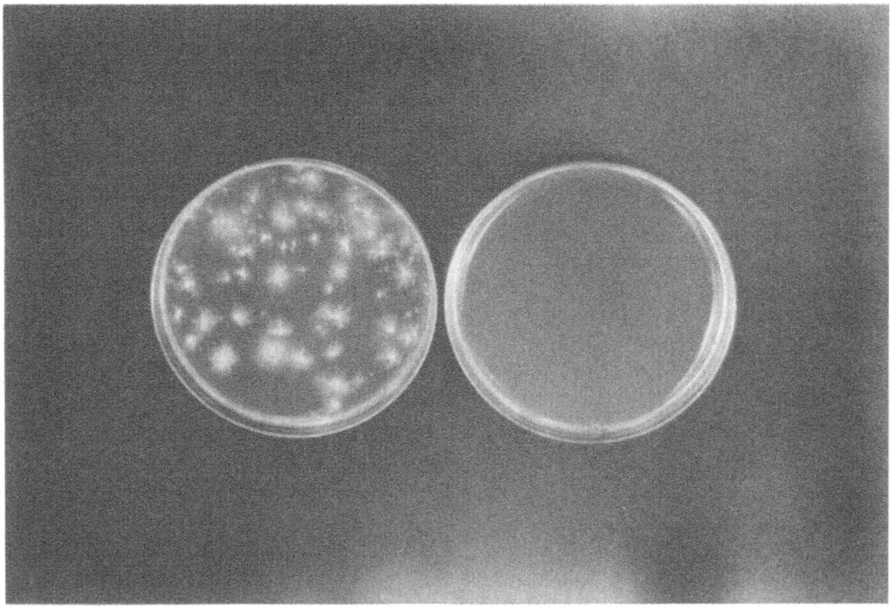

Figure 1. Transformation of *F. oxysporum* with plasmid pLD (Kistler and Benny, 1992). Left: transformant colonies developing on hygromycin B-containing medium after 4 d incubation at 25°C; right: untransformed control.

2.2. Mitotic Stability of pAN7-1 and pLD Transformants *in vitro* or after Recovery from Soil Microcosms

During 4 rounds of *in vitro* growth in the absence of hygromycin B or after 4 weeks from release in soil microcosms, eight over nine pAN7-1 transformants showed stable inheritance of the hygromycin B resistance phenotype. As expected, pLD transformants were no more able to grow in the presence of hygromycin B after one or two rounds of cultivation on non-selective medium. Southern analysis revealed that only one pAN7-1 transformant underwent no obvious DNA rearrangement, while the other transformants showed different degrees of rearrangement of the transforming DNA, mostly consisting of loss of plasmid copies (Fig.2) (Migheli *et al.*, 1996). Large scale release of genetically manipulated organisms requires careful evaluation of their behaviour in natural and agricultural environments (Cairns and Orvos, 1992; Day and Fry, 1992; Gullino *et al.*, 1995). The genetic stability of transformed biocontrol fungi is crucial in evaluating the environmental risk of their release: genetic instability in contained preliminary experiments may lead to erroneous conclusion in risk evaluation (Leslie and Dickman, 1991). The high level of mitotic instability observed in our studies is not surprising in the case of multicopy transformants, where homologous recombination events between plasmid copies scattered around the genome might be responsible for the observed rearrangement of transforming DNA. Thus, alternatives should be investigated in order to increase the level of stability in this fungus, such as the identification of trapping DNA sequences in the *Fusarium* genome and their use in constructing new plasmid vectors, restriction enzyme-mediated integration-REMI (Schliest and Petes, 1991) of plasmid DNA, and the use of gene tagging and transformation systems based on the *Fot1* and *Impala* transposable elements, recently discovered in *F. oxysporum* (Daboussi, 1996).

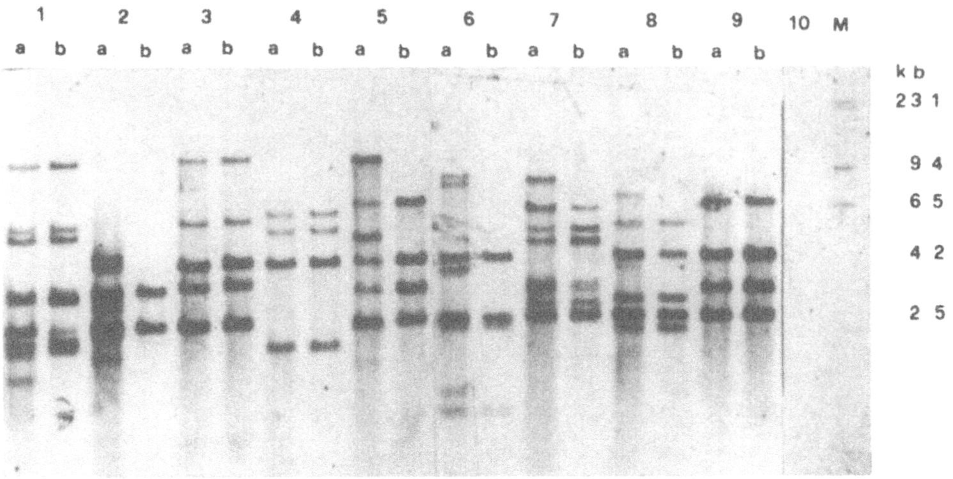

Figure 2. Southern hybridization analysis of EcoRI-digested DNAs obtained from *Fusarium oxysporum* transformants T26.1, T26.6, T26.19, T26.40, T27.7, T27.11, T27.28, T28.9, and T28.26 (1-9, respectively) grown under selective pressure (lanes a) or recovered four weeks after release in soil microcosms (lanes b); lane 10: EcoRI-digested DNA from the recipient strain 233/1 C5; M: HindIII-digested bacteriophage l DNA, digoxygenin-labeled. Molecular sizes (kb) are indicated in the right margin (Migheli *et al.*, 1996).

3. BIOLOGICAL CONTROL OF *FUSARIUM* WILT OF CARNATION

Most of the nine pbenA33 transformants tested in the first series of experiments were able to contain significantly the pathogen. The biocontrol efficacy was particularly evident in the first two trials (disease index reduced from 45-70 to 0-20), while in the third trial the transformants were less effective, probably due to the high incidence of *Fusarium* wilt (Table-1). In the second series of experiments, the wild type reference strain 233 WT, the mutant 233/1 and the recipient strain 233/1 C5 effectively reduced the incidence of disease from 65-84 to 0-24. Among the nine transformants tested, only T26.40 consistently reduced the severity of *Fusarium* wilt, rating a disease index of 9-20 (Table 2). Our results show that genetically manipulated strains originating from the same transformation experiment, and deriving from the same recipient strain, can perform differently as biocontrol agents against *Fusarium* wilt of carnation. Such a high variability in the antagonistic potential against *F. oxysporum* f. sp. *dianthi* was observed between fusants of *F. oxysporum* deriving from the same inter-strain cross (Migheli *et al.*, 1992) and even between monoconidial isolates of the same benomyl resistant mutant. None of the transformants, however, had completely lost its antagonistic potential against *Fusarium* wilt of carnation. These results, and the fact that all the tested strains were characterised by multicopy integration of pAN7-1 and pbenA33, make these two series of transformants unsuitable for the identification of tagged mutations in biocontrol controlling genes.

Table 1. Effect of different antagonistic strains of *Fusarium oxysporum* (nine benomyl resistant 141 WT-derived transformants and their recipient strain) on *Fusarium* wilt of carnation (cv Indios).

Strain	Disease index (0-100)		
	Experiment 1	Experiment 2	Experiment 3
-	45.0 b[1]	70.0 d	82.5 c
141 WT	10.0 a	7.5 ab	10.0 a
T8.1	2.5 a	0 a	20.0 ab
T8.2	5.0 a	20.0 c	37.5 b
T8.3	10.0 a	5.0 ab	20.0 ab
T8.4	2.5 a	5.0 ab	17.5 ab
T8.5	7.5 a	0 a	35.0 b
T8.6	5.0 a	5.0 ab	12.5 a
T8.7	12.5 a	7.5 ab	27.5 ab
T8.8	2.5 a	10.0 abc	32.5 b
T8.9	2.5 a	15.0 bc	40.0 b

[1]Values in a column followed by the same letter are not significantly different at p=0.05 based on Tukey's multiple comparison test.

Table 2. Effect of different antagonistic strains of *Fusarium oxysporum* (nine hygromycin B resistant transformants deriving from mutant 233/1 C5, isolate 233 WT, mutant 233/1 and the recipient strain 233/1 C5) on *Fusarium* wilt of carnation (cv Indios; Migheli *et al.*, 1996).

Strain	Disease index (0-100)		
	Experiment 1	Experiment 2	Experiment 3
-	83.7 d[1]	65.0 e	65.0 c
233 WT	16.2 ab	2.5 ab	0 a
233/1	21.2 abc	1.2 a	10.0 ab
233/1 C5	23.7 abc	10.0 abc	21.2 b
T26.1	10.0 a	26.2 cd	16.2 ab
T26.6	15.0 ab	22.5 bcd	15.0 ab
T26.19	43.7 c	26.2 cd	17.5 ab
T26.40	20.0 abc	16.2 abcd	8.7 ab
T27.7	17.5 ab	15.0 abcd	28.7 b
T27.11	30.0 abc	50.0 e	8.7 ab
T27.28	38.7 bc	21.2 abcd	27.5 b
T28.9	17.5 ab	23.7 bcd	16.2 ab
T28.26	35.0 abc	31.2 d	23.7 b

[1] See Table 1.

4. CONCLUSIONS

Different strains of *Fusarium oxysporum*, antagonistic against phytopathogenic formae speciales of *F. oxysporum*, were transformed by the insertion of the *benA* or *hph* genes. The frequency of transformation ranged from about 3-8 to 800-1000 transformants per microgram of DNA. Plasmids pbenA33 and pAN7-1 usually integrated in the genome, but most of the pAN7-1 transformants tested underwent loss of plasmid copies during *in vitro* growth without selective pressure or after release to soil microcosms, by retaining the hygromycin B resistance character. On the contrary, the plasmid vector pLD did not integrate in the chromosomes of *F. oxysporum* and transformants lost the resistant phenotype on non-selective medium. Biocontrol experiments against *F. oxysporum* f. sp. *dianthi* revealed that genetically manipulated strains can perform very differently as biocontrol agents. Results are discussed in relation to risk assessment of the release of transgenic antagonists and to the possible identification of tagged mutations in the biocontrol potential of this fungus.

ACKNOWLEDGEMENTS

Research supported by National Research Council of Italy (Binational Project Italy-U.S.A. N. 97.00197.CT06) and by Ministero delle Risorse Agricole, Alimentari e Forestali (M.I.R.A.A.F., Piano Nazionale "Biotecnologie vegetali", Area 10 - I diagnostici, Programma n° 451).

REFERENCES

Alabouvette, C. 1986, *Fusarium* wilt suppressive soils from Châteaurenard region: review of a 10 years study, Agronomie 6: 273-284.

Cairns, J. and Orvos, D.R. 1992, Establishing environmental hazards of genetically engineered microorganisms, Rev. Environ. Contam. Toxicol. 124: 19-39.

Daboussi, M.J. 1996, Fungal transposable elements: generators of diversity and genetic tools, J. Genetics 75: 325-339.

Day, M.J. and Fry, J.C. 1992, Microbial ecology, genetics and risk assessment, in : Release of Genetically Engineered and other Micro-organisms, M.J. Day and J.C. Fry, eds., Cambridge University Press, Cambridge, Great Britain pp. 160-167.

Garibaldi, A., Brunatti, F. and Gullino, M.L. 1986, Suppression of *Fusarium* wilt of carnation by competitive non-pathogenic strains of Fusaria, Med. Fac. Landbouw, Rijksuniv. Gent 51/2b: 633-638.

Garibaldi, A., Gullino, M.L. and Aloi, C. 1990, Biological control of *Fusarium* wilt of carnation, Proc. BCPC 1990 3A-1: 89-95.

Garibaldi, A., Minuto, G. and Gullino, M.L. 1994, Biological control of *Fusarium* wilt of basil. Proc. BCPC 1994 6D-18: 811-815.

Gullino, M.L., Migheli, Q. and Mezzalama, M. 1995, Risk analysis for biological control agents: antagonistic *Fusarium* spp. as a case study, Pl. Dis. 12: 1193-1201.

Kistler, H.C. and Benny, U. 1992, Autonomously replicating plasmids and chromosome rearrangement during transformation of *Nectria haematococca*, Gene 117: 81-89.

Langin, T., Daboussi, M.J., Gerlinger, C. and Brygoo, Y. 1990, Influence of biological parameters and gene transfer technique on transformation of *Fusarium oxysporum*, Curr. Genet. 17: 313-319.

Lemanceau, P. and Alabouvette, C. 1991, Biological control of *Fusarium* diseases by fluorescent *Pseudomonas* and non-pathogenic *Fusarium*, Crop Prot. 10: 279-286

Leslie, J.F. and Dickman, M.B. 1991, Fate of DNA encoding hygromycin resistance after meiosis in transformed strains of *Gibberella fujikuroi* (*Fusarium moniliforme*), Appl. Environ. Microbiol. 57: 1423-1429.

May, G.S., Tsang, M.L.S., Smith, H., Fidel, S. and Morris, N.R. 1987, *Aspergillus nidulans* β-tubulin genes are unusually divergent, Gene 55: 231-243

Migherli, Q., Del, Tedesco, D., Musso, M.R. and Gullino, M.L. 1995, High efficiency transformation of antagonistic *Fusarium oxysporum* (Abstr), in : Proc. International Seminar on *Fusarium*, Mycotoxins, Taxonomy and Pathogenicity, May 9-13, 1995, Martina Franca, Italy, p. 204.

Migheli, Q., Friard, O., Del, Tedesco, D., Musso M.R. and Gullino, M.L. 1996, Stability of transformed antagonistic *Fusarium oxysporum* strains *in vitro* and in soil microcosms, Mol. Ecol. 5: 641-649.

Migheli, Q., Piano, S., Enrietti, S. and Gullino, M.L. 1992, Protoplast fusion in antagonistic *Fusarium* spp.

IOBC/WPRS Bull. 1992/XV/1: 196-198.
Minuto, A., Migheli, Q. and Garibaldi, A. 1995, Evaluation of antagonistic strains of *Fusarium* in the biological and integrated control of *Fusarium* wilt of cyclamen, Crop Prot. 14: 221-226
Minuto, A., Minuto, G., Migheli, Q., Mocioni, M. and Gullino, M.L. 1997, Effect of antagonistic *Fusarium* spp. and of different commercial biofungicide formulations on *Fusarium* wilt of basil (*Ocimum basilicum* L.), Crop Prot. 16: 765-769.
Punt, P.J., Oliver, R.P., Dingemanse, M.A., Pouwels, P.H. and Van den Hondel, CAMJJ. 1987, Transformation of *Aspergillus* based on the hygromycin B resistance marker from *Escherichia coli*, Gene 56: 117-124.
Sambrook, J., Fritsch, E.F. and Maniatis, T. 1989, Molecular cloning, a laboratory manual, Cold Spring Harbor Laboratory Press, Cold Spring Harbor, USA.
Schliestl, R.H. and Petes, T.D. 1991, Integration of DNA fragments by illegitimate recombination in *Saccharomyces cerevisiae*, Proc. Natl. Acad. Sci. USA 88: 7585-7589.
Tramier, R., Pionnat, J.C. and Tebibel, N. 1983, Role of the fungi in the induction of suppressiveness into substrates to *Fusarium* wilt of carnation, Acta Hortic. 141: 55-59.

THE APPLICATION OF *lux*-GENE TECHNOLOGY IN THE CONTROL OF SOIL-BORNE DISEASES

Duncan White[1], Oliver G.G. Knox[2], Ken Killham[2] and Carlo Leifert[2]

[1]Department of Agriculture, University of Aberdeen, MacRobert Building, Aberdeen AB24 5UA,
[2]Department of Plant and Soil Science, University of Aberdeen, Cruickshank Building, Aberdeen AB24 3UU.

1. INTRODUCTION

Soil-borne plant diseases, caused by both fungi and bacteria, are responsible for considerable economic losses to the agricultural and horticultural industries (Whipps, 1992). Soil acts as a reservoir for many of the major diseases which afflict plants ranging from attack on the root system such as *Pythium* damping-off, root and crown rots to spoilage of above ground tissue including leaves and fruit. There is currently much awareness in the potential damage caused by chemical means of controlling these diseases and in the residues which may be left in the soil after treatment. Biological control using fungal and bacterial antagonists is now seen in some cases as a viable alternative to chemical means of control.

There are a number of bacterial antagonists which are active against fungal plant pathogens and are currently being developed for commercial use. To accurately predict the behaviour of current biological control agents, prior to release, an efficient marker system must be used to track the introduced antagonist and monitor its behaviour towards the pathogen and environment. This will provide a firm foundation for the design of biological control of soil-borne fungal diseases for future antagonists, and to realise the full potential of biological control either within existing or future pathogen management strategies. Information regarding the mode of action of the antagonism in the environment to which the antagonist will be introduced and, the behaviour of the antagonist in terms of the interaction with the indigenous flora and fauna must be known prior to full-scale release.

This review presents a detailed account of the current work in the use of bioluminescence as a technology and how it can be used in the biological control of soil-borne diseases. Biological control using bacterial antagonists is currently constrained by the lack of accurate knowledge of the behaviour of the introduced antagonists in the environment, and how they interact with the indigenous microbial flora. Bioluminescence is one of a number of marker systems which have been used to track microbial inocula and this review considers

Biotechnological Approaches in Biocontrol of Plant Pathogens
Edited by Mukerji *et al.*, Kluwer Academic / Plenum Publishers, New York, 1999

the general requirements of biological control for a particular marker system and focuses on the benefits and disadvantages of the use of bioluminescence in biological control. The current uses for bioluminescence are considered in the context of biological control as are future uses. Current methodology in the introduction of bioluminescence genes into antagonists and the constraints to the visualisation of bioluminescent-marked inocula are also considered.

2. MARKER TECHNOLOGY IN BIOLOGICAL CONTROL

There are a number of microbial marker systems which can be used to track bacterial biocontrol agents. Each one possesses distinct advantages and disadvantages. These include the use of genetically introduced traits which express detectable phenotypes and include bioluminescence (*lux*), *xyl*E (de Leij *et al.*, 1994), *lac*ZY (Ryder, 1994) and ß-glucuronidase (*GUS*) (Jefferson *et al.*, 1987; Wilson, 1995). Other markers include the use of resistance markers, for example heavy metals and antibiotics (Ryder, 1995). Molecular probes based on PCR detection of specific sequences have also been used (Hartung, 1998).

To indicate how useful a particular marker system would be, several general requirements can be made of any marker technology. These can be summarised as:

1. Possess low environmental background signal
2. Have a high information content regarding the behaviour of the antagonist in the environment, preferably in the one to which the antagonist will be introduced
3. Relatively high sensitivity
4. Use of existing methodologies to introduce marker genes into host antagonist and the ease with which the antagonist can be marked.

The current main focus for the use of marker technology with biological control agents is determining the spatial relationships between antagonist and pathogen by tracking the antagonist once released into the environment, for example colonisation of the root surface. To design better and more effective disease management strategies for biological control agents, this information must ideally be linked to microbial and biological activity. From this, a further requirement regarding the use of marker technology can be made;

5. Provide information coupling directly the mode of antagonism (i.e. biological activity) with metabolic activity/gene expression in the environment.

Specific advantages relating to the use of bioluminescence as a marker technology are given in Table 1.

Table 1. Advantages of the use of *lux* technology with biological control agents

Advantage	Comments
Real time sensing of microbial activity	Use on either whole cell populations (e.g.as part of an on line system of monitoring biological activity) or focussing on localisation studies in e.g. plant root systems.
Coupling of metabolic activity with biological activity	e.g. for the study of nutrient gradients.
High information content	Can be both quantitiative and qualitative.
Detection systems are flexible	CCD and photographic monitoring can accommodate large samples e.g. whole plants.
Not limited to terrestrial samples	Used in plant and animal tissues as well as aquatic systems.
Low background signal	No free living indigenous light producing soil microorganisms (with the same detection wavelength).
Uses established cloning methods	

3. THE NATURE OF BIOLUMINESCENCE

Natural bioluminescence has been known for over 100 years and occurs in very few organisms and those that do occupy a narrow niche, occurring as symbionts in marine fish and squids or as planktonic forms. The marine bacteria *V. fischeri* and *V. harveyi* (*Photobacterium phosphoreum*) are the most widely studied bioluminescent organisms. No terrestrial bacteria, have however, been found capable of producing light, although bioluminescent fungi do exist as well as a luminescent bacterial nematode symbiont, *Xenorhabdus luminescens* which is not present free-living in soil. (Schmidt *et al.*, 1989).

The molecular biology and enzymology of the bacterial luciferase genes have been extensively characterised and are reviewed by Meighen (1988, 1991) and Koncz *et al.* (1990) with DNA sequences available on international databases (Table 2). The *lux* operon consists of five structural genes *lux*ABCDE. *lux*AB are responsible for the coding of the luciferase αß-subunits (Cohn *et al.*, 1985; Foran and Brown, 1988; Johnsen *et al.*, 1986, 1990) and, *lux*CDE which are responsible for the enzymes which recycle an aldehyde substrate (Engebrecht and Silverman, 1984) (Figure 1). Since this early work two other *lux* genes, *lux*G and *lux*H, have also been identified which are responsible for coding for proteins involved in the synthesis of the reduced flavin (Bassler *et al.*, 1993). Regulation of expression is through autoinduction (*lux*R and *lux*I) and also through a variety of cellular mechanisms, for example, via the cAMP-CRP catabolite repression mechanism (Dunlap and Greenberg, 1985). The aldehyde substrate (*n*-tetradecyl aldehyde in *V. fischeri*) can be applied exogenously (usually as *n*-decyl aldehyde) in which case only those genes coding for the luciferase enzyme need to be introduced into the target organism. With these strains, careful optimisation of the bioluminescence assay needs to be carried out, for example with respect to the timing of the addition of the substrate, substrate concentration and careful choice of pH buffers (Blouin *et al.*, 1996). The inclusion of all genes from the *lux* operon for the constitutive production of light may place an energetic burden on the host cell, and comparative tests will have to be made with the parent strain to ensure there is no loss in overall biological activity (e.g. biocontrol).

Table 2. Selection of *lux* gene sequences available from the GenBank Database

Organism	*lux* genes	Accession Number
Vibrio fischeri	A and B	X06758
	G and 3' of E	M62812
	R, I and 5' of C	M19039
V. harveyi	A and B	X5879
	C	X07084
	D	J03950
	G and H	M27139
	L, M and N	L13940
Photobacterium leiognathi	C, D, A, B, E and G	M63594
Xenorhabdus luminescens	A, B, C, D and E	M90092

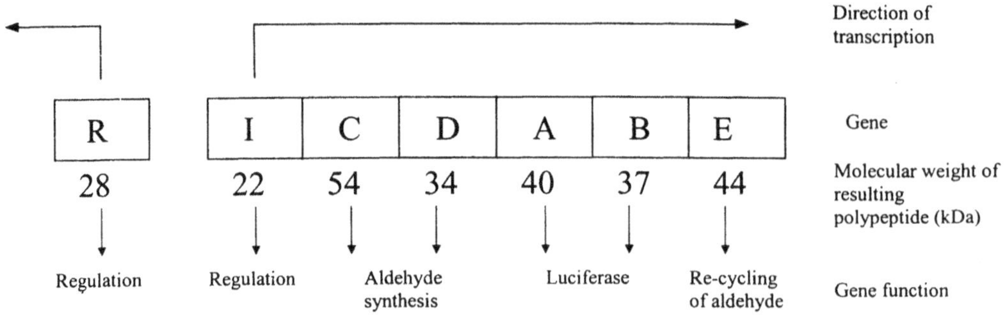

Figure 1. Organisation of *lux* genes from *Vibrio fischeri*.

The production of light using bacterial luciferase is a three step process (Blouin *et al.*, 1996);

1. Formation of oxidised $FMNH_2$ using luciferase catalysed reaction;
   $E_1 + FMNH_2 \rightarrow E_1 - FMNH_2$
   $E_1 - FMNH_2 + O_2 \rightarrow E_1 - FMNHOOH$
2. Formation of a high energy intermediate;
   $E_1 - FMNHOOH + RCHO \rightarrow [E_1 - FMNHOH]^* + RCHOOH$
3. Production of light and re-cycling of FMN
   $[E_1 - FMNHOH]^* \rightarrow E_1\text{-}FMNHOH + h\nu$ (light) (at 490 nm)
   $E_1 - FMNHOH \rightarrow E_1 + FMN + H_2O$
4. NAD(P)H:FMN oxidoreductase recycles the FMN and NAD(P);
   $FMN + NAD(P) + H^+ + E_2 \rightarrow NAD(P)^+ + FMNH_2 + E_2$

The wavelength of light emitted is a function of the folding of the luciferase enzyme-subunits with emission spectra between 490 to 535 possible with different structural mutations (Knocz *et al.*, 1990). Differences in the wavelength of light produced has opened up the possibility of investigating the simultaneous expression of different genes in the same organism (Koncz *et al.*, 1990) and the monitoring of different species within a consortium (Burlage and Kuo, 1994) or when multiple antagonists are to be applied for biological control (Pierson and Weller, 1994).

4. THE INTRODUCTION OF *lux* GENES INTO BIOCONTROL AGENTS

Genes coding for luciferase production can be introduced into recipient biocontrol strains by plasmid transfer or by the uptake of exogenous DNA. Conjugative plasmids, protoplast transformation conjugative mating and electroporation have all been used with varying degrees of success. Transposon mediated introduction of *lux* genes has been used in many studies using Tn5 and Tn7 as has the use of phage (Waterhouse *et al.*, 1993a), but in most cases this is dependent upon a plasmid delivery system such as pUT (de Lorenzo *et al.*, 1990). Differences in the ease of transformation between Gram negative and Gram-positive bacteria are highlighted in the number of Gram-negative strains which have been cloned with *lux* and the relatively few Gram-positive strains for use in environmental studies (Table 3).

Table 3. Selected examples of *lux*-marked bacteria used in biocontrol and environmental studies.

Lux-marked strain	Plant association	Investigation	Reference
P. fluorescens	Wheat	Colonisation	Rattray *et al.* (1993); de Weger *et al.* (1997);
	Soy bean	Colonisation	de Weger *et al.* (1991)
	Barley	Phosphate starvation/ colonisation	Kragelund *et al.* (1997)
	Brassica campestris L cv. Tobin (canola)	Colonisation	Boelens *et al.* (1993)
	Capsicum annuum	Colonisation	Marschner and Crowley (1996)
		Soil survival studies phosphorous starvation	Meikle *et al.* (1995)
		(soil survival)	Kragelund *et al.* (1995)
Enterobacter cloacae	Lettuce (*Lactuca sativa* L.)	Colonisation (root pressings)	Fravel *et al.* (1990)
	wheat	Colonisation	Rattray *et al.* (1995)
P. syringae pv *phaseolicola*	Chinese cabbage	Colonisation	Waterhouse *et al.* (1996)
		soil colonisation	Silcock *et al.* (1992)
E. coli	*Phaseolus vulgaris*	Colonisation/competition	Kozdroj (1996)
Rhizobium leguminsoarum bv. *phaseoli*	maize and lettuce	Colonisation/survival	Chabot *et al.* (1996)
Erwinia amylovora	Apple	Colonisation	Helfert and Zeller (1997); Bogs *et al.* (1998)
Er. carotovora	Potato	Location in Tubers	McLennan *et al* (1992)
		soil survival	Grant *et al.* (1991)
P. putida	Spinach	Colonisation	Sakai *et al.* (1997)
B. subtilis	*Phaseolus vulgaris*	Colonisation/competition	Kozdroj (1996)
several strains	Several plants	Colonisation	Shaw and Kado (1986)

The uptake of DNA by competent cells is the most widely used method of transformation, although at present for *lux*, it is restricted to transforming *E. coli* or *B. subtilis* strains. The production of competent cells can be easily achieved using a combination of $CaCl_2$ and a brief heating treatment. The effect of the $CaCl_2$ treatment is uncertain at present and although it is thought to influence DNA binding by either affecting the membrane binding site or DNA attachment to the membrane (Brown, 1990). The effect of heating is to render the membrane more fluid in readiness for preparation for DNA transfer. Other binding agents including the use of DMSO and combinations of divalent cations have been tried to improve transformation efficiency. Table 4 shows a simple protocol for the preparation of competent *E. coli* cells.

Table 4. Protocol for the preparation and transformation of *Escherichia coli.*

Preparation of competent *E. coli*
Grow cells overnight (100ml culture) to OD_{560} of ~0.4.
Harvest by centifugation at 2500g and 4°C.
Remove supernatant and suspend cell pellet in 1ml of TSS buffer. (TSS: 10ml LB, 2ml 10% PEG 6000, 1.5ml 5% DMSO and 1.5ml 50mM $MgCl_2$)
Chill on ice for 10 minutes.
Use immediately or store as small aliquots at -70°C with 10% glycerol.
To Transform
Remove aliquot and thaw slowly.
Add plasmid DNA (10-100ng per 100µl of cells), vortex mixture and place on ice for 30 minutes.
Incubate at 42°C (water bath) for 90 seconds and then place back on ice for 5 minutes.
Add 500 µl Luria Bertani broth to every 100 µl of cells and incubate at 37°C for 30 minutes.
Plate out cells onto selective agar and culture overnight.
Select for transformed colonies.

Gram-negative biocontrol strains can be transformed easily with di- and tri-parental mating systems using *Tn*5 mediated transfer of *lux* genes (de Lorenzo *et al.*, 1990). For example, in the tri-parental mating, three strains are involved; the recipient to be transformed, the host bearing the *lux* genes and a helper strain which provides the nucleases required for the transfer.

The introduction of *lux* genes into Gram-positive strains, particularly *Bacillus* spp. can be achieved through the transformation of competent cells. The early pioneers in this field were Anagnostopoulos and Spizizen (1961) whose methods were later refined by Dubnau and Davidoff-Abelson (1971). Competent cells are prepared by initiating growth in rich media and then transferring to media less rich in nutrients and devoid of the amino acids required by the auxotrophic mutants usually used. Competence of *Bacillus* is therefore achieved as a function of growth. The transformation of *Bacillus* does not however work efficiently for monomeric closed circular forms of plasmid DNA. Multimeric forms of plasmids (usually generated by cutting the plasmid at only one site with an endonuclease prior to re-ligation to form multimers) when used to transform *Bacillus* generally lead to higher transformation efficiencies (Hardy, 1985). In such cases only a small proportion of the cells are able to take up exogenous DNA and not all strains have auxotrophic requirements.

To overcome many of the problems associated with the use of competent cells, Chang and Cohen (1979) proposed a protocol for the production of protoplasts. This approach has the advantage that a greater proportion of cells are able to take up DNA and that they can accept monomeric form of plasmids and are therefore suitable for the range of bacteria for which competent cell preparation has not been developed (Hardy, 1985). However, protoplasts are difficult to store and the selection after transformation can not be done directly but requires recovery in a rich media and then replica plating. The methods usually involves growing the bacteria in liquid culture to roughly mid exponential phase, harvesting the cells prior to suspending them in a solution containing lysozyme. The formation of the protoplasts has to be carefully monitored to ensure the majority of cells are in the required state (Chang and Cohen, 1979). At this point the DNA is added and the recovery of the protoplasts initiated.

A more radical solution to the transformation of otherwise difficult strains is the use of electroporation which has been shown to be successful with *Bacillus* sp. (Brigidi *et al.*, 1990). This technique was originally designed for use with eukaryotic cells but has since been developed for use with many genera of bacteria. Electroporation of bacteria uses a pulse of electricity (Dower *et al.*, 1988), which is delivered through a chilled sample of cells and plasmid in a low ionic strength buffer between two electrodes in a specially designed cuvettte.

Cells are harvested at mid-exponential phase and washed in a low salt containing buffer. At this point the cells can be frozen and stored for several months. The cells are mixed with dialysed plasmid and transferred to chilled cuvettes. The reaction is carried out on ice to keep the efficiency of the transformation as high as possible. After a short incubation, the cells and plasmid are then exposed to a 2500V charge. Afterwards, the cells are quickly recovered into fresh media prior to selecting for the required phenotype. For each strain, the field strength and the pulse delivery time need to be characterised. Transformation efficiencies for electroporation can be very high but differences in membrane width, composition and the electroporation solution can account for differences in the transformation frequency between different strains.

5. DETECTING BIOLUMINESCENT-MARKED BIOCONTROL AGENTS IN THE ENVIRONMENT

There are several methods of detecting light producing strains (Table 5). The choice of method depends on the level of information that is required, the nature of the *lux* construct that is being used (i.e. how bright it is for a given concentration of cells) and the number of active cells in the sample. The latter is important if soil or the rhizosphere/rhizoplane is being studied where populations of the introduced organism may be very low.

Table 5. Detection methods for bioluminescent bacteria

Method	Advantages	Disadvantages
Charge Coupled Device Imagery	Very sensitive, exposure and time for obtaining results can be rapid, long time exposure possible, quantitative and qualitative. Detection of single cell possible	Expensive, very long exposures (> 20 mins) may require use of liquid nitrogen for cooling, thermal "noise" may limit detection sensitivity.
Photographic Emulsion	Inexpensive, quantitative and qualitative	Long time exposure may cause insensitivity due to nature of emulsion used, photographic film development time (polaroid film may not have sufficient sensitivity)
Visualy	Good for colony detection if colonies are young; can enable rapid screening of culture plates, qualitative	Limited to bright strains, period of adjustment to darkness required prior to viewing.

Charge coupled device (CCD) enhanced imagery at present remains the most sensitive, but most expensive detection system and can detect weakly bioluminescent colonies or single cells. CCD-imagery has been used to visualise; bacterial microcolonies (Rattray *et al.*, 1995; Waterhouse *et al.*, 1993b), sites of pathogenic infection in plants (McLennan *et al.*, 1992; Shaw & Kado, 1986), animals (Stewart & Williamson, 1992) and single cells (Mahro *et al.*, 1992; Masuko *et al.*, 1991 a,b; Silcock *et al.*, 1992; Waterhouse *et al.*, 1993). Differences in cellular light output can give information regarding cell activity which can be estimated through the associated computer software.

CCD technology relies on the use of photon sensitive silicon chips which convert photon energy to an electrical signal through the generation of "potential wells" (Aikens, 1994). Many of the CCD-enhanced images of bacterial microcolonies and single cells are formed after long exposure times. Typical exposures are around 2 min. for some culture plates to around 30 to 60 min. for single cells and microcolonies (Waterhouse *et al.*, 1993b), depending on the age of the cells/colonies and promoter strength. To reduce background "noise" after long exposure times and improve the image quality, slow scan cooled CCD cameras. Two methods of cooling are found in current cameras; peltier based devices and those cooled by liquid nitrogen. For long exposures (30 min. or more), the liquid nitrogen cooled cameras generally give better images. The cooling of the CCD camera is also necessary to remove the "dark current" formed from thermally generated charge (Aitkins, 1994). Improvements in the design of CCD technology and software has led to the development of rapid sequential imaging systems which are capable of imaging up to the equivalent of 1000 frames per second. These systems are being used to image, for example accumulation of intracellular ions such as Ca^{2+} (Bright, 1994). This technology is therefore capable of imaging real-time luminescence, although it is currently restricted by the relative light output of current marked organisms.

In some cases photographic film can replace CCD instruments to obtain an image of, for example, root colonisation and, give semi-quantitative information about microbial activity. de Weger *et al.* (1991) used a fibre optic "light-pipe" as an alternative to CCD based detection to visualise bioluminescent colonies. Adaptation of this technique may be used *in situ* similar to the use of mini-rhizotrons which can be used to visualise roots in soil (Pan *et al.*, 1998).

Plating methods form the principal method of monitoring biocontrol agents. The culture plates can be imaged to verify the presence of the marked strain which is useful where the background of indigenous organisms is high, through either low soil dilution levels or if specific colonies are covered by fungal mycelium or larger colonies or in cases where selective antibiotics cannot be used (Prosser, 1994), for example when antibiotic resistance markers are not included for the cloning selection methods, for example if the marked strain is ultimately to be released into the environment.

The inclusion of antimicrobials into plating media should ideally be unnecessary with the use of bioluminescence but they are often included to limit the background of indigenous microorganisms which can be present at far greater numbers than the introduced inocula. This is often the case when the released marked strain has been in prolonged contact with the environment and low populations of the introduced strain exist. In these cases bioluminescence is used to confirm the presence of the marked organism on the culture plate. The sensitivity of plate count detection and enumeration can be increased using standard microbiological techniques, for example through the use of enrichment cultures (Grant *et al.*, 1991; Shaw *et al.*, 1992) and MPN-microtitre assays, coupled to light imaging detection (Flemming *et al.*, 1994). This latter method was able to detect 5 cells per g soil. Colony identification is possible by the use of either naked eye if the strains are sufficiently bright.

6. USE OF BIOLUMINESCENCE TO MONITOR BIOLOGICAL CONTROL AGENTS

There are currently three main uses of bioluminescence for the control of soil borne diseases:

1. As a tracking system to provide population and spatially relevant information of marked inocula.
2. To provide metabolic information of overall cell activity.
3. To provide information regarding the expression of other genes (reporter system).

6.1. Use of Bioluminescence to Track Biological Control Agents

The tracking of microbial inocula is normally achieved by monitoring population dynamics through the use of plate counts which often incorporate a selective antimicrobial to limit the growth of indigenous microorganisms. Confirmation of *lux*-marked colonies can then be made using CCD-detection or exposure to photographic film. Very bright colonies can often be imaged directly by eye. The antimicrobial resistance used can either be from resistance genes incorporated during the introduction of the bioluminescent genes or from spontaneous resistant mutants. In either case the stability of the resistance factor needs to be assessed. Resistance to rifampicin (spontaneous resistance), ampicillin (transformation) and kanamycin (transformation) are common antibiotics.

To provide spatial information about the colonisation of the rhizosphere, direct assessment of active bacteria on plant roots can be made. This identifies one of the main constraints to biological control by assessing likely sites for introducing the antagonist and can be achieved through the use of enhanced light imaging techniques or photographic emulsions of roots directly or using root pressings on agar. Bacteria present on the phyloplane (Waterhouse *et al.*, 1993b) and rhizoplane (root surface) (Brennerova and Crowley, 1994; Rattray *et al.*, 1995) can be easily detected. Population dynamics must also be assessed for an accurate assessment of colonisation on plants. Bacteria present on the phyloplane are likely to be more easily detected than those on the root surface. This is due to the higher numbers which can be applied to the phyloplane and environmental quenching of the light signal due to soil particles associated with the root surface.

Bioluminescence can be used to track *lux*-marked inocula associated with soil invertebrates and monitor their activity. Earthworms have been shown to translocate microbial inocula such as *Pseudomonas* spp. (Madsen and Alexander, 1982; Thorpe *et al.*, 1993) and plant pathogens (Toyota and Kimura, 1994) through the soil. This is therefore an important route in the dissemination of biocontrol agents through soil. This is achieved by the earthworms either carrying soil microorganisms on the outside of their body (Heijnen and Marinissen, 1995), ingesting soil material or through their foraging activities through litter (Thorpe *et al.*, 1996).

6.1.1. Use of Bioluminescence in Post-harvest Control: *lux*-marking of Pathogens

The tracking of bacteria and fungi in the environment using bioluminescence is not only limited to the use of antagonists. The deliberate release of *lux*-marked pathogens in a controlled situation will indicate likely routes of infection and pathogenesis. This could be important in the control of post harvest spoilage when the pathogens responsible are either derived from soil associated material or from insect vectors. McLennan *et al.* (1992) introduced *lux*-marked *Erwinia carotovora* on to potato tubers and could observe the spread of tissue degradation without resorting to destructively harvesting the tuber to enumerate the pathogen. Helfert and Zeller (1997) were able to follow the translocation of *lux*-marked *Er. amylovora* through apple leaves from the point of infection to the leaf tip by exposing the infected leaves to X-ray film. In a more recent study, again using apple seedlings, Bogs *et al.* (1998) were able to follow the translocation of different strains of *lux*- and green fluorescent protein (*gfp*)-marked *Er. amylovora* through different plant tissues from the xylem to the parenchyma, and even to the roots after stem inoculation. The tracking of pathogens in this way could be coupled to measurements of microbial activity to obtain information regarding the association between the production of metabolic signals from the pathogen within plant tissue and the spread of plant pathogens through the plant. Bogs *et al.* (1998) suggested that the levels of nutrients found in the xylem sap could account for the increased activity of the infecting cells, perhaps aiding their translocation and colonisation capacity and subsequent ability to cause disease.

Future developments in the management of pathogen and pathogen activity could involve *lux*-marking as a reporter of those genes responsible for the production of elicitors and hypersensitivity factors (Vivian and Gibbon, 1997). Conversely, it could be possible, once the mechanisms responsible for the response to the elicitors were identified, to introduce bioluminescent genes (*luc*) to monitor the plant defence response.

6.1.2. Use of Bioluminescence in Protected Crops

The use of protected horticultural crops, for example in glasshouses, provides an ideal opportunity to use bioluminescence to track pathogens and antagonists in a controlled but essentially enclosed environment. Protected crops, grown in pots or "grow-bags" are often the choice of the grower as they afford a controlled environment for a standardised crop. The diversity of pathogens is also likely to be less since the soil/peat substrate can be made relatively free of pathogens by sterilisation/sanitation methods. Irrigation systems provide an ideal opportunity to disperse both antagonists and pathogens through the soil (Rattink, 1992). The risks of spreading pathogens in this way could be easily monitored by tracking *lux*-marked cells. Such environments provide an ideal testing ground by providing a link between laboratory scale environments and commercial practices.

6.2. Use of *lux* to Monitor Microbial Activity

Microbial activity can be defined at two levels; metabolic and biological. To ensure fitness of the biocontrol agent, both must be considered in relation to the environment and how the introduced antagonist behaves once it is released into the environment.

Bioluminescence is directly coupled to the cells' energy requirements through the pool of cellular NAD(H) and dehydrogenase activity (Meikle *et al.*, 1992). Measurement of the bioluminescence (as actual bioluminescence) relates the light output with the present energy status of the cell and can be determined directly using a luminometer and a suspension of the soil containing the *lux*-marked organisms.

In soil, bioluminescence is coupled to metabolic activity rather than the amount of microbial biomass. The addition of nutrients such as glucose (Rattray *et al.*, 1992) to the soil samples indicates the potential for strains to recover from periods of starvation or nutrient stress (measured as potential bioluminescence). The BIOLOG metabolic profile system, originally designed to identify microbial isolates, identifies the metabolism of a range of C and N sources (Garland and Mills, 1991). The BIOLOG system has been adapted to measure changes of microbial diversity in the soil and rhizosphere after perturbation (Ellis *et al.*, 1995) and, could be further adapted to monitor microbial activity as bioluminescence. This would provide a direct link between microbial activity and nutrient availability in the soil and could provide information about colonisation and survival in the soil.

Root exudates stimulate the associated microbial flora attached to or near the root by providing a ready supply of assimilatable nutrients. The behaviour and amount of stimulation of the colonising biocontrol agent toward these exudates may determine how successful the control will eventually be. Rattray *et al.* (1995) introduced *lux*-marked *Enterobacter cloacae* into the rhizosphere of wheat and into non-rhizosphere arable soil and measured the microbial activity by luminometry after the addition of aldehyde substrate. Microbial activity (luminescence per cell) of introduced *lux*-marked *Ent. cloacae* was found to be greater in the rhizosphere than non-rhizosphere and those cells introduced into non-rhizosphere soil were only active after addition of growth substrate.

Biological control agents are often associated with plants through original contact with the germinating seeds. Monitoring the activity of cells by measuring bioluminescence and how *lux*-marked inocula respond to the germination of seeds will determine how effective seed coating is at establishing rhizoplane microorganisms and, may even give some information of how germinating seed initially respond to the environment. Germinating seeds are known to produce metabolites which can stimulate germination of fungal pathogens (Nelson and Hsu, 1994; Stanghellini and Hancock, 1971). Optimisation of seed coating methods and, how fast the inocula may use up nutrients included with the seed coat formulation can also be assessed using bioluminescence. This information will be of use in assessing the shelf life of the coating preparation (Cassidy *et al.*, 1997). *Bacillus subtilis* is often formulated as a spore suspension for storage and ease of application. However, endospores of *lux*-marked vegetative cells are phenotypically dark (Hill *et al.*, 1994). Monitoring of light production will therefore show when the endospores germinate and are likely to become available for biological control as well as providing information about nutrient fluxes produced by the seeds.

Monitoring of the microbial inoculum using bioluminescence will give information about the persistence (survival) and activity of the microbial population. Soil matric potential (related to water availability) has been shown using stained inocula to affect inoculum placement (White *et al.*, 1994) and survival (Wright *et al.*, 1993). In luminometric studies using a *lux*-marked *E. coli*, Rattray *et al.* (1992) have shown that matric potential markedly affects microbial activity. Reduction of microbial activity was found with increasing matric stress. This was thought to be due to the reduction of substrate diffusion/translocation through the soil pore system.

Microbial activity in soil is often constrained by the lack of available nutrients. Bioluminescence-based biosensors which determine nutrient availability in soil and the rhizosphere can give information about specific nutrient responses, for example during cell starvation. Bacteria introduced into soil are thought to undergo a starvation phase and perhaps enter into a viable but non culturable state (VBNC) (McDougald, 1998) and may therefore be unavailable for biocontrol activity. Duncan *et al.* (1994), using luminometry, monitored changes to microbial activity in soil of *lux*-marked *P. fluorescens* and *lux*-marked *E. coli* as the cells entered into the VBNC state. Reactivation of starved cells can be achieved from the input of C (Campbell, 1991).

6.3. Use of Reporter Gene Fusions

To understand more fully and to optimise biological control using introduced microbial inocula, information is required about the relationship between microbial colonisation and survival in the soil/rhizosphere and biological activity. Metabolic activity of the inocula gives information about the cell population reacts to nutrient fluxes. However, to identify specific constraints relating to biological activity in the environment, information is needed which will relate biological behaviour with specific gene activity. One way of doing this is to use a gene-reporter system which use a readily detectable marker allowing the activity of unrelated gene(s) to be determined.

Reporter gene fusions use the promoter and regulatory gene sequences of the gene(s) of interest for example inducible promoters, coupled to the structural genes of the marker system (Lindow, 1995). Expression of the marker phenotype thereby indicating expression of the environmental gene of interest which would otherwise be very hard to detect (Lindow, 1995). Bioluminescence is an ideal reporter gene by providing an easily measurable and quantifiable method of detection. Several successful reporter gene fusions have been constructed to monitor using on-line systems the bioavailability of heavy metals (Condee and Summers, 1992) and, the extent of degradation of soil contaminants, such as toluene and naphthalene (Burlage *et al.*, 1990; 1994). It may be possible in the future to adapt these on-line systems to monitor biological activity relevant to the control of soil borne diseases. The major constraints to the use of reporter genes is knowledge of the gene sequence of the gene to be reported and availability of the appropriate transformation vector.

Bioavailability of essential elements in the soil can be monitored using *lux*-gene fusions. Nitrogen availability has been assessed by Nybroe and his colleagues using a *lux*-based (Tn*5*::*lux*AB) nitrogen-starvation-inducible gene in *P. fluorescens* DF57. Expression of the light phenotype indicated N starvation in pure culture and in non-sterile soil (Jensen *et al.*, 1998). Nitrogen fixation in legumes by *lux*-marked rhizobia has been studied through the use of *nif*DH promoter genes of nitrogenase genes and *lux*AB (O'Kane *et al.*, 1988). Phosphorous availability has also been assessed using P-inducible starvation genes fused with *lux*AB genes (Kragelund *et al.*, 1995; 1997).

Many of the mechanisms which are thought to contribute to biological control in soil can be identified and optimised by the use of reporter genes. The expression of antifungal metabolites such as cyanide and pyrrolnitrin, siderophores and, enzyme activity such as chitenases can be monitored using *lux* as a reporter gene probe, having once identified target gene sequences. The use of reporter genes in those genes which determine biological adaptation to the environment for example factors responsible for root colonising, may identify likely constraints to the environment. Many of the factors associated with biological control have only been identified to work *in vitro* (Kraus and Loper, 1992), reporter gene fusions may indicate which ones will function *in vivo*.

Regulation of nutrient fluxes within the rhizosphere are thought to regulate the size of microbial populations associated with the root (Jones and Darrah, 1993). The response of starved cells as they come into contact with the root exudates could be monitored using cells containing a ribosomal-promoter driven *lux* reporter gene (Marschner and Crowley, 1996). These particular genetic constructs could be used to optimise how antagonists respond to rhizosphere derived nutrient influxes and also perhaps optimise microbial competition (e.g. between introduced antagonist and indigenous microbial flora) in the rhizosphere.

7. REAL-TIME MONITORING OF BIOLOGICAL CONTROL

One of the main features of bioluminescence is that under the right conditions real time monitoring of microbial activity and spatial colonisation can be measured. Constitutive light output by the *lux*-marked strains can be monitored continuously which could be useful, for example, in determining the action of pathogens on antagonists and *vice versa* or in monitoring when cells come into contact with the plant root exudates. With these strains, problems may arise, however, due to the energy drain on the cells which may affect its competitiveness in the environment.

8. QUORUM SENSING AND BIOLOGICAL CONTROL

An important aspect of microbial interactions, including pathogen-antagonist interactions, is the sensing of metabolic signals produced by specific strains. Microbes naturally coexist with others in the environment and it would be advantageous for a cell to be able to monitor gene activity from another species/individual cell. The *lux*I regulatory gene from *V.fischeri* of the *lux*-operon codes for the synthesis of the autoinducible N(3-oxohexanoyl) homoserine lactone (Eberhard *et al.*, 1981; Fuqua *et al.*, 1996) which is diffusible across the membrane and operates at high cell densities in a positive feedback loop. The *lux*R/*lux*I system is one of a number of regulatory genes which are responsible for the regulation of genes that are involved in microbial interactions (Pierson, 1998; Pierson *et al.*, 1998).

Several important genes for biological control are those responsible for the production of the phenazine-based family of antimicrobials (phenazine-1-carboxylic acid, 2-hydroxy-phenazine-1-carboxylic acid and 2-hydroxy-phenazine) which are active against a large number of organisms and act by uncoupling their oxidative phosphorylation (Turner and Messenger, 1986). For example, the fungus responsible for take-all disease in wheat, *Gaeumannomyces graminis* var. *tritici* is sensitive to the phenazine antimicrobials produced from several *Pseudomonas* spp. *Phz*I, which codes for a N-acyl HSL synthetase, is similar to LuxI (Wood and Pierson, 1996) and is involved in the regulation of the production of the phenazine antibiotics. These two operons (perhaps from different organisms) can therefore be complementary in their transcription. A similar mechanism was found in the production of carbapenem from *Er. carotovora* (Bainton *et al.*, 1992). Inserting the *lux* genes into antibiotic producing strains and monitoring light output, real time information can be obtained of the production of an antimicrobial in the environment in response to pathogen activity and in response to assay conditions (Leifert *et al.*, 1995).

9. BIOLUMINESCENCE AND INTEGRATED BIOLOGICAL CONTROL

The integration of biological control with existing chemical control methods is a likely step for future plant disease control programs. Advantages of this include reduced environmental impact of chemical residues, an extended environmental window of activity and reduced likelihood of resistance (Tronsmo and Hjeljord, 1998). Understanding how the microbial antagonist behaves towards chemical residues is important in assessing which combinations of antagonist (and its formulation) and chemical formulation will give the best outcome for good control. The various possible combinations of fungicide and biocontrol agent and their introduction to the plants must be considered prior to use.

Monitoring of cell populations and bioluminescence of *lux*-marked antagonists in the presence of the chemical fungicides will indicate how an inoculum may behave during integrated control. Populations of *lux*-marked *P. corrugata* and *lux*-marked *P. fluorescens* were treated with Apron Combi 453FS commercial fungicide (Ciba Geigy Agrochemicals, Cambridge, UK) and their population dynamics and bioluminescence monitored (Figure 2a). There was a significant difference in the population dynamics between those cells exposed to the fungicide and the control samples with fewer cells surviving after 2 weeks exposure. In addition, there was a significant difference between the two biocontrol strains indicating a difference in sensitivity towards the fungicide. This was also seen by measurements of bioluminescence (Figure 2b) where activity significantly decreased with time with *lux*-marked *P. corrugata*, but generally increased with *lux*-marked *P. fluorescens*. The differences between the two antagonists clearly indicates a need for careful choice in establishing combinations of antagonist and fungicide.

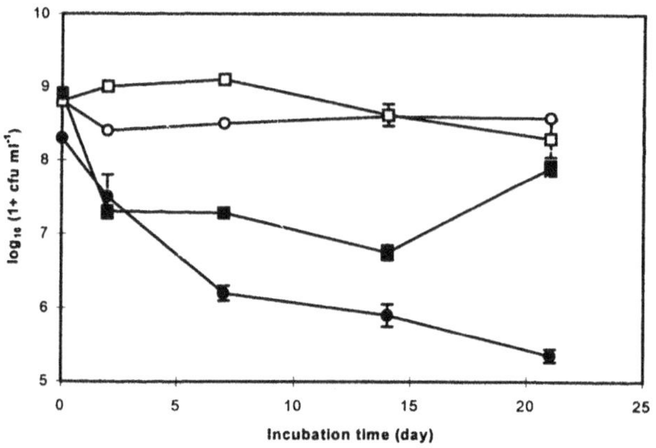

Figure 2a. Toxicity effect of Apron Combi fungicide (Ciba Geigy Agrochemicals, Cambridge, UK) on the population dynamics of *lux*-marked *Psudomonas corrugata* and *lux*-marked *P.fluorescens* (both spontaneous resistance mutants to rifampicin). Washed cells from an 18 h old culture suspension were initially challenged with fungicide (10^{-3} dilution (d. H_2O)). Aliquots (10 ml) were removed at intervals and centrifuged (3,000 x g) before being finally re-suspended to the same volume in 1/4-strength Ringers solution to remove excess fungicide. Viable cell counts were determined using Luria Bertani media (in gl^{-1}; Tryptone (10), Yeast Extract (5), NaCl (5), pH 7). Fungicide treated samples are; *lux*-marked *P. corrugata* (●) and *lux*-marked *P. fluorescens* (■). Control (d.H_2O) samples are; *lux*-marked *P. corrugata* (O) and *lux*-marked *P. fluorescens*(□). Error bars are ± SEM, n=3 replicates.

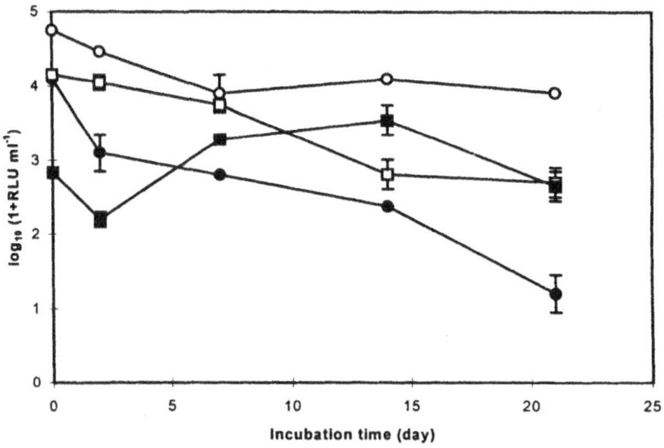

Figure 2b. Toxicity effect of Apron Combi fungicide (Ciba Geigy Agrochemicals, Cambridge, UK) on the activity(bioluminescence) of *lux*-marked *Pseudomonas corrugata* and *lux*-marked *P. fluorescens* (both spontaneous resistance mutants to rifampicin). Washed cells from an 18 h old culture suspension were initially challenged with fungicide (10^{-3} dilution (d.H_2O). Aliquots (10 ml) were removed at intervals and centrifuged (3,000 x g) before finally re-suspended to the same volume in 1/4-strength Ringers solution to remove excess fungicide. Bioluminescence from 1 ml samples (5 replicates per sample) was measured after the addition of 1 µl *n*-decyl aldehyde (33%) v/v) over a 10 sec. period in an LKB luminometer. Fungicide treated samles are; *lux*-marked *P. corrugata* (●) and *lux*-marked *P. fluorescens*(■). Control (d.H_2O) samples are; *lux*-marked *P. corrugata* (O) and *lux*-marked *P. fluorescens* (□). Error bars are ± SEM, n=3 replicates.

10. PRACTICALITIES TOWARDS THE USE OF *lux*

There are two main constraints regarding the use of *lux*-marked biological control agents which need to be addressed; legislative considerations associated with the containment and any subsequent release into the environment and, problems associated with the detection sensitivity. The general problems and disadvantages associated with using *lux* as a marker system are outlined in Table 6. Many of these problems can be overcome by the use of appropriate strains and experimental conditions.

Table 6. Disadvantages of the use of *lux* for biological control

Disadvantage	Comments
Cost of some specialised equipment	Relevant to Charge Coupled Devices and luminometry.
Pomoter availability	The lack of promoter or vector in some antagonists may preculde their current use.
Quenching of signal	The soil environment may quench the light signal.
Signal could be related to inducer concentration than to activity	This is also related to promoter strength. High cell number may be required for sufficient light output due to inducer availability.
Soil properties	Soil pH and chemistry may affect light signal and detection and microbial activity.
Use of foreign genes	Careful selection of clones required to ensure biological activity is not affected by gene insertion.
n-decyl aldehyde	Toxic to cell after prolonged exposure, could be adsorbed to soil particles; potential health hazard; concentration may be critical.
Use of plasmids	Could be undesirable in certain circumstances for example due to mobility to the indigenous soil population.
Detection limit	Could be high for some constructs (e.g. greater than 10^3 cfu g^{-1} soil)
Oxygen sensitivity	Bioluminescence dependent upon critical level of oxygen

10.1. Legislative Considerations

As in the use of any genetically modified organism, specific guidelines for the use of *lux*-marked biocontrol agents must be adhered to prior to experimental use. Current legislation concentrates on ensuring adequate containment and the prevention of foreign genes from entering the natural ecosystem (Ryder, 1994; Tiedje *et al.*, 1989). For this reason, laboratory scale experiments are normally initially carried out, although a limited field release of *lux*-marked *Xanthomonas campestris* has taken place under specific conditions (Shaw *et al.*, 1992). The use of strains out-with containment level I and the use of non-indigenous microorganisms may pose significant problems of, for example, containment and safety. In these cases, such strains are unlikely to be *lux*-marked.

10.2. Detection of light emitting cells

The method chosen to detect light output is largely determined by cost considerations and the level of sensitivity required. These constraints are found both by light imaging of microbial colonisation/growth and quantification of light output by luminometry. The cost of expensive detection systems has been a major criticism against the use of *lux* imaging although cheaper but perhaps less sensitive systems can be used if desired (Table 5). Dedicated luminometers for the quantification of light output can be replaced by the use of ATP detection instruments or scintillation counters (Burlage and Kuo, 1994).

In soil, one of the main constraints which affects detection sensitivity is the quenching of the light signal by soil particles interacting with photons of light prior to detection. For example, light from cells adhering to roots which are covered in soil particles may not be detected. The use of pH-unbuffered samples (particularly for luminometric studies) may alter the light emission spectra since the activity of the luciferase enzyme is closely pH-regulated (Prosser *et al.*, 1996).

Adsorption of the substrate (e.g. *n*-decyl aldehyde) required for light emission to soil particles will limit its availability to the *lux*-marked strain and, in some cases, the addition of the substrate may be unpractical or hazardous to health. Under very dry soil conditions, some soils become hydrophobic preventing the diffusion of *n*-decyl aldehyde through the soil pore space. The use of constitutive light producing strains will overcome these problems and have the advantage that in growing cells, the amount of light produced is directly proportional to the microbial biomass concentration (Prosser, 1994).

The activity of the inoculum and its behaviour in the environment may affect the detection sensitivity and limit the application of bioluminescence in some cases. In unamended, non-sterile soil, light output as measured by luminometery is dependent upon cell number (White *et al.*, 1996). During prolonged soil incubation in non-sterile soil, cell populations can decline by approximately 1 log unit per week (Postma *et al.*, 1989), with the light signal decreasing to the detection limit very rapidly (White *et al.*, 1996). Enrichment of the soil may therefore be necessary to raise the detection limit after prolonged incubation.

Problems of detection sensitivity may arise due to the nature of the *lux* construct and the interaction of the marked biocontrol agent with the environment. The light signal is dependent upon the strength of the promoter and if *lux*-genes are maintained on a plasmid, the plasmid copy number. In pure culture light production is dependent upon the level of inducer present (Prosser *et al.*, 1996). Too few metabolically active cells may limit the amount of available inducer present, thereby decreasing the sensitivity of detection.

Visualisation of colonies on a culture plate can be influenced by the age of the growing colony. Older colonies, produced by prolonged incubation on dilute media which are often used to enumerate soil bacteria, can produce less light for a given *lux*-marked strain than younger ones. Most light from the older colonies is limited to the colony periphery and longer exposures may be necessary to ensure all *lux*-marked colonies can be accounted for.

The *lux* system also requires the presence of oxygen which may become limiting under either very wet or under very dry soil conditions. The solubility of oxygen is also affected by extremes of temperature. This may therefore place a disadvantage on the widespread use of bioluminescence as a marker system, for example, its use with strict anaerobes. Phillips-Jones (1993) was however, able to demonstrate bioluminescence in the aerotolerant *Clostridium perfringens*. Clostridia have been implicated in carrot and potato soft rots, particularly when the tubers are stored in poorly aerated containers (Lund, 1982). However, careful choice of the origin of the luciferase enzyme may overcome this constraint. Oxygen concentration is critical for the activity of the luciferase enzyme from *V. harveyi* but not *V. fischeri* (Knocz *et al.*, 1990).

11. THE FUTURE OF BIOLUMINESCENCE AND BIOLOGICAL CONTROL OF SOIL-BORNE DISEASES

The use of bioluminescence in biological control is at present in its infancy. As cheaper detection systems are being made and, better and more efficient cloning systems are being developed, a greater use of marker technology will be used to answer specific questions in biological control over a greater range of biological control agents and soil-borne plant diseases. Reporter gene technology, perhaps using the simultaneous production of light at different wavelengths, will play an important part in identifying biological mechanisms which occur in the environment. The use of strains emitting different wavelengths of light will allow the use of microbial consortia as a biological control method.

The real-time monitoring of biological activity, for example in response to microbial colonisation of the root and of the interaction with pathogens will indicate the optimum time for biocontrol. This could be coupled with metabolic studies of how both the antagonist and pathogen respond to root exudates. The integration of *lux*-marked biocontrol agents in long term studies could identify how natural biocontrol mechanisms operate and perhaps suggest which biocontrol agents are capable of activity under specific circumstances.

12. CONCLUSIONS

The transformation of biological control antagonists using genes responsible for bioluminescence is an efficient method with which to optimise biological control behaviour in the environment. Many of the problems associated with the optimisation of microbial biological control agents can be solved using marker technology. Bioluminescence using *lux* genes from *Vibrio fischeri* or *V. harveyi* can be easily introduced into bacterial antagonists, and using this unique soil marker system, the spatial distribution of marked strains can be visualised directly in soil and on the rhizoplane. Metabolic activity, biological activity (i.e. biocontrol) by the use of reporter gene technology strains can be monitored using *lux*. The introduction of the light-emitting phenotype into plant pathogens can be used to monitor post harvest spoilage and spoilage occurring on the plant and provide information to optimise integrated control of soil borne diseases. The linking of *lux* genes with those responsible for chemical signalling in the environment will give information of how biological control agents behave toward not only signals produced by the plant for example, root exudates but also by plant pathogens.

ACKNOWLEDGEMENTS

We thank the Ministry of Agriculture, Fisheries and Food (MAFF) UK for their support of this work (MAFF Open Contract Grants CSA 2767 and AE 9038).

REFERENCES

Aikens, R. 1994, Properties of low-light-level slow scan detectors, in : Biological Techniques: Fluorescent and Luminescent Probes for Biological Activity, A Practical Guide to Technology for Quantitative Real-Time Analysis, W.T. Mason, ed., Academic Press, London pp. 277-286.

Anagnostopoulos, C. and Spizizen, J. 1961, Requirements for Transformation in *Bacillus subtilis*, J. Bacteriology 81: 741-746.

Bainton, N.J., Bycroft, B.W., Chhabra, S.R., Stead, P., Gledhill, L., Hill, P.J., Rees, C.E.D., Winson, M.K., Salmond, G.P.C., Stewart, G.A.S.B. and Williams, P. 1992, A general role for *lux* autoinducer in bacterial cell signalling: control of antibiotic biosynthesis in Erwinia, Gene 116, 87-91.

Bassler, B.L., Wright, M., Showalter, R.E. and Silverman, M.R. 1993, Intercellular signalling in *Vibrio harveyi*: sequence and function of genes regulating expression of luminescence, Molecular Microbiology 9: 773-786.

Blouin, K., Walker, S.G., Smit, J. and Turner, R.F.B. 1996, Characterisation of *in vivo* reporter systems for gene expression and biosensor applications based on *lux*AB luciferase genes, Appl. Biochem. Microbiol. 62, 6: 2013-2021.

Boelens, J., Zoutman, D., Campbell, J., Verstraete, W. and Paranchych, W. 1993, The use of bioluminescence as a reporter gene to study the adherence of the plant growth promoting rhizopseudomonads 7NSK2 and ANP15 to canola roots, Can. J. Microbiol. 39: 329-334.

Bogs, J., Bruchmüller, I., Erbar, C. and Geider, K. 1998, Colonisation of host plants by the fire blight pathogen *Erwinia amylovora* marked with genes for bioluminescence and fluorescence, Phytopathology 88: 416-421.

Brennerova, M.V. and Crowley, D.E. 1994, Direct detection of rhizosphere-colonizing *Pseudomonas* sp. using an *Escherichia coli* rRNA promoter in a Tn7-*lux* system, FEMS Microbiology Ecology 14: 319-330.

Bright, G.R. 1994, Multiparameter Imaging of Cellular Function, in : Biological Techniques: Fluorescent and Luminescent Probes for Biological Activity, A Practical Guide to Technology for Quantitative Real-Time Analysis, W.T. Mason, ed., Academic Press, London pp. 204-215.

Brigidi, P., De Rossi, E., Berterini, M.L., Riccardi, G. and Matteuzzi, D. 1990, Genetic transformation of intact cells of *Bacillus subtilis* by electroporation, FEMS Letters, 67: 135-138.

Brown, T.A. 1990, Gene-cloning - An introduction, Chapman and Hall, Second Edition pp. 84-102.

Burlage, R.S., Sayler, G.S. and Larimer, F. 1990, Monitoring of naphthalene catabolism by bioluminescence with nah-*lux* transcriptional fusions. J. Bacteriology 172: 4749-4757.

Burlage, R.S. and Kuo, C.T. 1994, Living biosensors for the management and manipulation of microbial consortia, Annual Review of Microbiology 48: 291-309.

Burlage, R.S., Palumbo, A.V., Heitzer, A. and Sayler, G. 1994, Bioluminescent reporter bacteria detect contaminants in soil samples, Appl. Biochem. Biotech. 45-6: 731-740.

Campbell, F.A. 1991, Luminescence-based detection of genetically modified *Erwinia carotovora* subsp. *carotovora*, PhD Thesis, University of Aberdeen.

Cassidy, M.B., Lee, H. and Trevors, J.T. 1997, Survival and activity of lac-*lux* marked *Pseudomonas aeruginosa* UG2Lr cells encapsulated in k- carrageenan over four years at 4 C, J. Microbiological Methods 30: 167-170.

Chabot, R., Antoun, H., Kloepper, J.W. and Beauchamp, C.J. 1996, Root colonisation of maize and lettuce by bioluminescent *Rhizobium leguminosarum* biovar. *phaseoli*, Appl. Environ. Microbiol. 62: 2767-2772.

Chang, S. and Cohen, S.N. 1979, High frequency transformation of *Bacillus subtilis* protoplasts by plasmid DNA, Molecular and General Genetics 168: 111-115.

Cohn, D.H., Mileham, A.J., Simon, M.I., Nealson, K.H., Raush, S.K., Bonam, D. and Baldwin, T.O. 1985, Nucleotide sequence of the *lux*A gene of *Vibrio harveyi* and the complete amino acid sequence of the alpha subunit of luciferase, J. Biol. Chem. 260: 6139-6146.

Condee, C.W. and Summers, A.O. 1992, A mer-*lux* transcriptional fusion for real-time examination of *in vivo* gene expression kinetics and promoter response to altered superhelicity, J. Bacteriology 174: 8094-8101.

de Leij, F.A.A.M., Sutton, E.J., Whipps, J.M., and Lynch, J.M. 1994, Effect of a genetically modified *Pseudomonas aureofaciens* on indigenous microbial population of wheat. FEMS Microbiology Ecology 13: 249-258.

de Lorenzo, V., Herrero, M., Jakubzik, U. and Timmis, K.N. 1990, Mini-Tn5 transposon derivatives for insertion mutagenisis, promoter probing, and chromosomal insertion of cloned DNA in Gram-negative Eubacteria, J. Bacteriology 172: 6568-6572.

de Weger, L.A., Dunbar, P., Mahafee, W.F., Lugtenberg, B.J.J. and Sayler, G.S. 1991, Use of bioluminescence markers to detect *Pseudomonas* spp. in the rhizosphere, Appl. Biochem. Microbiol. 57: 3641-3644.

de Weger, L.A., Kuiper, I., van der Bij, A.J. and Lugtenberg, B.J.J. 1997, Use of a *lux*-based procedure to rapidly visualise root colonisation by *Pseudomonas fluorescens* in the wheat rhizosphere, Antonie van Leeuwenhoek Int., J. Gen and Molecular Microbiology 72: 365-372.

Dower, W.J., Miller, J.F. and Ragsdale, C.W. 1988, High efficiency transformation of *E. coli* by high voltage electroporation, Nucleic Acids Research 16: 6127-6145.

Dubnau, D. and Davidoff-Abelson, R. 1971, Fate of transforming DNA following uptake by competent *Bacillus subtilis*, I, Formation and properties of the donor-recipient complex, J. Molecular Biology 56: 209-221.

Duncan, S., Glover, L.A., Killham, K. and Prosser, J.I. 1994, Luminescence-based detection of starved and non-culturable but viable bacteria, Appl. Environ. Microbiol. 60: 1308-1316.

Dunlap, P.V. and Greenberg, E.P. 1985, Control of *Vibrio fischeri lux* genes, Transcription by a cyclic AMP receptor protein-*lux* R protein regulatory circuit, J. Bacteriology 170: 4040-4046.

Eberhard, A., Burlingame, A.L., Eberhard, C., Kenyon, G.L., Nealson, K.H. and Oppenheimer, N.J. 1981, Structural identification of autoinducer of *Photobacterium fischeri luciferase*, Biochemistry 20: 2444-2449.

Ellis, R.J., Thompson, I.P. and Bailey, M.J. 1995, Metabolic profiling as a means of characterising plant-associated microbial communities, FEMS Microbiology Ecology 16: 9-18.

Engebrecht, J. and Silverman, M. 1984, Identification of genes and gene products necessary for bacterial bioluminescence, Proceedings National Academy Sciences, USA 81: 4154-4148.

Flemming, C.A., Lee, H. and Trevors, J.T. 1994, Bioluminescent most-probable-number method to enumerate *lux*-marked *Pseudomonas aeruginosa* UG2Lr in soil, Appl. Environ. Microbiol. 60: 3458-3461.

Foran, D.R. and Brown, W.M. 1988, Nucleotide sequence of the *lux*A and *lux*B genes of the bioluminescent marine bacterium *Vibrio fischeri*, Nucleic Acid Research 16: 777.

Fravel, D.R., Lumsden, R.D. and Roberts, D.P. 1990, *In situ* visualisation of the biocontrol rhizobacterium *Enterobacter cloacae* with bioluminescence, Plant and Soil 125: 233-238.

Fuqua, C., Winans, S.C. and Greenberg, E.P. 1996, Census and consensus in bacterial ecosystems: The *lux*R-*lux*I family of quorum-sensing transcriptional regulators, Annual Review of Microbiology 50, 727-751.

Garland, J.L. and Mills, A.L. 1991, Classification and characterisation of heterotrophic microbial communities on the basis of patterns of community-level sole-carbon source utilisation, Appl. Environ. Microbiol. 57: 2351-2359, Geoderma 56: 633-640.

Grant, F.A., Glover, L.A., Killham, K. and Prosser, J.I. 1991, Luminescence based viable cell enumeration of *Erwinia carotovora* in the soil, Soil Biol. Biochem. 23: 1021-1024.

Hardy, K.G. 1985, *Bacillus* cloning methods, in : DNA cloning - A practical approach, D.M. Glover, ed., IRL Press, Oxford. pp. 1-17.

Hartung, J.S. 1998, Molecular probes and assays useful to identify plant pathogenic fungi, bacteria, and marked biocontrol agents, in : Plant-Microbe Interactions and Biological Control, G.J. Boland and L.D. Kuykendall, eds., Marcel Dekker Inc., New York, pp 393-413.

Heijnen, C.E. and Marinissen, J.C.Y. 1995, Survival of bacteria introduced into soil by means of transport by *Lumbricus rubellus*, Biology and Fertility of Soils 20: 63-69.

Helfert, S. and Zeller, W. 1997, Luminescence based detection of the spread of fire blight (*Erwinia amylovora*) in plant tissue after the treatment with plant extracts, J. Plant Diseases and Protection, 104: 93-94.

Hill, P.J., Hall, L., Vinicombe, D.A., Soper, C.J., Setlow, P., Waites, W.M., Denyer, S. and Stewart, G.S.A.B. 1994, Bioluminescence and spores as biological indicators of inimical processes, J. Applied Bacteriology, Symposium Supplement 76: 129S-134S.

Jefferson, R.A., Kavanagh T.A. and Bevan, M.W. 1987, GUS fusions: ß-glucuronidase as a sensitive and versatile gene fusion marker in higher plants, The EMBO Journal 6: 3901-3907.

Jensen, L.E., Kragelund, L. and Nybroe, O. 1998, Expression of a nitrogen regulated *lux* gene fusion in *Pseudomonas fluorescens* DF57 studied in pure culture and in soil, FEMS Microbiology Ecology 25: 23-32.

Johnston, T.C., Thompson, R.B. and Baldwin, T.O. 1986, Nucleotide sequence of the *lux*B gene of *Vibrio harveyi* and the complete amino acid sequence of the ß-subunit of bacterial luciferase, J. Biol. Chem. 261: 4805-4811.

Johnston, T.C., Rucker, E.B., Cochrum, L., Hruska, K.S. and Vandegrift, V. 1990, The nucleotide sequence of the *lux*A and the *lux*B genes of *Xenorhabdus luminescens* HM and a comparison of the amino acid sequences of luciferase from four species of bioluminescent bacteria, Biochemical and Biophysical Research Communications 170: 407-415.

Jones, D.L. and Darrah, P.R. 1993, Influx and efflux of amino acids from *Zea mays* L. roots and their implications for N nutrition and the rhizosphere, Plant and Soil 155/156: 87-90.

Koncz, C., Langridge, W.H.R., Olsson, O., Schell, J. and Szalay, A.A. 1990, Bacterial and firefly luciferase in transgenic plants: Advantages and disadvantages of a reporter gene, Developmental Genetics 11: 224-232.

Kozdroj, J. 1996, Survival of *lux* marked bacteria introduced into soil and the rhizosphere of bean (*Phaseolus vulgaris* L.), World Journal of Microbiology and Biotechnology 12: 261-265.

Kragelund, L., Christoffersen, B., Nybroe, O. and Debrun, F.J. 1995, Isolation of *lux* reporter gene fusions in *Pseudomonas fluorescens* DF57 inducible by nitrogen or phosphorous starvation, FEMS Microbiology Ecology 17: 95-106.

Kragelund, L., Hosbond, C. and Nybroe, O. 1997, Distribution of metabolic activity and phosphate starvation response of *lux*-tagged *Pseudomonas fluorescens* reporter bacteria in the barley rhizosphere, Appl. Environ. Microbiol. 63: 4920-4928.

Kraus, J. and Loper, J.E. 1992, Lack of evidence for a role of antifungal metabolite production by *Pseudomonas*

fluorescens Pf-5 in biological control of *Pythium* damping-off of cucumber, Phytopathology 82: 264-271.
Leifert, C., Li, H., Chidburee, S., Hampson, S., Workman, S., Sigee, D., Epton, H.A.S. and Harbour, A. 1995, Antibiotic production and biocontrol activity by *Bacillus subtilis* CL27 and *Bacillus pumilus* CL45. J. Applied Bacteriology 78: 97-108.
Lindow, S.E. 1995, The use of reporter genes in the study of microbial ecology, Molecular Ecology 4: 555-566.
Lund, B.M. 1982, Clostridia and Plant Disease: New Pathogens? in : Phytopathogenic Prokaryotes, Vol. I, M.S. Mount and G.H. Lacy, eds., Academic Press, London. pp 263-283.
Madsen, E.L. and Alexander, M. 1982, Transport of *Rhizobium* and *Pseudomonas* through soil, Soil Science America Journal 46: 557-560.
Mahro, B., Schleinkofer, L. and Kasche, V. 1991, *In situ* detection of bioluminescent *Streptomyces coelicolor* in liquid and soil by photon imaging, Acta Biotechnologica 11: 541-546.
Marschner, P. and Crowley, D.E. 1996, Root colonisation of mycorrhizal and non-mycorrhizal pepper (*Capsicum annuum*) by *Pseudomonas fluorescens* 2-79RL, New Phytologist 143: 115-122.
Masuko, M., Hosoi, S. and Hayakawa, T. 1991a, A novel method for detecting and counting of single bacteria in a wide field using an ultra high sensitivity TV camera without a microscope, FEMS Microbiology Letters 81: 287-290.
Masuko, M., Hosoi, S. and Hayakawa, T. 1991b, Rapid detection and counting of a single bacteria in a wide field using a photon counting TV camera, FEMS Microbiology Letters 83: 231-238.
McDougald, D., Rice, S.A., Weichart, D. and Kjelleberg, S. 1998, Nonculturability: adaptation or debilitation, FEMS Microbiol Ecology 25: 1-9.
McLennan, K., Glover, L.A., Killham, K., Prosser, J.I. 1992, Luminescence based detection of *Erwinia carotovora* associated with rotting potato tubers, Letters in Applied Microbiology 15: 121-124.
Meighen, E.A. 1988, Enzymes and genes from the *lux* operons in bioluminescent bacteria, Annual Review of Microbiology 42: 151-176.
Meighen, E.A. 1991, Molecular biology of bacterial bioluminescence, Microbiological Reviews 55: 123-142.
Meikle, A., Glover, L.A., Killham, K. and Prosser, J.I. 1992, Potential luminescence as an indicator of activation of genetically modified *Pseudomonas fluorescens* in liquid culture and in soil, Soil Biol. Biochem. 26: 747-755.
Meikle, A., Amin-Hanjani, S., Glover, L.A., Killham, K. and Prosser, J.I. 1995, Matric potential and the survival and activity of a *Pseudomonas fluorescens* inoculum in soil, Soil Biol. Biochem. 27: 881-892.
Nelson, E.B. and Hsu, J.S.T. 1994, Nutritional factors affecting responses of sporangia of *Pythium ultimum* to germination stimulants, Phytopathology 84: 677-683.
O'Kane, D.J., Legocki, M., Legocki, R.P., Lingle, W.L., Wampler, J.E. and Szalay, A.A. 1988, Visualisation of bioluminescence as a marker of gene expression in *Rhizobium* infected soybean root nodules, Plant Molecular Biology 10: 387-399.
Pan, W.L., Bolton, R.P., Lundquist, E.J. and Hiller, L.K. 1998, Portable rhizotron and color scanner system for monitoring root development, Plant and Soil 200: 107-112.
Phillips-Jones, M.K. 1993, Bioluminescent (*lux*) expression in the anaerobe *Clostridium perfringens*, FEMS Microbiology Letters 106: 265-270.
Pierson, E.A. and Weller, D.M. 1994, Use of mixtures of fluorescent pseudomonads to suppress take-all and improve the growth of wheat, Phytopathology 84: 940-947.
Pierson, L.S. 1998, Genetic analysis of selected antifungal metabolites produced by *Pseudomonas aureofaciens*, in : Plant-Microbe Interactions and Biological Control, G.J. Boland and L.D. Kuykendall, eds., Marcel Dekker Inc., New York. pp 355-392.
Pierson III, L.S., Wood, D.W., Pierson, E.A. and Chancey, S.T. 1998, N-acyl-homoserine lactone-mediated gene regulation in biological control by fluorescent pseudomonads: Current knowledge and future work, European J. Plant Pathology 104: 1-9.
Postma, J., van Veen, J.A., and Walter, S. 1989, Influence of different initial soil moisture conditions on the distribution and population dynamics of introduced *Rhizobium leguminosarum* biovar. *trifoli*, Soil Biol. Biochem. 21: 437-442.
Prosser, J.I. 1994, Molecular marker systems for detection of genetically engineered micro-organisms in the environment, Microbiology 140: 5-17.
Prosser, J.I, Killham, K., Glover, L.A. and Rattray, E.A.S. 1996, Luminescence-based detection systems for detection of bacteria in the environment, CRC Critical Reviews in Biotechnology 16: 157-183.
Rattink, H.1992, Targets for pathology research in protected crops, Pesticide Science 36: 385-388.
Rattray, E.A.S., Prosser, J.I., Glover, L.A. and Killham, K. 1992, Matric potential in relation to survival and activity of a genetically modified inoculum in soil, Soil Biol. Biochem. 5: 421-425.
Rattray, E.A.S., Tyrrell, J.A., Prosser, J.I., Glover, L.A. and Killham, K. 1993, Effect of soil bulk density and

temperature on wheat rhizosphere colonisation by *lux*-marked *Pseudomonas fluorescens*, European J. Soil Biology 29: 73-82.
Rattray, E.A.S., Prosser, J.I., Glover, L.A. and Killham, K. 1995, Characterisation of rhizosphere colonisation by luminescent *Enterobacter cloacae* at the population and single cell levels, Appl. Environ. Microbiol. 61: 2950-2957.
Ryder, M. 1994, Key issues in the deliberate release of genetically-manipulated bacteria, FEMS Microbiology Ecology 15: 139-146.
Ryder, M.H. 1995, Monitoring of biocontrol agents and genetically engineered microorganisms in the environment: biotechnological approaches, in : Molecular Methods in Plant Pathology, U.S. Singh and R.P. Singh, eds., CRC Lewis Publishers, pp 475-492.
Sakai, M., Futamata, H., Ozawa, H., Sueguchi, T., Kim, J.S. and Matsuguchi, T. 1997, Use of bacterial bioluminescence for monitoring the behaviour of rhizobacteria introduced to plant rhizosphere, Soil Science and Plant Nutrition 43: 395-404.
Schmidt, T.M., Kopecky, K. and Nealson, K.H. 1989, Bioluminescence of the insect pathogen *Xenorhabdus luminescens*, Appl. Environ. Microbiol. 55: 2607-2612.
Shaw, J.J. and Kado, C.I. 1986, Development of a *Vibrio bioluminescence* gene-set to monitor phytopathogenic bacteria during on going process in a non-disruptive manner, Biotechnology 4: 560-564.
Shaw, J.J., Dane, F., Geiger, D. and Kloepper, J.W. 1992, Use of bioluminescence for detection of genetically engineered microorganisms released into the environment, Appl. Environ. Microbiol. 58: 267-273.
Silcock, D.J., Waterhouse, R.N., Glover, L.A., Prosser, J.I. and Killham, K. 1992, Detection of a single genetically modified bacterial cell in soil by using Charged Coupled Device-enhanced microscopy, Appl. Environ. Microbiol. 58: 2444-2448.
Stanghellini, M.E. and Hancock, J.G. 1971, Radial extent of the bean spermosphere and its relation to the behaviour of *Pythium ultimum*, Phytopathology 61: 165-168.
Stewart, G.S.A.B. and Williamson, P. 1992, *lux* genes and the application of bacterial bioluminescence, J. General Microbiology 138: 1289-1300.
Thorpe, I.S., Killham, K., Prosser, J.I. and Glover, L.A. 1993, Novel method for the study of population dynamics of a genetically modified microorganism in the gut of *Lumbricus terrestris*, Biology and Fertility of Soils 15: 55-59.
Thorpe, I.S., Prosser, J.I., Glover, L.A. and Killham, K. 1996, The role of the earthworm *Lumbricus terrestris* in the transport of bacterial inocula through soil, Biology and Fertility of Soils 23: 132-139.
Tiedje, J.M., Colwell, R.K., Grossman, Y.L., Hodson, R.E., Lenski, R.E., Mack, R.N. and Regal, P.J. 1989, The planned introduction of genetically engineered organisms: ecological considerations and recommendations, Ecology 70: 298-315.
Toyota, K. and Kimura, M. 1994, Earthworms disseminate a soil borne plant pathogen, *Fusarium oxysporum* f. sp. *raphani*, Biology and Fertility of Soils 18: 32-36.
Tronsmo, A. and Hjeljord, L.G. 1998, Biological control with *Trichoderma* species, in : Plant-Microbe Interactions and Biological Control, G.J. Boland and L.D. Kuykendall, eds., Marcel Dekker Inc., New York, pp 111-126.
Turner, J.M. and Messenger, A.J. 1986, Occurrence, biochemistry and physiology of phenazine pigment production, Advances in Microbial Physiology 27: 211-275.
Vivian, A. and Gibbon, M.J. 1997, Avirulence genes in plant-pathogenic bacteria: signals or weapons? Microbiology 143: 693-704.
Waterhouse, R.N., Silcock, D.J. and Glover, L.A. 1993a, Use of charge-coupled device (CCD) image-enhancement for rapid screening and monitoring of prokaryotic promoter expression, Letters Applied Microbiology 16: 136-141.
Waterhouse, R.N., Silcock, D.J., White, H.L., Buhariwalla, H.K. and Glover, L.A. 1993b, The cloning and characterisation of phage promoters, directing high expression of luciferase in *Pseudomonas syringae* pv. *phaseolicola*, allowing single cell and microcolony detection, Molecular Ecology 2: 285-294.
Waterhouse, R.N., Buhariwalla, H., Bourn, D., Rattray, E.A.S. and Glover, L.A. 1996, CCD detection of *lux* marked *Pseudomonas syringae* pv. *phaseolicola* L-forms associated with Chinese cabbage and the resulting disease protection against *Xanthomonas campestris*, Letters in Applied Microbiology 22, 4: 262-266.
Whipps, J.M. 1992, Status of biological disease control in horticulture, Biocontrol Science and Technology 2: 3-24.
White, D., FitzPatrick, E.A. and Killham, K. 1994, Use of stained bacterial inocula to assess spatial distribution after introduction into soil, Geoderma 63: 245-254.
White, D., Leifert, C., Ryder, H.M. and Killham, K. 1996, *Lux* gene technology - a strategy to optimise biological control of soil-borne diseases, New Phytologist 133: 173-181.
Willets, N. and Wilkins, B. 1984, Processing of plasmid DNA during bacterial conjugation, Microbiological Reviews 48: 24-30.

Wilson, K.J., Sessitsch, A., Corbo, J.C., Giller, K.E., Akkermans, A.D.L. and Jefferson, R.A. 1995, ß-Glucuronidase (GUS) transposons for ecological and genetic studies of rhizobia and other Gram-negative bacteria, Microbiology 141: 1691-1705.

Wood, D.W. and Pierson III, L.S. 1996, The *phzI* gene of *Pseudomonas aureofaciens* 30-84 is responsible for the production of a diffusable signal for phenazine antibiotic production, Gene 168: 49-53.

Wright, D.A., Killham, K., Glover, L.A. and Prosser, J.I. 1993, The effect of location in soil on protozoal grazing of a genetically modified inoculum, Geoderma 56: 633-640.

INDEX

Batch number: 10755809

Printed by Printforce, the Netherlands